AF508662

Advances in Textile Structural Composites

Advances in Textile Structural Composites

Editor

Rajesh Mishra

MDPI • Basel • Beijing • Wuhan • Barcelona • Belgrade • Manchester • Tokyo • Cluj • Tianjin

Editor
Rajesh Mishra
Czech University of Life
Sciences Prague
Suchdol
Prague

Editorial Office
MDPI
St. Alban-Anlage 66
4052 Basel, Switzerland

This is a reprint of articles from the Special Issue published online in the open access journal *Polymers* (ISSN 2073-4360) (available at: hhttps://www.mdpi.com/journal/polymers/special_issues/Adv_Text_Struct_Compos).

For citation purposes, cite each article independently as indicated on the article page online and as indicated below:

LastName, A.A.; LastName, B.B.; LastName, C.C. Article Title. *Journal Name* **Year**, *Volume Number*, Page Range.

ISBN 978-3-0365-6594-1 (Hbk)
ISBN 978-3-0365-6595-8 (PDF)

Contents

Preface to "Advances in Textile Structural Composites"

The direction of fiber orientation plays a crucial role in deciding the mechanical performance of textile structural composites. Unlike conventional composite materials, geometrically oriented textile structures, e.g., woven, knitted, and braided constructions, can be designed and developed for load bearing in a particular direction. Their properties can be enhanced by modifying the geometry and material composition. One major challenge in producing textile structural composites with superior mechanical properties at a reasonably lower price is cost effective prepreg. Composites constructed from reinforcement having a well-defined geometry perform better than randomly oriented fibers at a reasonable cost. Their flex fatigue is superior to conventional preforms in specific applications. This Special Issue invites research as well as review articles dealing with different types of (2D, 3D, multiaxial) woven, knitted, and braided structures for load bearing structural composite applications. Use of industrial multifilament yarns of pure and hybrid composition in textile geometrical reinforcement structures can also be included. The methods of impregnation of such structures by thermoplastic and thermoset resins should be described. The superior performance in such structural composites must be highlighted. Methods of characterizing woven, knitted, and braided textile reinforced composites is the focus of this issue. Current and future applications of advanced textile structural composites are summarized in the submitted articles. Theoretical (computational, numerical simulation etc.) as well as experimental work are submitted with sufficient scientific innovation.

Rajesh Mishra
Editor

Editorial

Advances in Textile Structural Composites

Rajesh Kumar Mishra

Department of Material Science and Manufacturing Technology, Faculty of Engineering, Czech University of Life Sciences Prague, Kamycka 129, 165 00 Prague, Czech Republic; mishrar@tf.czu.cz

Citation: Mishra, R.K. Advances in Textile Structural Composites. *Polymers* **2023**, *15*, 808. https://doi.org/10.3390/polym15040808

Received: 28 January 2023
Accepted: 3 February 2023
Published: 6 February 2023

Textile-reinforced structural composites are a major discipline of modern-day research and development. The geometry of the reinforcing textile plays a crucial role in deciding the mechanical performance of textile-reinforced composites. Unlike conventional materials, textile geometrical structures, e.g., yarns or fabrics, can be designed and developed for load bearing in a particular direction [1]. Their properties can be enhanced by modifying the structure and material composition. The challenge in producing textile geometry-based structural composites with superior mechanical properties at a reasonably lower price is cost effective prepreg.

There are many ways of making fabrics from textile fibers. The most common and most complex category comprises fabrics made from interlaced yarns. These are the traditional methods of manufacturing textiles. Great scope lies in choosing fibers with particular properties, arranging fibers in the yarn in several ways, and organizing interlaced yarn in multiple ways within the fabric. This provides the textile designer with greater freedom and variation for controlling and modifying the fabric. The most common form of interlacing is weaving, where two sets of threads cross and interweave with one another. The yarns are held in place due to the inter-yarn friction. Another form of interlacing, where the thread in one set interlocks with the loops of a neighboring thread by looping, is called knitting. The interloping of yarns results in positive binding. Knitted fabrics are widely used in apparel, home furnishings, and technical textiles. Lace, crochet, and different types of net are other forms of interlaced yarn structures. The basic unit of a knitted structure is called a loop. A stitch is formed when one loop is drawn through another loop. Stitches may be formed in horizontal or in a vertical direction. Weft knitting is a method of forming a fabric by means of the intermeshment of horizontal loops in a circular or flat form on a course wise basis. In this method, one or more number of yarns are fed to a group of needles placed in either a lateral or circular fashion. Warp knitting is a method of forming a fabric by the intermeshment of loops made in a vertical way from each warp yarn. In this method, a number of end of yarns are fed simultaneously to individual needles placed in a lateral fashion [2].

Braiding is another way of thread interlacing for fabric formation. Braided fabric is formed by the diagonal interlacing of yarns. Braided structures are mainly used for industrial composite materials. Other forms of fabric manufacture use fibers or filaments laid down, without interlacing, in a web where they are bonded together mechanically or by using an adhesive. The former are needle-punched nonwovens and the later are spun-bonded. The resulting fabric after bonding normally produces a flexible and porous structure. These find use mostly in industrial and disposable applications. All these fabrics are broadly used in three major applications such as apparel, home furnishings, and for industrial uses. The traditional methods of weaving and hand weaving will remain supreme for high-cost fabrics with a rich design content [3]. The woven structures provide a combination of strength with flexibility. The flexibility at small strains is achieved by yarn crimp due to the freedom of yarn movement, whereas at high strains, the threads take the load together, providing a high strength. A woven fabric is produced by interlacing two sets of yarns, the warp, and the weft, which are at right angles to each other in the plane of the cloth. The warp is along the length and the weft is along the width of the fabric. Individual warp and weft yarns are called the ends and picks [4]. The interlacement of the ends and picks with each other produces a coherent and stable structure. The repeating unit of interlacement is called the weave. The structure and properties of a woven fabric are dependent upon the constructional parameters, such as the thread density, yarn fineness, crimp, weave, etc. [5].

The forming of textile reinforcements is an important stage in the manufacturing of textile composite parts. Fiber orientations and part geometry obtained from this stage have a significant impact on the subsequent resin injection and final mechanical properties of the composite part [6]. The numerical simulation of the forming of textile reinforcement is in strong demand as it can greatly reduce the time and cost in the determination of the optimized processing parameters, which is the foundation of the low-cost application of composite materials. The literature presents the state-of-the-art for forming, modeling methods for textile reinforcement and the corresponding experimental characterization methods developed in this field [7].

Textile-reinforced composites and fabric-reinforced mortars have emerged as a promising solution for lightweight construction elements and structural retrofit and strengthening. These novel composites consist of two- or three-dimensional textile structures, alkali-resistant glass, or polymer multi-filament yarns, etc. Given the high tensile strength of the textile yarns and their durable nature under normal service conditions, construction elements and strengthening covers reinforced with textile structures feature small thicknesses and offer a higher flexibility in terms of fabrication and application technology as well as the element's shape [8].

The Special Issue provides an overview of the available textile geometrical reinforcements possible for use in composite reinforcement. It dealt with the different types of textile structures for load bearing applications. The uses of industrial multifilament yarns of a pure and hybrid composition in textile structures, e.g., woven, knitted, braided, and multiaxial structures, were summarized [9].

In this Special Issue, the microscopic, mesoscopic, and macroscopic models were discussed since this is the most common defect occurring in the manufacturing of textile reinforcement structures. The characterization and analysis of their performance is addressed in detail in this issue.

This Special Issue focused on the application of textile structural composites. The significance and potential of textile composites was described briefly. Given the regularity in the textile preforms, of which textile composites are made, there are plenty of applications reported. Textile-based composites are fast becoming key in many industries, such as automotive, military, aeronautical and aerospace, and construction industries, mainly due to their very attractive specific properties (e.g., strength to weight ratio). Particularly, 3D woven composites (a three-dimensional weaving pattern) have appeared in many new applications requiring high mechanical properties. Evidently, the increasing interest in these materials has generated a high demand for the proper characterization of the methods, accurate FE simulations, suitable nondestructive testing techniques, and adequate visualization techniques. The structured nature of textile-reinforced composites, as described by their geometrical pattern, allows for a new type of analysis, namely one that takes into account their topology [10]. The Special Issue details the various applications of advanced textile structural composites into collective groups.

The Editor is thankful to all the contributors and editorial staff for preparing this Special Issue successfully and effectively. These specialized contributions will lead to new ways of research and development in this area.

Conflicts of Interest: The author declares no conflict of interest.

Polymers **2023**, *15*, 808

References

1. Tufail, M.R.; Jamshaid, H.; Mishra, R.; Hussain, U.; Tichy, M.; Muller, M. Characterization of hybrid composites with polyester waste fibers, olive root fibers and coir pith micro-particles using mixture design analysis for structural applications. *Polymers* **2021**, *13*, 2291. [CrossRef] [PubMed]
2. Jamshaid, H.; Mishra, R.; Zeeshan, M.; Zahid, B.; Basra, S.A.; Tichy, M.; Muller, M. Mechanical performance of knitted hollow composites from recycled cotton and glass fibers for packaging applications. *Polymers* **2021**, *13*, 2381. [CrossRef] [PubMed]
3. Kolář, V.; Müller, M.; Tichý, M.; Mishra, R.K.; Hrabě, P.; Hanušová, K.; Hromasová, M. Experimental investigation of wavy-lap bonds with natural cotton fabric reinforcement under cyclic loading. *Polymers* **2021**, *13*, 2872. [CrossRef] [PubMed]
4. Vorhof, M.; Sennewald, C.; Schegner, P.; Meyer, P.; Hühne, C.; Cherif, C.; Sinapius, M. Thermoplastic composites for integrally woven pressure actuated cellular structures: Design approach and material investigation. *Polymers* **2021**, *13*, 3128. [CrossRef] [PubMed]
5. Mishra, R.; Petru, M.; Novotna, J. Bio-composites reinforced with natural fibers: Comparative analysis of thermal, static and dynamic-mechanical properties. *Fibers Polym.* **2020**, *21*, 619–627. [CrossRef]
6. Kamble, Z.; Mishra, R.K.; Behera, B.K.; Tichý, M.; Kolář, V.; Müller, M. Design, development, and characterization of advanced textile structural hollow composites. *Polymers* **2021**, *13*, 3535. [CrossRef] [PubMed]
7. Qu, Z.; Gao, S.; Zhang, Y.; Jia, J. Analysis of the mechanical and preforming behaviors of carbon-kevlar hybrid woven reinforcement. *Polymers* **2021**, *13*, 4088. [CrossRef] [PubMed]
8. Mishra, R.K.; Petru, M.; Behera, B.K.; Behera, P.K. 3D woven textile structural polymer composites: Effect of resin processing parameters on mechanical performance. *Polymers* **2022**, *14*, 1134. [CrossRef] [PubMed]
9. Blonder, A.; Brocato, M. Layered-fabric materiality fibre reinforced polymers (L-FMFRP): Hysteretic behavior in architectured FRP material. *Polymers* **2022**, *14*, 1141. [CrossRef] [PubMed]
10. Pais, V.; Silva, P.; Bessa, J.; Dias, H.; Duarte, M.H.; Cunha, F.; Fangueiro, R. Low-velocity impact response of auxetic seamless knits combined with non-newtonian fluids. *Polymers* **2022**, *14*, 2065. [CrossRef] [PubMed]

Article

Design, Development, and Characterization of Advanced Textile Structural Hollow Composites

Zunjarrao Kamble [1], Rajesh Kumar Mishra [2,*], Bijoya Kumar Behera [1], Martin Tichý [2], Viktor Kolář [2] and Miroslav Müller [2]

1 Department of Textile & Fiber Engineering, Indian Institute of Technology Delhi, New Delhi 110016, India; ttz178482@textile.iitd.ac.in (Z.K.); B.K.Behera@textile.iitd.ac.in (B.K.B.)
2 Department of Material Science and Manufacturing Technology, Faculty of Engineering, Czech University of Life Sciences Prague, Kamycka 129, 165 00 Prague, Czech Republic; martintichy@tf.czu.cz (M.T.); vkolar@tf.czu.cz (V.K.); muller@tf.czu.cz (M.M.)
* Correspondence: mishrar@tf.czu.cz

Abstract: The research is focused on the design and development of woven textile-based structural hollow composites. E-Glass and high tenacity polyester multifilament yarns were used to produce various woven constructions. Yarn produced from cotton shoddy (fibers extracted from waste textiles) was used to develop hybrid preforms. In this study, unidirectional (UD), two-dimensional (2D), and three-dimensional (3D) fabric preforms were designed and developed. Further, 3D woven spacer fabric preforms with single-layer woven cross-links having four different geometrical shapes were produced. The performance of the woven cross-linked spacer structure was compared with the sandwich structure connected with the core pile yarns (SPY). Furthermore, three different types of cotton shoddy yarn-based fabric structures were developed. The first is unidirectional (UD), the second is 2D all-waste cotton fabric, and the third is a 2D hybrid fabric with waste cotton yarn in the warp and glass multifilament yarn in the weft. The UD, 2D, and 3D woven fabric-reinforced composites were produced using the vacuum-assisted resin infusion technique. The spacer woven structures were converted to composites by inserting wooden blocks with an appropriate size and wrapped with a Teflon sheet into the hollow space before resin application. A vacuum-assisted resin infusion technique was used to produce spacer woven composites. While changing the reinforcement from chopped fibers to 3D fabric, its modulus and ductility increase substantially. It was established that the number of crossover points in the weave structures offered excellent association with the impact energy absorption and formability behavior, which are important for many applications including automobiles, wind energy, marine and aerospace. Mechanical characterization of honeycomb composites with different cell sizes, opening angles and wall lengths revealed that the specific compression energy is higher for regular honeycomb structures with smaller cell sizes and a higher number of layers, keeping constant thickness.

Keywords: textile structural composite; 3D weaving; hollow structure; spacer fabric; woven honeycomb; sandwich; waste cotton; impact; compression; flexural rigidity

Citation: Kamble, Z.; Mishra, R.K.; Behera, B.K.; Tichý, M.; Kolář, V.; Müller, M. Design, Development, and Characterization of Advanced Textile Structural Hollow Composites. *Polymers* **2021**, *13*, 3535. https://doi.org/10.3390/polym13203535

Academic Editor: Andrea Zille

Received: 12 September 2021
Accepted: 11 October 2021
Published: 14 October 2021

Publisher's Note: MDPI stays neutral with regard to jurisdictional claims in published maps and institutional affiliations.

1. Introduction

Textile structures have shown remarkable performance in advanced composites for aerospace, automotive, marine, civil engineering, wind energy, protective clothing, and many other applications. Unidirectional (UD) and two-dimensional (2D) woven textile-reinforced composites have exhibited clear advantages over the traditional metallic materials in terms of performance-to-weight ratio. Various three-dimensional (3D) woven textile structures have started to receive serious attention for structural composites due to better structural integrity, high delamination resistance, etc. The modern low-cost manufacturing methods of single and multilayer non crimp woven preform have created research interest in these new reinforcement structures [1–6]. Modern preform manufacturing technology (weaving, braiding, warp knitting, and nonwoven) also facilitates the development of a variety of complex geometrical shapes [7,8].

In 3D woven fabric-reinforced composites, adjusting parameters of the internal geometry of the preform leads to efficient optimization of the performance of the final product. The fiber orientation in an engineered preform determines the direction of the best possible stiffness and strength performance, while the matrix is responsible for stress transfer and load redistribution in a textile structural composite [9–12]. The internal architecture of the material governs the mechanical properties of the part and hence offers enormous space for designers to match the ultimate criteria for a specific application. Over and above, modern computational tools help to predict, and hence design, special textile architectures of desired mechanical performance [13–18]. Fiber architectures of 3D woven preforms can be adjusted in a wide range by changing the weaving parameters, such as warp/weft density and weaving patterns. The introduction of fiber in thickness direction improves the interlaminar properties. Fiber architectures directly affect the formability of the preforms [19–22].

An advantage of 3D weaving is that preforms can be made on standard industrial weaving looms used for producing 2D fabrics by making minor modifications to the machinery [23–28]. A specialized 3D woven fabric is spacer or distance fabric. This material consists of two parallel 2D woven fabrics integrally connected by a low density of the through-thickness yarns. Spacer fabric composites are an alternative to honeycomb or foam material to make sandwich structures because they exhibit superior mechanical properties [29–32]. These composites are primarily used to manufacture double-walled tanks or the wall lining for chemical storage tanks, car and truck spoilers/fairings, lightweight walls, dome structures and composite tooling [33]. Sandwich structures constitute a thick and light-weight core sandwiched between two relatively thin face sheets and offer high bending stiffness while being light-weight. Sandwich structures reinforced with integrally produced 3D spacer preforms have very high delamination resistance compared to the conventional sandwich composites [34,35]. The characterization of compressive and bending properties of corrugated core sandwich structures with different core thickness, corrugation angle, and bonding length between core-face sheets have been reported [36–38]. In order to produce spacer structures with different cell geometrical parameters, i.e., with different cell wall opening angles and with different cell widths (at almost constant cell heights) and different cell heights (at almost constant cell widths), the required number of picks in different sections of the cross-sections need to be calculated. Using these calculated number of picks, the generalized weave designs for each type of structure can be modified to obtain the actual weave designs [39–41]. Sandwich structures with integrated woven core piles have higher skin–core debonding resistance as compared to other sandwich composites [42]. Quasi-static and dynamic compression of such structures demonstrate ductile failure and very good energy-absorbing capability [43]. Increase in the height of core piles reduces the out-of-plane compression load [44], whereas thinner panels exhibit higher absorbed energy per unit volume in quasi-static compressive and three point-bending evaluation [45]. Though these spacer composites are better than traditional sandwich composites in some respect, these structures are not strong enough for flexural loading conditions [46].

In light of the above discussion, intensive research has been carried out to investigate and establish the relative mechanical advantage of some special textile structural composites using a wide range of preform architectures starting from simple chopped fiber to the most complex 3D structures, such as energy-absorbing hollow structures, honeycomb structures, spacers with augmented cores, profiled structures, stiffeners, and aerodynamic structures [47,48]. Woven spacer fabrics with woven cross-links and different cell geometries were produced. The sandwich composites were analyzed for their quasi-static lateral compression and flexural performance to compare their load-bearing capacity and energy absorbency [49–53]. Further, complex profiled 3D fabrics, e.g., I, U, + or X shapes are used in composites where superior joint strength is desired [54,55].

The manufacturing possibilities of woven spacers with woven cross-links have been reported by several researchers. However, to establish end-use based on structural characteristics, it is necessary to study and compare their mechanical behaviors, such as compressive and bending properties. In this research work, sandwich structures with different cell geometrical shapes were manufactured using 3D integrally woven spacer fabrics. These structures were subsequently evaluated for their compressive and flexural performance to reveal their load-bearing capacity, energy absorbency and failure mechanisms.

The current research mainly focuses on several UD, 2D and 3D woven textile hollow structures and profiled structures used in composite reinforcement. Several novel architectures have been designed and developed for applications in aircraft wings, wind turbine blades, etc. The mechanical performance of such hollow composites with respect to their impact, compression and flexural properties were evaluated. Novel sandwich structures were developed by using waste cotton fibers recycled from textile wastes. Hollow structural composites, namely spacer, honeycomb and sandwich, with special geometries were designed for optimal aerodynamic performance. Further, the junction strength of profiled geometries was analyzed. These innovative textile structural composites offer several advantages over chopped fiber or conventional 2D fabric-reinforced composites.

2. Materials and Methods

2.1. Materials

E-Glass multifilament yarn of 600 tex (Saint-Gobain, Paris, France) was used to produce various woven constructions. The high tenacity polyester multifilament yarn (Reliance Industries Ltd., Mumbai, India) was used to produce sandwich structures with a stiffener section and an augmented core. The physical and mechanical properties of the glass tow and polyester yarn are shown in Table 1. Epoxy resin LY 556 (Sigma-Aldrich Chemicals Private Limited, Bangalore, India) as a matrix and Aradur 22962 (Sigma-Aldrich Chemicals Private Limited, Bangalore, India) as a curing agent was used with a weight ratio of 10:1. The yarn produced from cotton shoddy (fibers extracted from waste textiles) was used to develop homogeneous and hybrid preforms.

Table 1. Physical and mechanical properties of E-glass and polyester yarn.

Material	E-Glass Yarn	Polyester Yarn	Waste Cotton Yarn
Linear density (tex)	600	333	476
Density (g/cm^3)	2.54	1.38	~1.5
Tenacity (gf/den)	5.78	4.88	0.36
Young's modulus (gf/den)	233	89.94	-
Strain at break (%)	4.5	12.16	15.88

2.2. Methods

2.2.1. Development of Various Textile Structures

In this study, unidirectional (UD), two-dimensional (2D) and three-dimensional (3D) fabric preforms were developed. The preform specifications are shown in Table 2. The UD, 2D and 3D fabric preforms were produced using 600 tex E-glass yarn (Figure 1). All the preforms were developed on a sample weaving machine with a multi-beam creel, and the 3D preform weaving technique has been explained in previous studies [5,6]. The spacer fabrics were produced using 600 tex E-glass yarns. The woven spacer fabric preforms with single-layer woven cross-links with four different geometrical shapes, namely rectangular (single wall structure = RECTSL, double wall structure = RECTDL), trapezoidal (single-level structure = TPZ45°, double-level structure = TDL45°), and triangular (TR47°), were produced. The performance of the woven cross-linked spacer structure was compared with the sandwich structure connected with the core pile yarns (SPY). The weaving specifications of spacer structures are shown in Table 3. The cross-sectional representation of different spacer structures is shown in Figure 2.

Table 2. Specifications of UD, 2D, and 3D woven preforms.

Preform	Unidirectional	2D Plain	3D Orthogonal	3D Angle Interlock	3D Warp Interlock
Stuffer/warp ends/m	708	394	158	98	394
Binder ends/m	-	-	315	492	-
Picks/m	78	275	315	315	275
Stuffer layers	-	-	3	3	-
Fabric thickness (mm)	0.4	0.52	1.45	1.46	1.49
Areal density (kg/m^2)	0.415	0.414	1.25	1.22	1.26
Fiber volume fraction	0.39	0.31	0.36	0.34	0.34

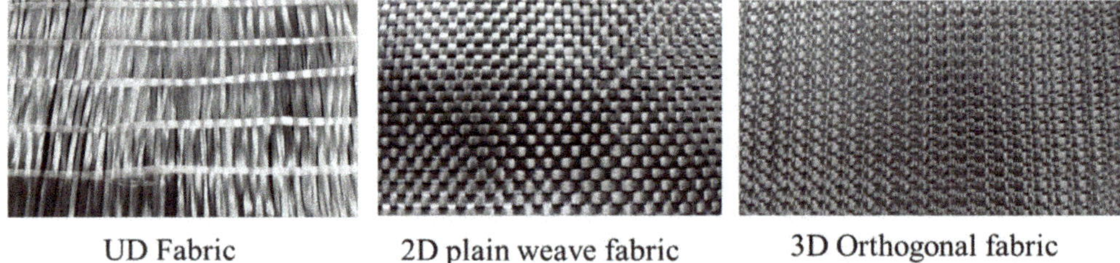

Figure 1. Optical images of UD, 2D and 3D woven preforms.

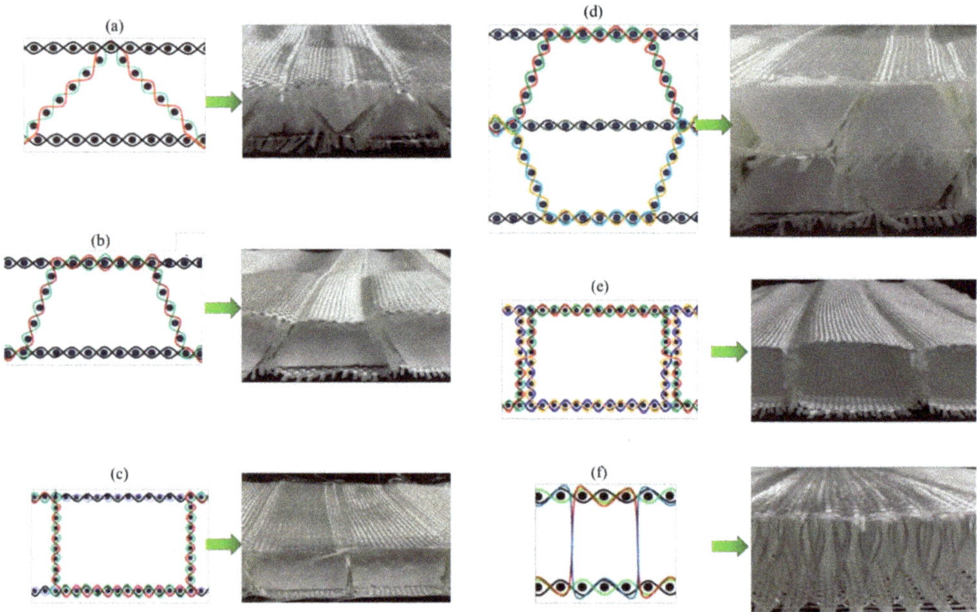

Figure 2. Cross-sectional representation of (**a**) TR, (**b**) TPZ45°, (**c**) RECTSL, (**d**) TDL45°, (**e**) RECTDL, and (**f**) SPY structures.

Further, an effort to weave sandwich structures with augmented core architecture was developed. Two different sandwich structures of this type were developed (Figure 3a,b). The fabric weaving specifications of these sandwich structures are shown in Table 4. The woven profiles with shapes U and + were produced on the double cloth weaving principle, which has two distinct layers separate from each other (Figure 3c,d). Both the layers of woven fabrics were integrated at a particular point to produce a crucial junction, resulting in an integrated profiled structure. The weaving specifications of these profile structures are shown in Table 5. Four different types of woven preforms (namely split tube, hull channel, inverted channel, and flat T bar) with integrated stiffener sections were developed using modified face to face weaving principle [15]. The line diagrams of these woven preforms and lifting patterns are shown in Figure 4. Based on the same weaving principle, 3D woven aerodynamic spacer structures were developed. The main objective of this invention was to design a woven fabric preform in a similar shape pertaining to a wing profile with the two outer skin sections and shear webs sections integrally woven (connected) so as to be used as reinforcement in developing a one-piece composite wing structure. The geometrical attributes of the airfoil are shown in Figure 5a. Airfoil structures with an 'I' and 'X' shear web profile were developed. The fabric geometry to be woven is chosen according to the required airfoil shape. The airfoil development process is detailed in Figure 5b. The plain weave was used to weave the preform. The ends and picks per meter were kept at 788 and 394, respectively. The initial information regarding wing profile and its characteristics (airfoil coordinates) was sourced from the NACA (National Advisory Committee for Airfoils) database, which provides the data for the construction of airfoils in the form of points. Each structural element of the 3D spacer structure was converted to an interlacement cross-section with equidistant pick spacing to create a weave design.

Table 3. Specifications of various woven spacer fabric preforms.

Structure	Cell Opening Angle (deg)	Structure Height (mm)	Top Side Length (mm)	Face Sheet Thickness (mm)	FVF [-]
TPZ45°	45	30	30	0.6	0.46
TDL45°	45	56.8	30	0.9	0.42
RECTSL	90	32	34	0.6	0.46
RECTDL	90	29	50	0.6	0.46
TR47°	47	28	-	0.6	0.47
SPY	-	30	-	0.6	0.40

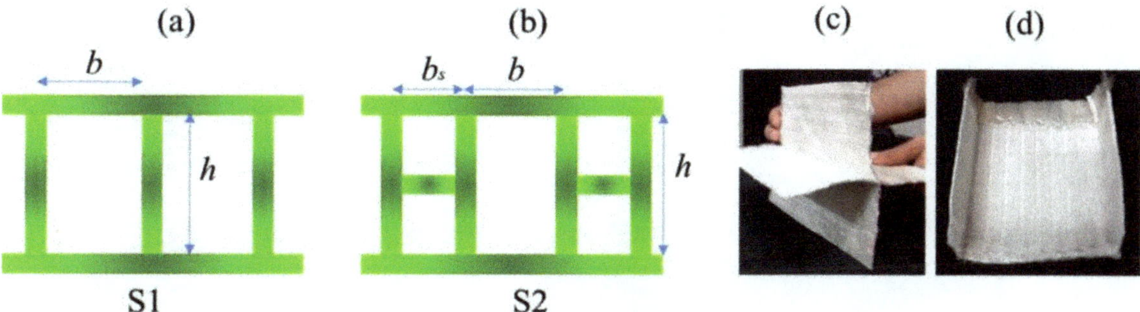

Figure 3. (**a,b**) Line diagram showing the geometry of the developed preform structures; (**c**) +-profiled and (**d**) U-profiled integrated woven profiles.

Table 4. Specifications of sandwich structures with an augmented core.

Structure	b (mm)	bs (mm)	h (mm)	Weaving Parameters
S1	45	30	30	Ends/m, Picks/m = 788,472
S2	45	56.8	30	

Advanced materials based on cellular solids have been used for decades in automotive, marine, and aerospace industries owing to their high energy-absorbing characteristics [18,19]. Metallic honeycombs have been explored as an energy-absorbing cellular structure. However, their strength-to-weight ratio is low, owing to a higher density compared to fiber-reinforced composite materials. E-glass yarn-based woven honeycomb composites with different cell shapes were developed in this research to study their compression properties. The geometrical parameters of the woven honeycomb cell are shown in Figure 6. The honeycomb fabrics with four different cell sizes keeping the opening angle constant were developed (Table 6). The honeycomb fabrics were woven using the double cloth weaving principle [21]. The honeycomb structure is denoted as (x,y)PzLθ, where x is the length of the free wall measured in the number of picks (P), y is the length of bonded measured in the number of picks (P), z is the number of fabric layers used to form a bonded wall, L denotes the layer, and θ is the cell opening angle [21].

Table 5. Weaving specifications of woven profile structures.

Profile Shape	U	H
Total no of ends	510	720
Jacquard capacity (hooks)	200	400
Ground ends	270	360
Ends/cm, Picks/cm	10,10	10,10
Profile ends	240	360
Reed count	20	20
Denting	5/10	5/10

(Note: profile ends indicate the number of ends required to weave the profile shape, where ground ends indicate the number of the ends needed to weave base structure).

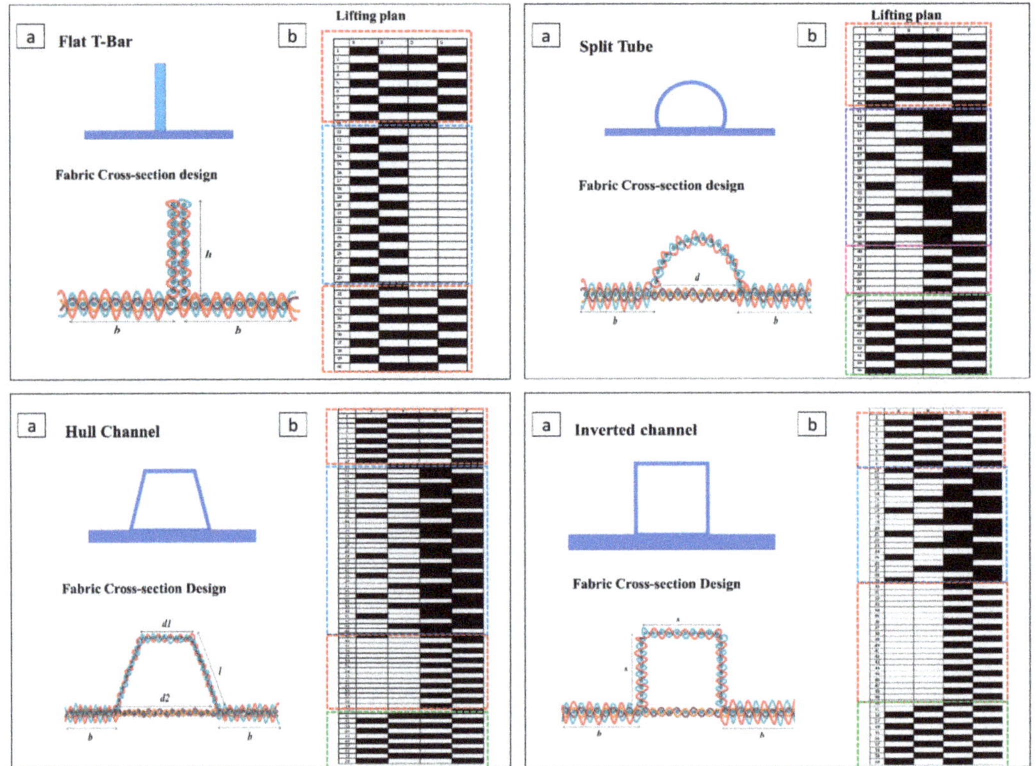

Figure 4. Line diagram of fabric cross-sections and lifting plans of various woven preforms.

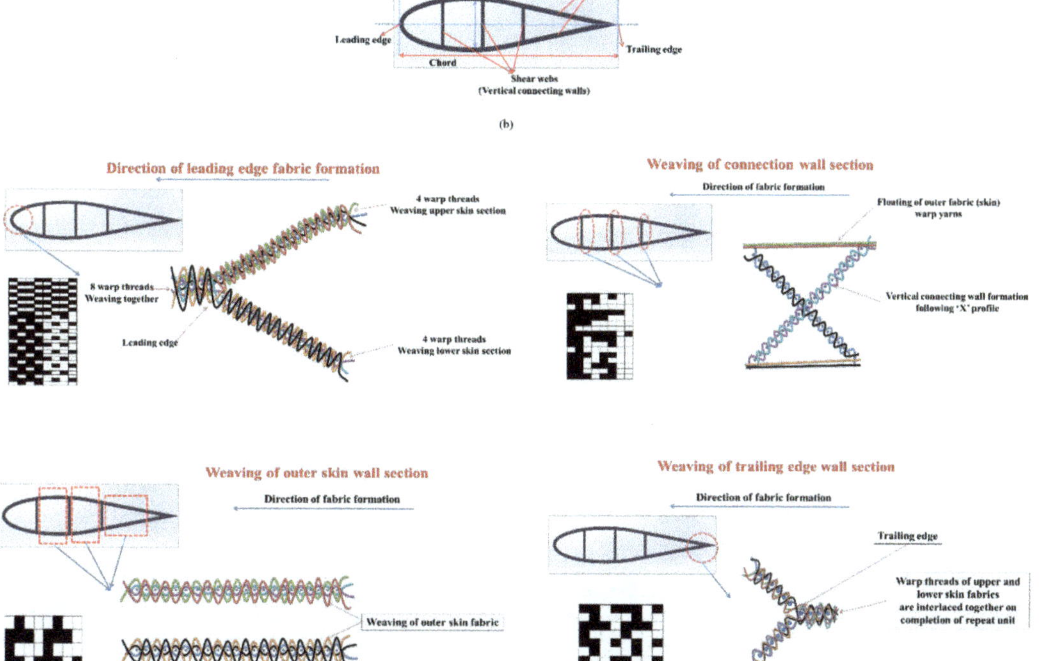

Figure 5. (**a**) Airfoil geometrical attributes; (**b**) development of woven airfoil structure.

Table 6. Structural parameters of woven honeycomb preforms.

Specimen	Bonded Wall Length (mm)	Free Wall Length (mm)	Cell Height (mm)	Opening Angle (°)
3P4L60	7.62	7.62	13.2	60
5P4L60	12.7	12.7	22	60
7P4L60	17.78	17.78	30.8	60
9P4L60	22.86	22.86	39.59	60

Furthermore, three different types of cotton shoddy yarn-based fabric structures were developed. The first is unidirectional (UD), the second is 2D all-waste cotton fabric, and the third is a 2D hybrid fabric with waste cotton yarn in the warp and glass multifilament yarn in the weft. Weaving specifications were decided to achieve approximately the same areal density (Table 7). Four different preform architectures were developed, as follows.

Table 7. Woven preform specifications.

Preform	Unidirectional	2D Fabric with All Waste Cotton Yarn	2D Hybrid Fabric (Warp—Waste Cotton Yarn, Weft—600 Tex Glass Yarn)
Ends/cm	10	5	5
Picks/cm	-	5	4
Areal density (g/m^2)	~450	~450	~450

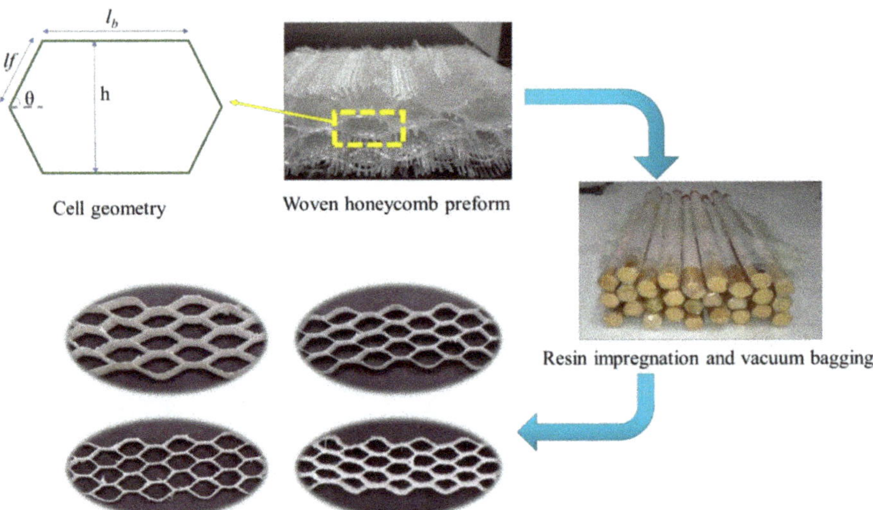

Figure 6. Woven honeycomb preforms, their cell geometry, and composite making process. (Note: *lf* and l_b are the lengths of a free and bonded wall).

1. Carded cotton shoddy web: The cotton shoddy, which is fibrous material obtained after mechanical shredding of waste cotton fabrics, was treated on a carding machine to produce a fiber web and used as a preform [28].

2. Cotton web sandwiched between woven fabrics: The weight of the preform desired in the composite was calculated based on a relationship between mass, volume, and density. The 2D woven all-waste cotton yarn fabric was cut in line with mold dimensions (30 cm × 30 cm) and weighed. The shoddy web weight was determined by subtracting the woven fabric weight from the total weight of the preform in the composite. The shoddy web has a size in line with mold dimensions. The shoddy web was then sandwiched between woven fabrics during composite fabrication.

3. Cotton web sandwiched between UD preform: this preform was developed by following a procedure similar to preform, with a cotton web sandwiched between woven fabrics. The preform has unidirectional waste cotton yarn fabric at the top and bottom, sandwiching shoddy web.

4. Cotton web sandwiched between hybrid woven fabrics: this preform was developed by following a procedure similar to preform, with a cotton web sandwiched between woven fabrics. The only difference is the hybrid fabric was used as the skin.

2.2.2. Development of Composites Reinforced with Various Textile Structures

The glass tows were chopped to the length of 25 mm uniformly. The stainless-steel mold (30 cm × 30 cm × 0.3 cm) was taken. The chopped glass fibers and matrix weight was calculated according to the relationship between volume, mass, and density, and desired fiber volume in the composite. The fibers were placed in the mold, and resin was applied to them. A uniform application of resin was ensured. The mold was covered with a Teflon sheet and placed in between preheated platens of the compression molding machine. The composites were cured for 60 min. The textile waste-based composite laminates were also developed using a similar methodology. The scheme of the experiment is shown in Table 8, according to which the composite specimens were developed.

Table 8. Scheme of composite laminate development.

Sample ID	Preform Type	Woven Fabric Weight (%) in the Composite	Web/Nonwoven Weight (%) in the Composite	Matrix Weight (%) in the Composite
SH	Carded cotton web	-	38	62
Wb	Cotton web sandwiched between woven fabrics	26.33	10.37	63.3
WbUD	Cotton web sandwiched between waste cotton yarn UD preform	23.24	14.45	62.43
WbH	Cotton web sandwiched between hybrid woven fabrics	26.79	8.93	64.28

The UD, 2D and 3D woven fabric-reinforced composites were produced using the vacuum-assisted resin infusion technique. The spacer woven structures were converted to composites by inserting wooden blocks with an appropriate size and wrapped with a Teflon sheet into the hollow space before resin application. A vacuum-assisted resin infusion technique was used to produce spacer woven composites. The samples were cured for 24 h at room temperature. A similar technique was used to produce sandwich composites, composites reinforced with integrated stiffener section, profiled composites, and airfoil structures (Figure 7). The process of woven honeycomb and aerodynamic spacer composite development is depicted in Figures 6 and 8, respectively.

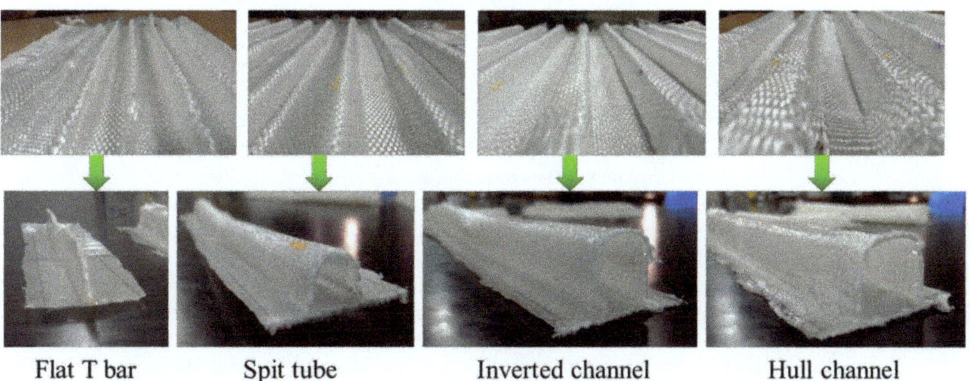

Figure 7. Woven preforms with integrated stiffener sections and their composites.

2.2.3. Characterization of Composite Materials

The lateral quasi-static compression and three-point bending of all spacer composites, sandwich composites with augmented core architecture and honeycomb composites were carried out according to ASTM C365 and ASTM C 393, respectively on an Instron 5982 universal testing machine. The junction strength of profiled preforms and their composites was characterized using a universal testing machine in tensile testing mode using a specially designed jaw. The flexural properties of composites reinforced with an integrally woven stiffener section were characterized according to ASTM D 790. The tensile, flexural, and izod impact properties of textile waste-based laminates were characterized according to ASTM D 3039, ASTM D 7264, and ISO 180:2000, respectively.

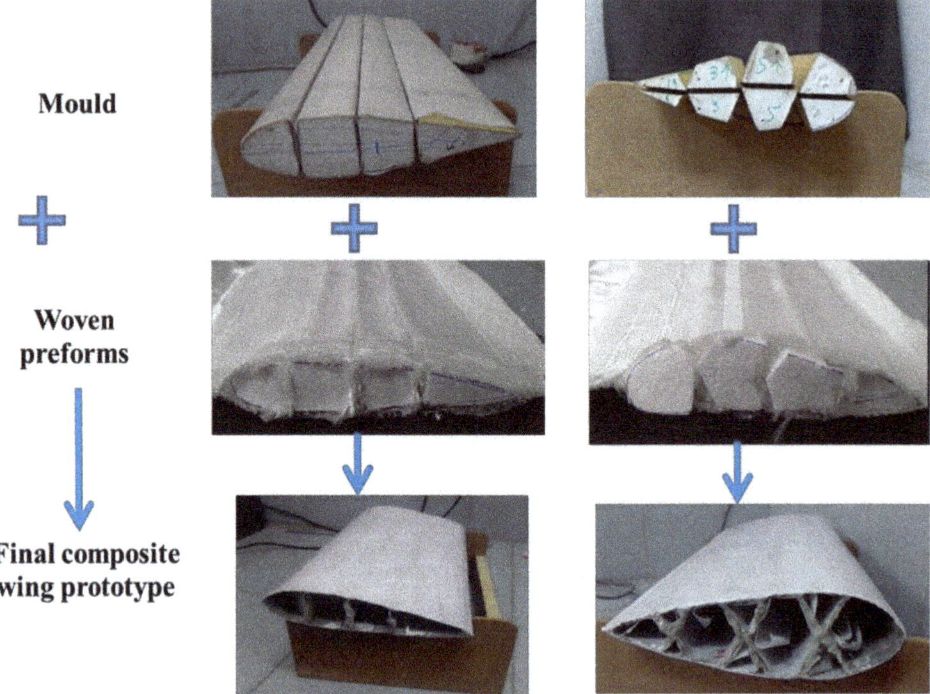

Figure 8. Flowchart for conversion of aerodynamic spacer fabric preforms to composite wing structures.

3. Results and Discussion

3.1. Mechanical Properties of Composites with UD, 2D and 3D Woven Reinforcement Structures

Three-dimensional woven composites are the materials of choice in many applications, such as aeronautic and astronautic, defense, automotive, construction, safety industry, etc. The fundamental advantage of 3D woven preforms over 2D laminate is the reinforcement in the thickness direction, which holds the yarn layers in place and provides structural stability [33]. This makes 3D woven composites delamination resistant. Further, the 3D weaving technique allows the production of near-net-shape and complex preforms. Three-dimensional woven composites have high tensile strain to failure values, high delamination, and high impact tolerance [34]. Various studies proved that 3D woven preforms produced using natural fiber yarn and their composites have mechanical properties comparable to high-performance fiber-reinforced composites.

Remarkable improvement in tensile strength and Young's modulus of textile structure-reinforced composites is observed compared to a neat matrix (Figure 9). While changing the reinforcement from chopped fibers to 3D fabric, its modulus and ductility increase substantially. Tensile test results clearly show that UD fabric-reinforced composite possesses the highest ultimate strength among all other composites. This is due to the higher fiber orientation in the loading direction, followed by 2D fabric-reinforced composite due to comparatively less fiber orientation in the loading direction, while in 3D fabric, reinforced composite fibers are disposed of in three perpendicular planes, leading to lesser strength in warp direction for the given fiber volume fraction and areal density compared to unidirectional reinforced (UD) 2D fabric-reinforced composites. A higher modulus in the composites with UD and 3D preform architecture is a negligible crimp in the warp yarns and zero crimp in the stuffer yarns.

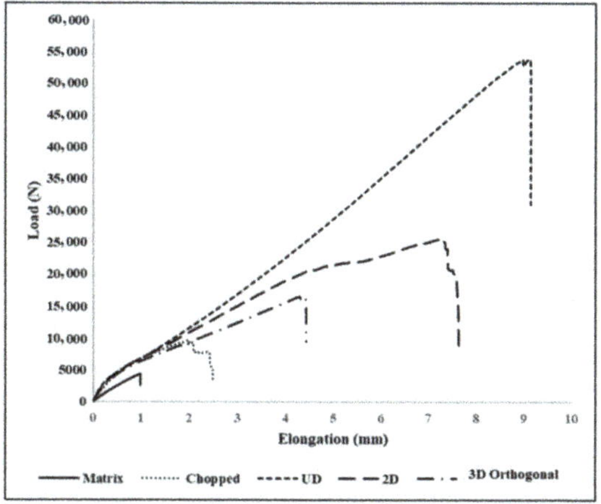
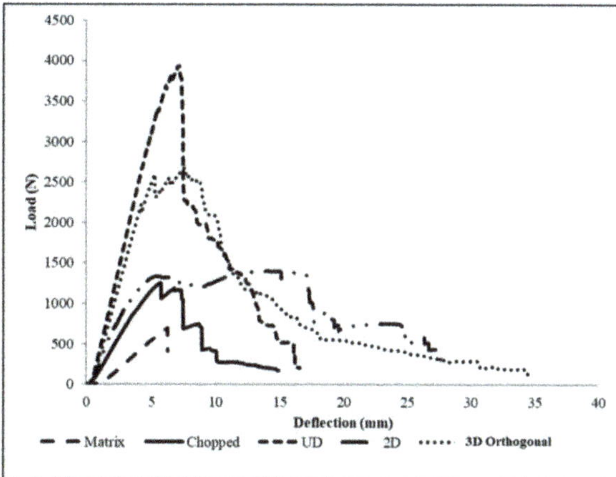

Figure 9. Load-elongation and load-deformation plots of composites reinforced with different reinforcement architecture.

These composites are translucent in nature. Hence, damaged regions of impacted samples become opaque, and internal damage can be visually identified. After testing, a composite's structural observation reveals that delamination is significantly higher in UD and 2D fabric-reinforced composites. The delamination in 3D is negligible due to through-thickness yarns, which will increase the interlaminar shear strength. The 3D fabric has an integrated architecture compared to all other preforms.

The microscopic analysis (side view) of tensile-tested specimens is shown in Figure 10. The side view near the rupture point of UD fabric layers and 2D fabric layers reinforced composites are shown in Figure 10b,c, respectively. It is clear from the images that delamination is the main reason for the failure of these composites. In these UD and 2D fabric-reinforced composites, the interlaminar connection is only by the matrix. This would form distinctive layers in the composites. When the composites are subjected to tensile loading, the interlaminar shear force will be exerted in the matrix region between the fabric layers. As the matrix has very poor shear strength, it will crack very quickly during loading. This crack in the matrix will increase in the loading direction with an increase in the tensile stress and ultimately lead to the composite's failure.

In contrast to these two composites, the 3D fabric-reinforced composite has a single integrated fabric in the reinforcement phase. The through-thickness yarns in the Z-direction have higher shear strength compared to the matrix. The microscopic image of tensile fractured 3D orthogonal fabric-reinforced composite is shown in Figure 10d. These two images show the significance of integrated fabric structure in the reinforcement phase. Hence, integrated 3D preform architecture could be majorly preferred for load-bearing applications.

A close observation reveals that a composite's flexural rigidity reinforced with chopped, 2D and 3D architecture is found to be 60%, 79% and 23% lower than that of a composite with UD fabric reinforcement. Similarly, a composite's flexural stress reinforced with chopped, 2D and 3D architecture is found to be 67%, 63%, and 25% lower, respectively, compared to a UD-reinforced composite. It indicates that strain energy is highest in UD, followed by 3D, 2D and chopped fiber-reinforced composites. This behavior is mainly because of the orientation of all the tows in the longitudinal direction, and also flexural testing is carried out in the warp direction. However, deflection at break is minimum for this UD composite, whereas in 3D fabric, the reinforced composite shows the second-highest energy absorption with a maximum deflection at the break. The 3D fabric composite shows comparatively less energy absorption than the UD fabric because the tows are oriented in three mutually perpendicular directions.

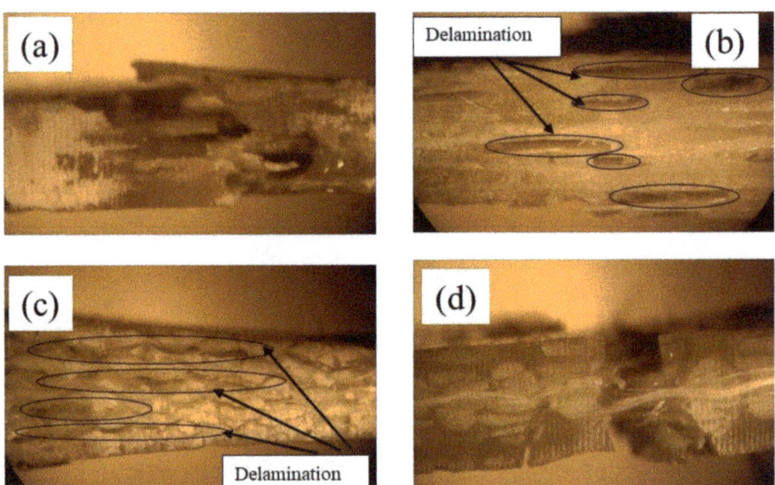

Figure 10. Optical microscope images of the side view of tensile fractured (**a**) chopped fibers; (**b**) UD; (**c**) 2D fabric; and (**d**) 3D orthogonal fabric reinforced composites.

The maximum deflection in the 3D-reinforced composite during three-point bending is shown in Figure 10. It could be observed from this that the 3D composite can withstand maximum load without a fail in the structure. The opaque region was observed around the ruptured zone indicating the delamination in the composite. As the load is applied in the transverse direction, the composite's top layer will undergo compression, and the bottom layer undergoes extension. Hence, interlaminar shear force will come into existence between the layers. Due to the poor strength of the matrix, composites reinforced with UD and 2D fabric layers are more prone to delamination than 3D fabric-reinforced composites. The initialization of matrix crack in the composite during flexural testing could be clearly seen from the microscopic images. The 3D fabric-reinforced composite shows a sharp break during transverse loading. This is because of the higher interlaminar strength between the fiber layers in the structure. The higher interlaminar strength is mainly the result of the yarns in the through-thickness direction. Hence, this composite reinforced with 3D fabric is the better choice in the places of load-bearing applications and crashworthiness.

3.2. Compressional and Flexural Properties of Sandwich Composites

The results of the compressive strength of the different composites are shown in Figure 11. It has been observed that the single-wall rectangular spacer structure shows the highest compressive force compared to TPZ and TR. This is due to the angle of load-bearing walls with respect to the direction of applied load. In the case of TPZ and TR, the effective load-carrying capacity of the connective wall reduces from applied load P to $P\sin\theta$ [38]. In the case of the RECTSL and RECTDL structure, the connecting wall is at a right angle to the face sheet, and therefore it exhibits high load-carrying capacity. However, the SPY structure shows the highest compressive load among all the spacer composites. This is attributed to the uniform distribution of core piles with a density of ~30 piles per square inch. However, in sandwich structures, only two walls take part in load bearing. The compressive strength of the double-wall RECTDL structure is multifold higher than RECTSL due to greater wall thickness. The compressive load-carrying capacity of the double-level TDL45° structure was lower than its single-level structure TPZ45°. Under the applied compressive load, the weak wall buckles sooner than that of the relatively stronger paired wall, which results in a moment at the junction. The mass of the specimen is considered in the calculation of specific compressive strength. However, it does not consider the different volumes of composite specimens, and therefore it cannot be a true representation of compressive performance. Therefore, the strain energy up to maximum compressive load (first peak load) was calculated from the load-deformation curves, and the values were normalized with the volume of the corresponding specimen. The compressional strain energy of the structures was found in order of SPY > RECTDL > TPZ45° > TDL45° > RECTSL > TR. The maximum flexural stress of the sandwich structures was calculated according to the equation below.

$$\text{Maximum flexural stress} = \frac{F_{max}Ly}{4I}$$

where *Fmax* indicates maximum bending load, L is supported span length, y is the distance from the neutral axis, and I is the area moment of inertia. The flexural stress of the sandwich composites was in the order of RECTDL > RECTSL > TR > TPZ > SPY. The flexural stress of the sandwich composite TDL was lower than its TPZ. In the case of the RECTDL and RECTSL structure, the connecting wall's alignment with the face sheet is at a right angle, which helps resist the bending deformation. However, in the case of TPZ and TR, the connecting wall is at an angle to the face sheet; thus, the stress experienced by the wall is less than RECTDL and RECTSL. The quasi-static compression test, results and compression force–displacement curves are shown in Figure 11.

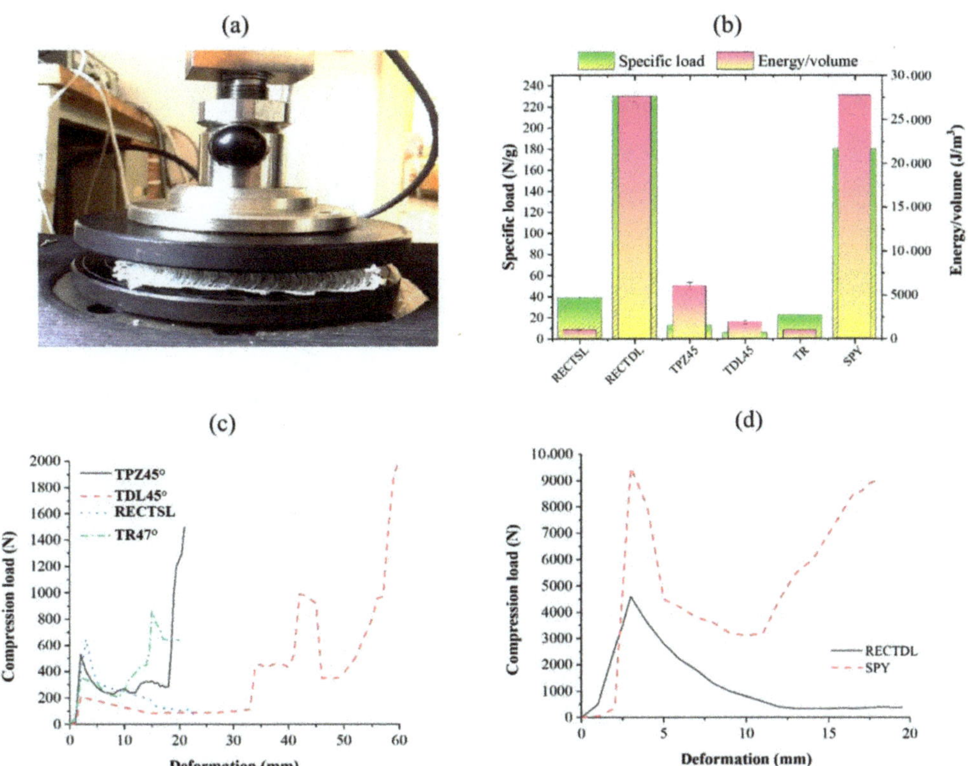

Figure 11. (**a**) Quasi-static compression test, (**b**) specific compressive load and energy/volume of different spacer composites, (**c,d**) compression load–deformation curves of different sandwich composites.

3.3. Compressional and Flexural Properties of Sandwich Composites with Augmented Core Architecture

Spacer fabrics with vertical connecting walls were selected with an intention to increase the equivalent thickness of their connecting walls (by replacing single connecting walls with double-layered connecting walls) in order to achieve enhanced mechanical performance. Under applied compressive stress, the vertical connecting wall's destruction, and thus structural deformation, occurs, which leads to core densification. During densification, the core becomes compacted, which indicates that the structure bears load even after core compaction. The peak load varies with core geometry [39]. In structure S1, the single vertical wall buckles or tilts under compressive load. In structure S4, the horizontally integrated section holds the connecting walls from buckling outwards during initial loading, and, therefore, its compressive strength is higher than structure S1 [40–42]. The compressional energy of S2 was found to be higher than S1. The compressional resistance is a function of core height, and it decreases with core height. The developed augmented structures exhibited better compressional properties than those of conventional materials [47]. The composites were characterized for flexural properties in three-point bending mode. The bending stiffness of the composite material depends on its elastic modulus, area moment of inertia of the cross-section, and length. The bending stiffness of composite structure S1 was found to be higher than that of S2, which is primarily due to additional load-bearing element in S1. The results have clearly shown that the face sheet acts as a weak point of structure under flexural loading, while its core architecture influences the flexural behavior. Figure 12 shows compression load–deformation and flexural load–deflection curves of sandwich composites with an augmented core.

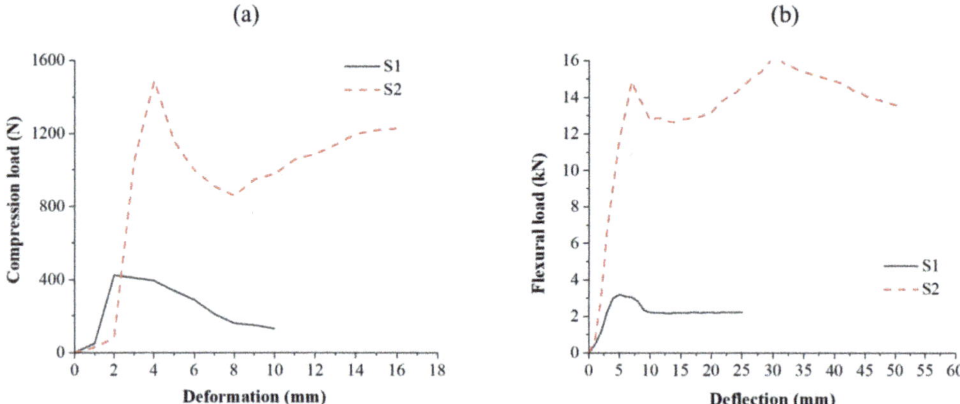

Figure 12. (**a**) Compression load–deformation and (**b**) flexural load–deflection curves of sandwich composites with an augmented core.

3.4. Flexural Properties of Composites Reinforced with an Integrally Woven Stiffener

The developed composites were tested in two modes: (1) stiffener section facing the indenter (SFI), (2) base section facing the indenter (BFI) (Figure 13c). Flexural load–deflection plots are shown in Figure 13a,b. The flexural properties of the developed composites were compared with 2D plain woven polyester fabric-reinforced composites. A higher peak load was observed in BFI mode than SFI. Under the SFI condition, the specimen fails due to the local indentation at the loading point and crippling of the stiffener sections. The stiffener sections, which have higher hollowness, cripple easily and deform under the indenter. A flat T bar with no hollowness exhibits higher flexural load-bearing capacity due to minor crippling and tilting away of stiffener section from the loading axis rather than being structurally deformed. Additionally, only the region of stiffener section which is under the indenter deform during loading and rest structure was observed undeformed. However, a remarkable increment in flexural load-bearing capacity was observed when the specimen was loaded in the BFI condition. This is because BFI condition allows stiffener section to work in coordination with base section, whereas in the SFI condition, the stiffener section deforms quickly. Further, the higher fabric areal density of the base section compared to the stiffener section may also be the reason for the higher flexural carrying capacity in the BFI condition.

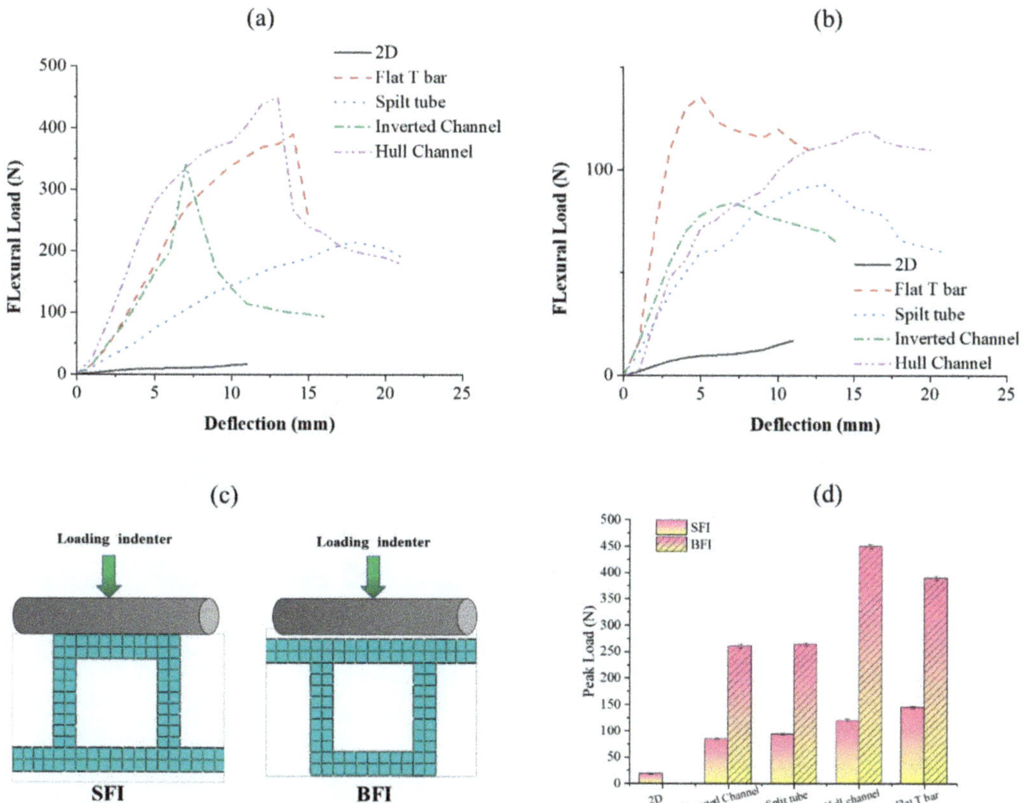

Figure 13. Flexural load–deflection plots of composites with different stiffener sections in (**a**) BFI and (**b**) SFI mode; (**c**) flexural loading configuration of composite samples; and (**d**) comparison of the peak load of different composites.

Figure 13d shows the peak loads of different stiffened structures characterized for flexural properties under SFI and BFI modes. The 2D fabric-reinforced composite shows little increase in flexural load with an increase in deformation. The peak flexural load of stiffened structures was higher than that of 2D fabric-reinforced composites due to the presence of integrated stiffener. The enhanced flexural performance of stiffened structures is due to an increase in the area moment of inertia of structure during bending. The peak flexural load in the BFI condition was 176, 173, 281, and 200% higher than SFI condition for flat T Bar, split tube, hull channel and inverted channel, respectively.

3.5. Junction Strength of Woven U and + Profiled Composites

Figure 14c shows the junction strength of U and + woven profiled structures and their composites. The junction strength of integrated woven U and + profiles is 72 and

43% higher than stitched profiles. The stitch line is a stress concentration point in stitched profiles, and the stitching thread is primarily responsible for load bearing at the junction. Additionally, the stitching causes fabric damage due to a higher needle cutting index (Figure 14a) [54]. However, in integrated woven profiles, the yarns within the structure are responsible for junction strength. Further, the junction strength of integrated woven U and + profile composites is 16 and ~39% higher than the junction strength of corresponding stitched profile composites. The stitched structure had a round corner and thick junction area, which results in a high-stress concentration at the junction. Due to the rounded corner, the stitched structure creates a hollow space around it when gripped in the tensile test jaw, which results in less junction strength. The stitched profile composites under tensile load fail when the applied external force exceeds the stitch strength. In this case, the yarn within the composite does not directly take part in load bearing at the junction. The integrated structure has a neat and clean junction with sharp edges. Further, the tensile stress applied on the composite is transferred to the reinforcement through the matrix, and yarns within the integrated woven composite bear the stress. The failure of the integrated woven composite's joint is primarily due to yarn fracture (Figure 14). Furthermore, it has been observed that the junction strength of stitched U and + profiles after converting them to the composites increases by 81 and 59%, respectively. However, the improvement in the junction strength of integrated woven U and + profiles upon converting to composites is 22 and 54%, respectively.

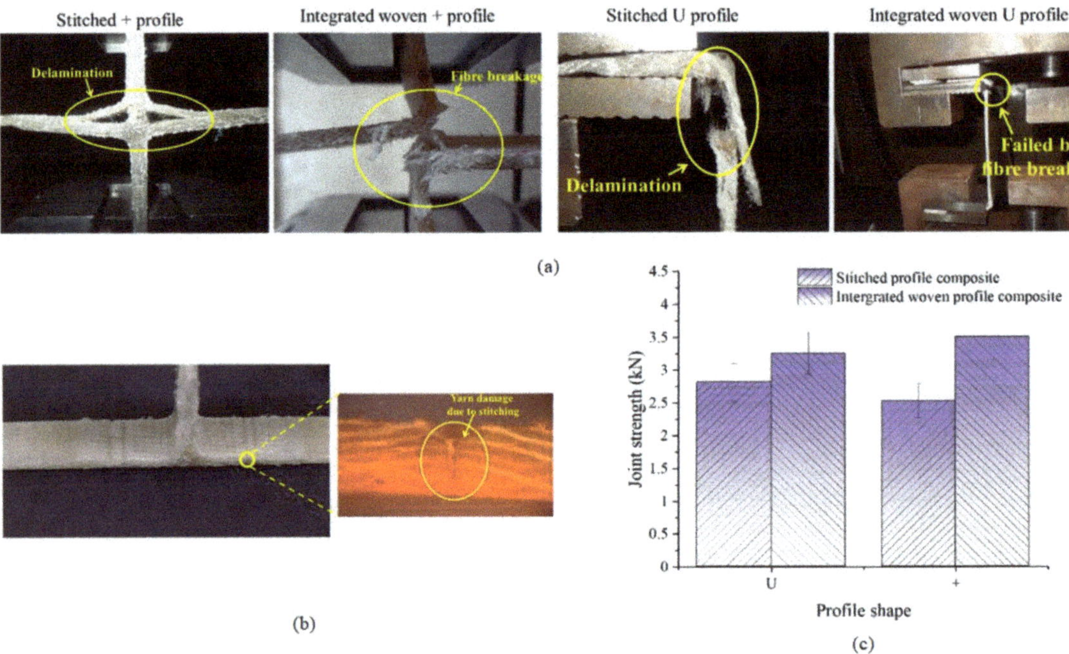

Figure 14. (**a**) Joint strength fractured stitched and integrated woven + and U profiled composite; (**b**) optical microscope image of stitched composite showing yarn damage due to stitching; and (**c**) comparison of joint strength of profiled structures and their composites.

3.6. Drag Force Analysis of Aerodynamic Spacer Structures

The flight conditions are assessed using wind tunnel to study aerodynamical efficiency of a prototype aircraft or wing structure. The wind tunnel is used by spacecraft and aircraft making companies namely Boeing, Northrop Gumman, and NASA, etc. The experimental measurement of the drag force generated on the surface of the airfoil was performed using a lab-scale wind tunnel. The measurement was carried out directly using the principle of the cantilever beam deflection (Figure 15a). The drag force measurement setup is described in Figure 15b.

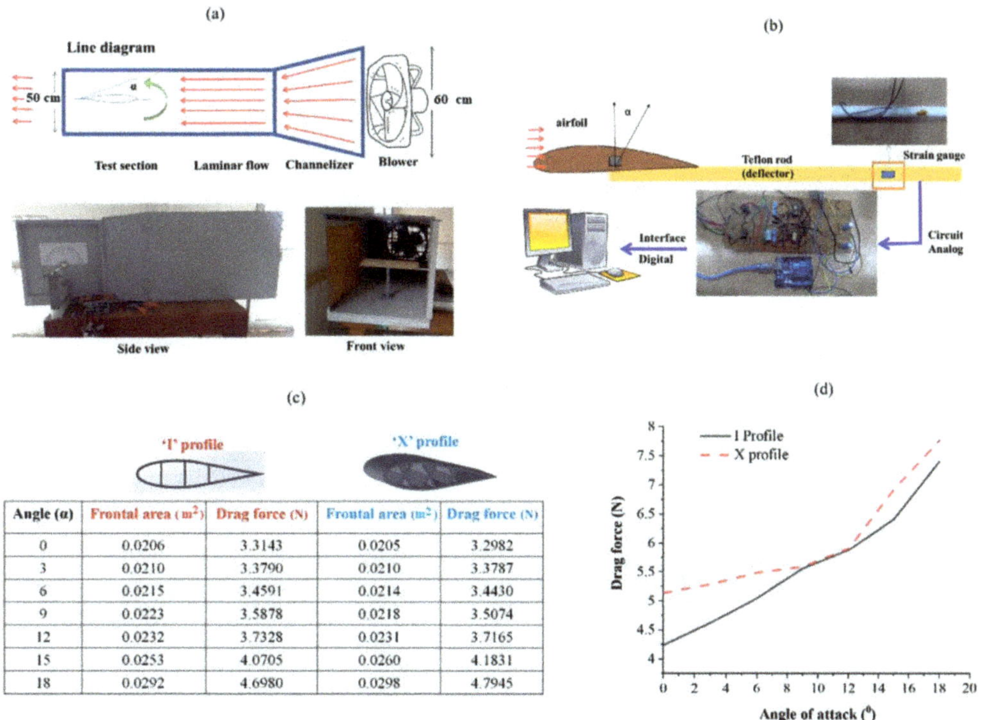

Angle (α)	Frontal area (m^2)	Drag force (N)	Frontal area (m^2)	Drag force (N)
0	0.0206	3.3143	0.0205	3.2982
3	0.0210	3.3790	0.0210	3.3787
6	0.0215	3.4591	0.0214	3.4430
9	0.0223	3.5878	0.0218	3.5074
12	0.0232	3.7328	0.0231	3.7165
15	0.0253	4.0705	0.0260	4.1831
18	0.0292	4.6980	0.0298	4.7945

Figure 15. (**a**) Line diagram and actual wing tunnel developed in the laboratory; (**b**) drag force measurement set up installed associated with wind tunnel; (**c**) frontal areas considered, and theoretical drag force calculated at their corresponding angle of attack; (**d**) drag force at a different angle of attack for 'I' and 'X' profiles.

The drag of an object moving in a fluid medium is a function of density, velocity, compressibility and viscosity of the air, the size and shape of the body, and inclination of the body to flow. Therefore, the measurement of drag becomes complex and thus it is necessary to characterize the dependence by a single variable. For drag, this variable is called the drag coefficient (C_d). The drag (D) is calculated as $0.5C_d A \gamma V^2$. Where γ and V are density and velocity of air, A is the reference area. The airfoil profile considered in this work is basically the symmetrical airfoils, and the coefficient of drag for a symmetrical airfoil is considered to be around 0.045 from the previous literature. The Area (A) given in the equation refers to the frontal area of the object that is perpendicular to the direction of the fluid flow at a particular angle of attack. The values of density and velocity of the air medium considered for the calculation are 1.223 Kg/m^3 and 43 m/s, respectively. The corresponding drag force of the wing structures calculated at various angles of attack is tabulated in Figure 15c. For airfoils, at small angles the value of drag is small. With an increase in angle of attack above 5 degrees, the frontal area increases, and thus the boundary layer thickness also increases. The drag force exponentially increases with the angle of attack due to an increase in the frontal area of the wing that tries to resist the flow (Figure 15d).

3.7. Compressive Performance of Woven Honeycomb Composites

The compressive load and energy per unit volume of different honeycomb structures are shown in Figure 16a,b. The compressive load-carrying capacity of the honeycomb composite increases with cell size. This is due to the different sizes of the specimen tested under flatwise compression of the composite. According to ASTM C365, a square shape specimen is required for flatwise compression of honeycomb. The honeycomb cell dimensions increase with a number of picks in the free and bonded wall. Thus, the specimen size increases with cell size, which results in increased load-carrying capacity. However, the strain energy per unit volume decreases linearly with an increase in honeycomb cell size. This is attributed to an increase in specimen volume with honeycomb cell size.

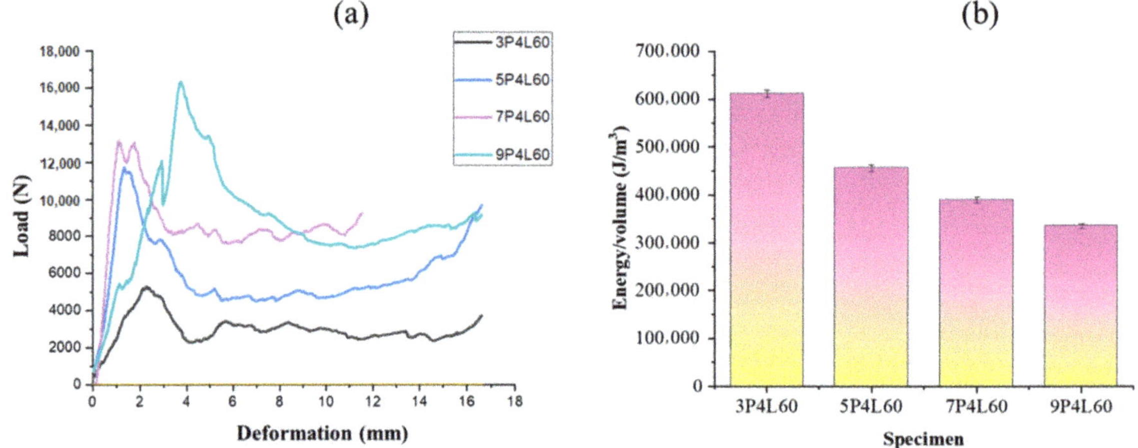

Figure 16. (**a**) Compressive load-deformation plots; and (**b**) energy/volume of honeycomb composites.

3.8. Mechanical Properties of Waste Cotton-Based Composite Laminates

Figure 17a,b shows the tensile stress–strain and flexural stress–deformation plots of textile waste-based composite laminates. It can be observed that the tensile strength and Young's modulus of composite specimen Wb are 43 and 17%, respectively, lower than SH. The lower tensile strength of composite specimen Wb is due to ~72% of the reinforcement's total weight within a composite being occupied by woven preform, and ~50% of yarns within the woven preform are not in the loading direction. However, the tensile strength of the WbUD composite is nearly the same as SH due to all the yarns within the skin layers are in the loading direction, and its Young's modulus is ~73% higher than SH. When tensile stress is applied on the WbUD composite, the outer layer initially bears the stress transferred by the matrix due to its high modulus (unidirectional yarn placement) compared to core material. Upon the outer layer fracture, the load is transferred to cotton fibers at the core, and the complete composite fails when the applied stress exceeds the bearing stress of the cotton fibers at the core. The composite's tensile strength improves upon stitching due to the enhanced interface between the layers.

The tensile strength and Young's modulus of the WbH composite is 74% and ~183% higher than the SH composite. This is attributed to high-modulus glass yarn in the loading direction. Further, under the tensile loading of composite specimen WbH, the glass filaments initially bear the stress transferred by the matrix due to its high modulus and low elongation. Upon fracture of glass filaments, the cotton fibers at the core experience the stress transferred by the matrix and fail when the applied stress exceeds its breaking stress. In contrast, tensile strength and Young's modulus of WbH composite is ~79% and ~63% higher than the WbUD composite.

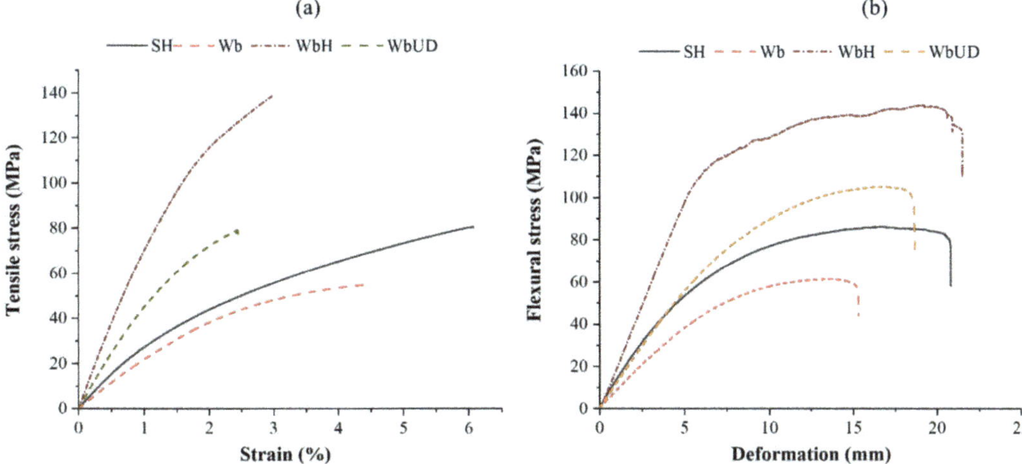

Figure 17. (**a**) Tensile stress–strain curves; and (**b**) flexural stress–deformation curves of composite laminates.

The flexural strength and flexural modulus of the composite specimen Wb are 40 and 66%, respectively, lower than composite specimen SH. According to sandwich panel theory, when the composite is under three-point bending, the top layer is put into the compressional load, and the bottom into tension, whereas the core is into shear. The laminated composite's flexural strength and stiffness are controlled by fiber type and its orientation at the composite skin [44,45]. The core is supposed to support the skin to reduce the maximum stress and deformation of the outer layer. The lower flexural strength of composite specimen Wb is due to the early failure of the woven fabric layer at the tension side [46]. However, when UD preform is used at the skin, the flexural strength and modulus of the WbUD composites increase by ~26% and ~74%, respectively, compared to SH. In the case of WbUD, all yarns within the skin layer take part in load-bearing. In contrast, when the 2D woven preform is used as skin, only half of the yarns within the preform take part in bearing tensile load generated at the tension side. The presence of high-strength glass fiber at the skin increases the load-bearing capacity of composite specimen WbH at the tension side, which results in its high flexural strength. The composite specimen WbH has ~68% higher flexural strength than SH.

The impact strength is the energy needed to fracture a composite specimen when subjected to impact loading [47,48]. The izod impact strength of Wb was ~40% higher than composite specimen SH. This was attributed to the high fracture toughness of the cotton yarns present at the skin layer. However, when all yarns within the skin layer are laid unidirectionally, as in composite specimen WbUD, the izod impact strength increases by ~72% than SH. This is due to the increased fracture toughness of the composite skin. The composite specimen WbH shows ~537% higher impact strength than SH. This is attributed to the high fracture toughness of glass filaments present within the skin of composite specimen WbH.

4. Conclusions

Fiber reinforced composites have emerged as viable structural materials due to their advantageous stiffness, thermal expansion, strength and density properties. These composites have a high modulus of elasticity, high resistance to fatigue failure, and good resistance to corrosion and they are increasingly used to replace traditional materials such as wood and metals such as steel, iron and aluminum. However, the strength of fiber-reinforced composites in a direction perpendicular to the fibers is extremely low compared with the strength along the length of fibers. The design of components made from these composites is complex and the manufacturing and testing of components are highly specialized. Conventional 2D woven fabrics have several disadvantages regarding the design of certain composite products which include anisotropy, limited conformability, poor in-plane shear resistance, difficulty in handling of open constructions, and reduced yarn to fabric tensile translation efficiency due to yarn crimp and crimp interchange. Three-dimensional weaving, on the other hand, can produce near-net-shaped preforms with complex geometry those are less expensive when converted into composites. Three-dimensional weaving allows the tailoring of properties for specific applications and the composites made out of them show better delamination resistance and damage tolerance, higher tensile strain-to-failure values and high interlaminar fracture toughness properties. Composites reinforced with net-shaped three-dimensional (3D) fabric preforms have emerged as a viable option for parts such as stiffeners and stringers. Three-dimensional weaving also made it possible to develop a wide range of air foils with desired aerodynamic behavior and high crossing strength. The driving forces for using 3D fabrics as reinforcement in composite materials includes the option of using different types of yarns in different directions, flexible fiber orientation and fabric architecture, higher impact tolerance and lower manufacturing costs due to reduced labor intensity in the manufacturing processes. It is established that the number of crossover points in the weave structures offered excellent association with the impact energy absorption and formability behavior which are important for many applications, including automobiles, wind energy, marine and aerospace. Mechanical characterization of 3D woven honeycomb composites with different cell sizes, opening angles and wall lengths revealed that the specific compression energy is higher for regular honeycomb structure with smaller cell sizes and a greater number of layers keeping constant thickness.

Author Contributions: Z.K., R.K.M., B.K.B., M.T., V.K. and M.M., experiment design; Z.K., R.K.M., B.K.B., M.T., V.K. and M.M., methodology; Z.K., R.K.M., M.T. and V.K., testing of mechanical properties and data analysis; Z.K., R.K.M. and B.K.B. wrote and edited the paper; B.K.B., R.K.M., M.T., V.K. and M.M., project administration; Z.K., R.K.M. and B.K.B., language correction; R.K.M., B.K.B., M.T., V.K. and M.M., resources; R.K.M., communication with the editors; R.K.M., B.K.B. and M.M., supervision. All authors have read and agreed to the published version of the manuscript.

Funding: Supported by the Internal grant agency of Faculty of Engineering no. 2021:31140/1312/3108 "Experimental research of hybrid adhesive bonds with multilayer sandwich construction, Czech University of Life Sciences Prague".

Institutional Review Board Statement: Not applicable.

Informed Consent Statement: Not applicable.

Data Availability Statement: Not applicable.

Conflicts of Interest: The authors declare no conflict of interest.

References

1. Bannister, M.K. Development and application of advanced textile composites. *Proc. Inst. Mech. Eng. Part L J. Mater. Des. Appl.* **2004**, *218*, 253–260. [CrossRef]
2. Karaduman, N.S.; Karaduman, Y.; Ozdemir, H.; Ozdemir, G. Textile Reinforced Structural Composites for Advanced Applications. In *Textiles for Advanced Applications*; Intech Open: London, UK, 2017. [CrossRef]
3. Behera, B.K.; Dash, B.P. Mechanical behavior of 3D woven composites. *Mater. Des.* **2015**, *67*, 261–271. [CrossRef]
4. Patel, D.K.; Waas, A.M.; Yen, C.F. Compressive response of hybrid 3D woven textile composites (H3DWTCs): An experimentally validated computational model. *J. Mech. Phys. Solids* **2019**, *122*, 381–405. [CrossRef]
5. Dash, A.K.; Behera, B.K. Weave design sspects of 3D textile preforms towards mechanical properties of their composites. *Fibers Polym.* **2019**, *20*, 2146–2155. [CrossRef]
6. Patel, D.K.; Waas, A.M.; Yen, C.F. Direct numerical simulation of 3D woven textile composites subjected to tensile loading: An experimentally validated multiscale approach. *Compos. Part B Eng.* **2018**, *152*, 102–115. [CrossRef]
7. Mishra, R.; Gupta, N.; Pachauri, R.; Behera, B.K. Modelling and simulation of earthquake resistant 3D woven textile structural concrete composites. *Compos. Part B Eng.* **2015**, *81*, 91–97. [CrossRef]

8.   Moshtaghian, Z.; Hasani, H.; Zarrebini, M.; Shirazi, M.P. Development and auxetic characterization of 3D composites produced with newly-designed multi-cell flat-knitted spacer fabrics. *J. Ind. Text.* **2020**, 1528083720971696. [CrossRef]

9.   Bunzel, F.; Wisner, G.; Stammen, E.; Dilger, K. Structural sandwich composites out of wood foam core and textile reinforced concrete sheets for versatile and sustainable use in the building industry. *Mater Today Proc.* **2020**, *31*, S296–S302. [CrossRef]

10.   Manjunath, R.N.; Behera, B.K.; Mawkhlieng, U. Flexural stability analysis of composite panels reinforced with stiffener integral woven preforms. *J. Text. Inst.* **2019**, *110*, 368–377. [CrossRef]

11.   Torre, L.; Kenny, J.M. Impact testing and simulation of composite sandwich structures for civil transportation. *Compos. Struct.* **2000**, *50*, 257–267. [CrossRef]

12.   Shin, K.B.; Lee, J.Y.; Cho, S.H. An experimental study of low-velocity impact responses of sandwich panels for Korean low floor bus. *Compos. Struct.* **2008**, *84*, 228–240. [CrossRef]

13.   Tripathi, L.; Neje, G.; Behera, B.K. Geometrical modeling of 3D woven honeycomb fabric for manufacturing of lightweight sandwich composite material. *J. Ind. Text.* **2020**, 152808372093147. [CrossRef]

14.   Li, D.; Zhao, C.; Jiang, N.; Jiang, L. Fabrication, properties and failure of 3D integrated woven spacer composites with thickened face sheets. *Mater. Lett.* **2015**, *148*, 103–105. [CrossRef]

15.   Li, D.; Zhao, C.; Jiang, L.; Jiang, N. Experimental study on the bending properties and failure mechanism of 3D integrated woven spacer composites at room and cryogenic temperature. *Compos. Struct.* **2014**, *111*, 56–65. [CrossRef]

16.   Kamble, Z.; Behera, B.K.; Mishra, R.; Behera, P.K. Influence of cellulosic and non-cellulosic particle fillers on mechanical, dynamic mechanical, and thermogravimetric properties of waste cotton fiber reinforced green composites. *Compos. Part B Eng.* **2021**, *207*, 1–10. [CrossRef]

17.   Mouritz, A.P.; Bannister, M.K.; Falzon, P.J.; Leong, K.H. Review of applications for advanced three-dimensional fiber textile composites. *Compos. Part A Appl. Sci. Manuf.* **1999**, *30*, 1445–1461. [CrossRef]

18.   Neje, G.; Behera, B.K. Investigation of mechanical performance of 3D woven spacer sandwich composites with different cell geometries. *Compos. Part B Eng.* **2019**, *160*, 306–314. [CrossRef]

19.   Pirouzfar, S.; Zeinedin, A. Effect of geometrical parameters on the flexural properties of sandwich structures with 3D-printed honeycomb core and E-glass/epoxy Face-sheets. *Structures* **2021**, *33*, 2724–2738. [CrossRef]

20.   Song, S.; Xiong, C.; Zheng, J.; Yin, J.; Zou, Y.; Zhu, X. Compression, bending, energy absorption properties, and failure modes of composite Kagome honeycomb sandwich structure reinforced by PMI foams. *Compos. Struct.* **2021**, *277*, 114611. [CrossRef]

21.   Manjunath, R.N.; Khatkar, V.; Behera, B.K. Influence of augmented tuning of core architecture in 3D woven sandwich structures on flexural and compression properties of their composites. *Adv. Compos. Mater.* **2020**, *29*, 317–333. [CrossRef]

22.   Fan, H.; Zhou, Q.; Yang, W.; Jingjing, Z. An experiment study on the failure mechanisms of woven textile sandwich panels under quasi-static loading. *Compos. Part B Eng.* **2010**, *41*, 686–692. [CrossRef]

23.   Zhao, C.; Li, D.S.; Ge, T.Q.; Jiang, L.; Jiang, N. Experimental study on the compression properties and failure mechanism of 3D integrated woven spacer composites. *Mater. Des.* **2014**, *56*, 50–59. [CrossRef]

24.   Vuure, A.W.; Ivens, J.A.; Verpoest, I. Mechanical properties of composite panels based on woven sandwich-fabric preforms. *Compos. Part A Appl. Sci. Manuf.* **2000**, *31*, 671–680. [CrossRef]

25.   Karahan, M.; Ulcay, Y.; Eren, R.; Karahan, N.; Kaynak, G. Investigation into the tensile properties of stitched and unstitched woven Aramid/Vinyl Ester composites. *Text. Res. J.* **2010**, *80*, 880–891. [CrossRef]

26.   Ahmed, K.S.; Vijayarangan, S. Tensile, flexural and interlaminar shear properties of woven jute and jute-glass fabric reinforced polyester composites. *J. Mater. Process. Technol.* **2008**, *207*, 330–335. [CrossRef]

27.   Kamble, Z.; Behera, B.K.; Kimura, T.; Haruhiro, I. Development and characterization of thermoset nanocomposites reinforced with cotton fibers recovered from textile waste. *J. Ind. Text.* **2020**, 1–27. [CrossRef]

28.   Lee, H.; Kureemun, U.; Ravandi, M.; Teo, W.S. Performance of interlaminar flax-carbon hybrids under bending. *Procedia Manuf.* **2020**, *43*, 658–665. [CrossRef]

29.   Jawaid, M.; Abdul Khalil, H.P.S.; Abu Bakar, A. Mechanical performance of oil palm empty fruit bunches/jute fibers reinforced epoxy hybrid composites. *Mater. Sci. Eng. A* **2010**, *527*, 7944–7949. [CrossRef]

30.   Dau, F.; Dano, M.L.; Vérone, B.; Girardot, J.; Aboura, Z.; Morvan, J.M. In-plane and out-of-plane characterization of a 3D angle interlock textile composite. *Compos. Part A Appl. Sci. Manuf.* **2021**, *149*, 106581. [CrossRef]

31.   Liu, T.; Fan, W.; Wu, X. Comparisons of influence of random defects on the impact compressive behavior of three different textile structural composites. *Mater. Des.* **2019**, *181*, 108073. [CrossRef]

32.   Ladani, R.B.; Wang, C.H.; Mouritz, A.P. Delamination fatigue resistant three-dimensional textile self-healing composites. *Compos. Part A Appl. Sci. Manuf.* **2019**, *127*, 105626. [CrossRef]

33.   Wondmagegnehu, B.T.; Paramasivam, V.; Selvaraj, S.K. Fabricated and analyzed the mechanical properties of textile waste/glass fiber hybrid composite materiál. *Mater. Today Proc.* **2021**, *46*, 7297–7303. [CrossRef]

34.   Zhang, J.; Zhang, W.; Huang, S.; Gu, B. An experimental–numerical study on 3D angle-interlock woven composite under transverse impact at subzero temperatures. *Compos. Struct.* **2021**, *268*, 113936. [CrossRef]

35.   Flores-Johnson, E.A.; Li, Q.M. Experimental study of the indentation of sandwich panels with carbon fibre-reinforced polymer face sheets and polymeric foam core. *Compos. Part B Eng.* **2011**, *42*, 1212–1219. [CrossRef]

36.   Conejos, F.; Balmes, E.; Tranquart, B.; Monteiro, E.; Martin, G. Viscoelastic homogenization of 3D woven composites with damping validation in temperature and verification of scale separation. *Compos. Struct.* **2021**, *275*, 114375. [CrossRef]

37. Yan, S.; Zeng, X.; Long, A. Effect of fibre architecture on tensile pull-off behaviour of 3D woven composite T-joints. *Compos. Struct.* **2020**, *242*, 112194. [CrossRef]
38. Jiao, W.; Chen, L.; Xie, J.; Yang, Z.; Fang, J.; Chen, L. Effect of weaving structures on the geometry variations and mechanical properties of 3D LTL woven composites. *Compos. Struct.* **2020**, *252*, 112756. [CrossRef]
39. Ruggles-Wrenn, M.B.; Alnatifat, S.A. Fully-reversed tension-compression fatigue of 2D and 3D woven polymer matrix composites at elevated temperature. *Polym. Testing* **2021**, *97*, 107179. [CrossRef]
40. Nayak, S.Y.; Shenoy, B.S.; Sultan, M.T.B.; Kini, C.R.; Shenoy, K.R.; Acharya, A.; Jaideep, J.R. Influence of stacking sequence on the mechanical properties of 3D E-glass/bamboo non-woven hybrid epoxy composites. *Mater. Today Proc.* **2021**, *38*, 2431–2438. [CrossRef]
41. Li, Z.; Li, D.; Zhu, H.; Guo, Z.; Jiang, L. Mechanical properties prediction of 3D angle-interlock woven composites by finite element modeling method. *Mater. Today Commun.* **2020**, *22*, 100769. [CrossRef]
42. Guo, Q.; Zhang, Y.; Guo, R.; Ma, M.; Chen, L. Influences of weave parameters on the mechanical behavior and fracture mechanisms of multidirectional angle-interlock 3D woven composites. *Mater. Today Commun.* **2020**, *23*, 100886. [CrossRef]
43. Zeng, C.; Liu, L.; Bian, W.; Leng, J.; Liu, Y. Bending performance and failure behavior of 3D printed continuous fiber reinforced composite corrugated sandwich structures with shape memory capability. *Compos. Struct.* **2021**, *262*, 113626. [CrossRef]
44. Zheng, T.; Li, S.; Wang, G.; Hu, Y.; Zhao, C. Mechanical and energy absorption properties of the composite XX-type lattice sandwich structure. *Eur. J. Mech.-A/Solids* **2022**, *91*, 104410. [CrossRef]
45. Chene, D.; Yan, R.; Lu, X. Mechanical properties analysis of the naval ship similar model with an integrated sandwich composite superstructure. *Ocean Eng.* **2021**, *232*, 109101. [CrossRef]
46. Reddy, C.N.; Rajeswari, C.; Malyadri, T.; Hari, S.N.S. Effect of moisture absorption on the mechanical properties of jute/glass hybrid sandwich composites. *Mater. Today Proc.* **2021**, *45*, 3307–3311. [CrossRef]
47. Khalili, S.; Khalili, S.M.R.; Farsani, R.; Mahajan, P. Flexural properties of sandwich composite panels with glass laminate aluminum reinforced epoxy facesheets strengthened by SMA wires. *Polym. Testing* **2020**, *89*, 106641. [CrossRef]
48. Flores-Johnson, E.A.; Li, Q.M. Structural behaviour of composite sandwich panels with plain and fibre-reinforced foamed concrete cores and corrugated steel faces. *Compos. Struct.* **2012**, *94*, 1555–1563. [CrossRef]
49. Caglayan, C.; Gurkan, I.; Gungor, S.; Cebeci, H. The effect of CNT-reinforced polyurethane foam cores to flexural properties of sandwich composites. *Compos. Part A Appl. Sci. Manuf.* **2018**, *115*, 187–195. [CrossRef]
50. Bharath, H.S.; Bonthu, D.; Gururaj, S.; Prabhakar, P.; Doddamani, M. Flexural response of 3D printed sandwich composite. *Compos. Struct.* **2021**, *263*, 113732. [CrossRef]
51. Nanayakkara, A.; Feih, S.; Mouritz, A.P. Experimental analysis of the through-thickness compression properties of z-pinned sandwich composites. *Compos. Part A Appl. Sci. Manuf.* **2011**, *42*, 1673–1680. [CrossRef]
52. Mishra, R.; Huang, J.; Kale, B.; Zhu, G.; Wang, Y. The production, characterization and applications of nanoparticles in the textile industry. *Text. Prog.* **2014**, *46*, 133–226. [CrossRef]
53. Wei, X.; Wu, Q.; Gao, Y.; Yang, Q.; Xiong, J. Composite honeycomb sandwich columns under in-plane compression: Optimal geometrical design and three-dimensional failure mechanism maps. *Eur. J. Mech.-A/Solids* **2022**, *91*, 104415. [CrossRef]
54. Hassanzadeh, S.; Hasani, H.; Zarrebini, M. Thermoset composites reinforced by innovative 3D spacer weft-knitted fabrics with different cross-section profiles: Materials and manufacturing process. *Compos. Part A Appl. Sci. Manuf.* **2016**, *91*, 65–76. [CrossRef]
55. Guo, J.; Wen, W.; Zhang, H.; Cui, H. Warp-loaded mechanical performance of 3D orthogonal layer-to-layer woven composite perforated structures with different apertures. *Compos. Struct.* **2021**, *278*, 114720. [CrossRef]

Article

3D Woven Textile Structural Polymer Composites: Effect of Resin Processing Parameters on Mechanical Performance

Rajesh Kumar Mishra [1,*], Michal Petru [2], Bijoya Kumar Behera [3] and Promoda Kumar Behera [2]

[1] Department of Material Science and Manufacturing Technology, Faculty of Engineering, Czech University of Life Sciences Prague, Kamycka 129, 16500 Prague, Czech Republic

[2] Department of Machinery Construction, Faculty of Mechanical Engineering, Technical University of Liberec, Studentska 1402/2, 46117 Liberec, Czech Republic; michal.petru@tul.cz (M.P.); promodabehera@gmail.com (P.K.B.)

[3] Department of Textile & Fiber Engineering, Indian Institute of Technology Delhi, New Delhi 110016, India; behera@textile.iitd.ac.in

* Correspondence: mishrar@tf.czu.cz

Abstract: This work presents the manufacture of polymer composites using 3D woven structures (orthogonal, angle interlock and warp interlock) with glass multifilament tows and epoxy as the resin. The mechanical properties were analyzed by varying the processing parameters, namely, add-on percentage, amount of hardener, curing time, curing temperature and molding pressure, at four different levels during the composite fabrication for three different 3D woven structures. The mechanical properties of composites are affected by resin infusion or resin impregnation. Resin infusion depends on many processing conditions (temperature, pressure, viscosity and molding time), the structure of the reinforcement and the compatibility of the resin with the reinforcement. The samples were tested for tensile strength, tensile modulus, impact resistance and flexural strength. Optimal process parameters were identified for different 3D-woven-structure-based composites for obtaining optimal results for tensile strength, tensile modulus, impact resistance and flexural strength. The tensile strength, elongation at break and tensile modulus were found to be at a maximum for the angle interlock structure among the various 3D woven composites. A composition of 55% matrix (including 12% of hardener added) and 45% fiber were found to be optimal for the tensile and impact performance of 3D woven glass–epoxy composites. A curing temperature of about 140 °C seemed to be optimal for glass–epoxy composites. Increasing the molding pressure up to 12 bar helped with better penetration of the resin, resulting in higher tensile strength, modulus and impact performance. The optimal conditions for the best flexural performance in 3D woven glass–epoxy composites were 12% hardener, 140 °C curing temperature, 900 s curing time and 12 bar molding pressure.

Keywords: textile structural composite; epoxy resin; add-on (%); amount of hardener (%); curing temperature; curing time; molding pressure; mechanical properties

Citation: Mishra, R.K.; Petru, M.; Behera, B.K.; Behera, P.K. 3D Woven Textile Structural Polymer Composites: Effect of Resin Processing Parameters on Mechanical Performance. *Polymers* **2022**, *14*, 1134. https://doi.org/10.3390/polym14061134

Academic Editor: Marcin Masłowski

Received: 14 February 2022
Accepted: 9 March 2022
Published: 11 March 2022

Publisher's Note: MDPI stays neutral with regard to jurisdictional claims in published maps and institutional affiliations.

1. Introduction

Composites can be defined as a selected combination of different materials formed with a specific geometrical structure and with a specific external shape or form. These are materials created of two or more different components (or phases) that are distinguishable and separated by an interface. One of these dissimilar materials is called the reinforcement and the other component is called the matrix. A textile structural composite is a polymer composite that is defined as the combination of a resin system with a textile-fiber-, yarn- or fabric-based reinforcement system. Textile-fabric-reinforced composites may be flexible or quite rigid [1,2]. Having a non-crimp 3D fabric as a composite reinforcement is obviously beneficial because of the significantly higher in-plane stiffness and strength. A single-layer preform or a few layers of such a preform combined with an advanced, automatically controlled closed-mold resin infusion system allows for avoiding a lot of flaws and irregularities. The fabric structural parameters, design and preparation of preforms, etc., influence the resin impregnation [3].

A single-layer, relatively thick, 3D woven composite is often better than a much thinner single layer 2D woven composite or a 2D woven laminate that does not match the thickness of the 3D woven composite. Quite often, composites made using vacuum-assisted resin transfer molding (VARTM) are compared to those made via compression molding under much higher pressure (resulting in significantly higher fiber volume fraction). Furthermore, the in-plane fiber architecture of the compared 3D weave and 2D weave composites may be significantly different. Moreover, factors such as the resin system, composite processing cycle and environmental conditions may be different. Any of these factors would produce a significantly quantifiable effect on the resulting properties of composite materials [4–6].

The mechanical properties of a laminated fiber-reinforced composite suffers significantly by a void content exceeding 2%. The presence of voids in a 3D reinforced composite materials, owing to the reinforcement in the thickness direction, would not affect their properties as much as in unidirectional or 2D composites, such as laminates. For woven fabrics, the out-of-plane orientations of the warp and weft yarns at their cross-over points are taken into account in an approximate way by representing the curved yarns as a set of straight yarn segments [7–9].

Three-dimensional woven fabrics exhibit a higher thickness and superior inter-laminar properties compared to laminated 2D woven fabrics because of their integrated internal geometry. While traditional laminated composites constitute the vast majority of applications, 3D woven composites have certain advantages over laminates because of the controlled orientation of the fibers in all three directions. Three-dimensional woven composites have superior through-thickness strength, stiffness and thermal conductivity compared to conventional 2D laminated composites. In addition, the interlocking nature of a 3D woven fiber preform leads to superior damage tolerance and fatigue behavior [10–12]. Liquid composite molding (LCM) processes, such as resin transfer molding (RTM), are identified as one of the potentially most advantageous manufacturing techniques. Textile preforms have several length scales, starting from the individual fibers making up the fiber bundles, which are woven or stitched together to create the fabric [13,14]. In effect, fabric reinforcement exhibits heterogeneous behavior, whereby the fiber bundles are porous entities with free open spaces between them.

The purpose of the polymer matrix is to bind the fibers or reinforcement together by virtue of its cohesive and adhesive characteristics, transfer the load to the reinforcing fibers and protect them from environments and handling. When unsaturated polyester resins are cured, the monomer reacts with the unsaturated sites on the polymer, converting it to a solid thermoset structure. The ester groups of vinyl ester are susceptible to water degradation via hydrolysis, which means that vinyl esters exhibit better resistance to water and many other chemicals than their polyester counterparts and are frequently found in applications such as pipelines and chemical storage tanks. The viscosity of the resin/hardener combinations limits the selection of a resin with a viscosity suited to application equipment and fabric weave and weight. A hardener that provides adequate curing time based on the ambient temperature in the environment must be selected [15–18].

The mechanical properties of polymer composites can be affected by resin infusion or resin impregnation. Resin infusion depends on many parameters, such as the processing conditions (temperature, pressure, viscosity and molding time), the structure of the reinforcement and the compatibility of the resin with the reinforcement. As viscosity increases, the resin infusion decreases, and as applied pressure increases, the infusion also increases. It reduces the add-on of the resin/matrix and removes air from inside the structure to reduce the void formation. It is therefore essential to optimize these parameters. The curing temperature also plays a very important role in resin impregnation, as it may cause degradation of the resin and fiber. The curing temperature and time needs to be optimized to achieve a requisite level of impregnation [19–25].

The principal objectives of this research were to produce 3D woven constructions using high-performance glass fiber; characterize the 3D structures with respect to thickness, fiber volume fraction and mechanical behavior; and optimize the resin application and curing parameters for each type of 3D woven fabric in order to achieve the desired mechanical properties.

2. Materials and Methods

2.1. Materials

Continuous multifilament glass tows (Saint-Gobain ADFORS Ltd., Litomyšl, Czech Republic) with a linear density of 600 tex were used to prepare various 3D woven fabrics. Araldite® LY 556 epoxy resin (SPOLCHEMIE in Ústí nad Labem, Czech Republic) was used as a matrix for composite preparation. It is an amine-cured laminating system without reactive diluents that shows excellent flexibility and high reactivity. It is a clear, pale yellow liquid in visual appearance. The density of this resin is 1.15–1.20 g/cm^3 with a viscosity of 10,000–12,000 cps at 25 °C. A hardener (CHS-HARDENER P11) was mixed with epoxy resin as per the manufacturer (SPOLCHEMIE in Ústí nad Labem, Czech Republic) guidelines with various ratios (8%, 10%, 12% and 14% of the matrix weight) and stirred well for uniform mixing.

2.2. Methods

2.2.1. Three-Dimensional Fabric Manufacturing

There are various methods of manufacturing 3D woven fabrics according to the fabric structure. The CCI sample weaving machine was used for this purpose. Some modifications in the conventional loom were made in order to weave a 3D fabric with a 2D loom. A separate negative let-off arrangement was made behind the loom to hold the binder beam while preparing 3D woven fabrics [26,27]. Three types of solid 3D woven structures, viz. orthogonal, angle interlock and warp interlock fabrics, were prepared on the same machine. The beams required to feed this loom were prepared on the CCI sample warping machine. The principle of 3D weaving on a 2D weaving machine is shown in Figure 1.

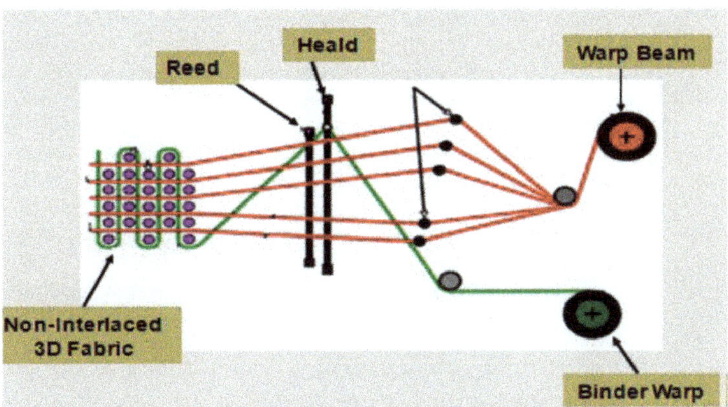

Figure 1. Principle of 3D weaving on a 2D weaving machine.

The weaving machine parameters are given in Table 1.

The schematic arrangement of constituent filaments/tows in the 3D woven fabrics is shown in Figure 2.

Table 1. Parameters of the weaving machine.

Sample Type	No. of Beams	No. of Healds	Warps (cm^{-1})	Wefts (cm^{-1})	Fabric Areal Density (g·m^{-2})	Machine Speed (rpm)	Machine Efficiency (%)
3D orthogonal	2	5	10	10	1200 ± 20	600	80
3D angle interlock	2	8	10	10	1200 ± 20	550	73
3D warp interlock	2	6	10	10	1200 ± 20	570	75

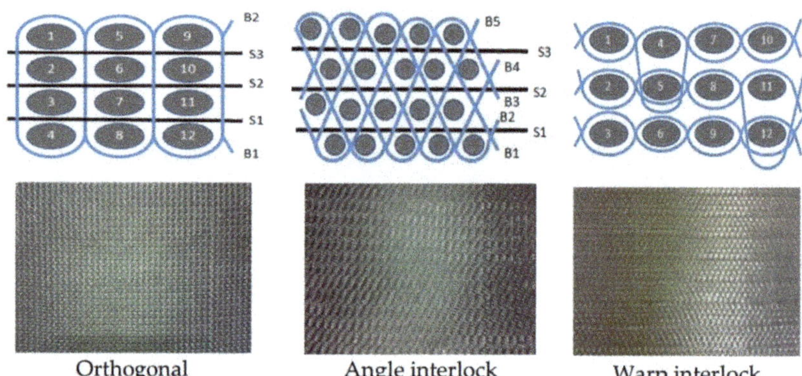

Figure 2. Schematics of the 3D woven structures.

2.2.2. Composite Manufacturing
Compression Molding

All composite samples were prepared using the compression molding technique. LY556 epoxy resin and hardener (CHS-HARDENER P11) from SPOLCHEMIE in Ústí nad Labem, Czech Republic, were used as a matrix component for all the various 3D fabric types. The resin and hardener were mixed and stirred thoroughly as per the manufacturer's guidelines. The principal advantage of compression molding is its ability to produce parts with a complex geometry in a short period. A SANTECH compression molding machine (Model SMC/DMC/FRP, Gurgaon, India) was used to prepare the composites. It operates at a high speed and with a high degree of precision with computer control. The slide moves with eight-point gibs with a lubrication arrangement that is a specially designed square gib construction that provides accurate guiding of the moving platen/slide with extra stability to resist any deflection under different load conditions. The structure of the press is monolithic or variable depending on requirements. The parameters set on the machine while preparing composites are mentioned below:

Machine used: SANTECH compression molding machine;
Curing time: 900 s;
Molding pressure: 6 bar;
Curing pressure: 12 bar;
Curing temperature: 120 °C;
Hardener/epoxy ratio: 1:10.

Specifications of Composite Samples

The composites were developed with different ratios of reinforcing 3D woven glass fabric and matrix (including resin and hardener). The final oven-dry weight of the composite samples was determined after curing and the add-on (%) of the matrix was calculated based on the oven-dry weight of the reinforcing glass fabric.

$$\mathrm{Add-on\ (\%)\ of\ matrix} = \frac{100 \times (\mathrm{Oven\ dry\ weight\ of\ composite} - \mathrm{Oven\ dry\ weight\ of\ fabric})}{\mathrm{Oven\ dry\ weight\ of\ reinforcing\ fabric}} \tag{1}$$

The specifications of the composite samples prepared are given in Table 2.

Table 2. Specification of various composite samples developed with variable parameters.

Variable Parameter	Reinforcement Fabric Type	Weight of Fabric (g)	Weight of Composite (g)	Weight of Matrix (g)	Weight Fraction of Fiber/Fabric in Final Composite
Add-on (%) of matrix					
45	Orthogonal	62.3	93.4	31.1	0.49
	Angle interlock	62.3	93.4	31.1	0.49
	Warp interlock	62.3	93.4	31.1	0.49
50	Orthogonal	63.5	97.2	33.7	0.47
	Angle interlock	63.5	97.2	33.7	0.47
	Warp interlock	63.5	97.2	33.7	0.47
55	Orthogonal	62.3	98.6	36.3	0.45
	Angle interlock	62.3	98.6	36.3	0.45
	Warp interlock	62.3	98.6	36.3	0.45
60	Orthogonal	64.4	103.3	38.9	0.44
	Angle interlock	64.4	103.3	38.9	0.44
	Warp interlock	64.4	103.3	38.9	0.44
Amount of hardener (%)					
8	Orthogonal	64.5	102.4	37.9	0.45
	Angle interlock	64.5	102.4	37.9	0.45
	Warp interlock	64.5	102.4	37.9	0.45
10	Orthogonal	65.4	106.1	40.7	0.44
	Angle interlock	65.4	106.1	40.7	0.44
	Warp interlock	65.4	106.1	40.7	0.44
12	Orthogonal	63.5	105.7	42.2	0.42
	Angle interlock	63.5	105.7	42.2	0.42
	Warp interlock	63.5	105.7	42.2	0.42
14	Orthogonal	62.9	93.4	30.5	0.47
	Angle interlock	62.9	93.4	30.5	0.47
	Warp interlock	62.9	93.4	30.5	0.47
Curing temperature (°C)					
100	Orthogonal	63.4	97.1	33.7	0.47
	Angle interlock	63.4	97.1	33.7	0.47
	Warp interlock	63.4	97.1	33.7	0.47
120	Orthogonal	54.6	86.2	31.6	0.45
	Angle interlock	54.6	86.2	31.6	0.45
	Warp interlock	54.6	86.2	31.6	0.45
140	Orthogonal	62.6	98.9	36.3	0.45
	Angle interlock	62.6	98.9	36.3	0.45
	Warp interlock	62.6	98.9	36.3	0.45

Table 2. *Cont.*

Variable Parameter	Reinforcement Fabric Type	Weight of Fabric (g)	Weight of Composite (g)	Weight of Matrix (g)	Weight Fraction of Fiber/Fabric in Final Composite
	Orthogonal	63.5	99.1	35.6	0.46
160	Angle interlock	63.5	99.1	35.6	0.46
	Warp interlock	63.5	99.1	35.6	0.46
Curing time (s)					
	Orthogonal	64.5	99.3	34.8	0.47
720	Angle interlock	64.5	99.3	34.8	0.47
	Warp interlock	64.5	99.3	34.8	0.47
	Orthogonal	63.2	97.2	34	0.47
900	Angle interlock	63.2	97.2	34	0.47
	Warp interlock	63.2	97.2	34	0.47
	Orthogonal	62.9	94.5	31.6	0.48
1080	Angle interlock	62.9	94.5	31.6	0.48
	Warp interlock	62.9	94.5	31.6	0.48
	Orthogonal	63.3	96.9	33.6	0.47
1260	Angle interlock	63.3	96.9	33.6	0.47
	Warp interlock	63.3	96.9	33.6	0.47
Molding pressure (bar)					
	Orthogonal	64.2	98.2	34	0.47
8	Angle interlock	64.2	98.2	34	0.47
	Warp interlock	64.2	98.2	34	0.47
	Orthogonal	63.8	98.5	34.7	0.46
10	Angle interlock	63.8	98.5	34.7	0.46
	Warp interlock	63.8	98.5	34.7	0.46
	Orthogonal	62.7	102.4	39.7	0.43
12	Angle interlock	62.7	102.4	39.7	0.43
	Warp interlock	62.7	102.4	39.7	0.43
	Orthogonal	63.1	99.7	36.6	0.45
14	Angle interlock	63.1	99.7	36.6	0.45
	Warp interlock	63.1	99.7	36.6	0.45

2.2.3. Testing

Tensile Testing

To measure the tensile properties, composite samples were cut in the size of 200 mm × 2.50 mm and the test was carried out on a universal testing machine Z05 (Zwick/Roell, Ulm, Germany) as per the ASTM-D 3039 standard [28]. The testing device works on the principle of a constant rate of elongation (CRE), which was set to 2 mm/min. A Vernier caliper was used before testing to measure the thickness of each specimen. The effective gauge length of the test samples was set as 100 mm. The load cell applied was 1000 kN [29,30]. The maximum load of the specimens was noted before fracture or failure. By monitoring the strain and load of the specimens, the stress–strain response was plotted. From this plot, the tensile modulus and ultimate tensile strength were calculated. For each sample, 30 measurements were carried out. The mean and standard deviation were calculated with a coefficient of variation (CV% < 5%). A Zwick/Roell Z05 Universal testing machine is shown in Figure 3a.

(a) Zwick Roell Z05 universal tensile testing machine

(b) Low-velocity impact testing

(c) Flexural test with three-point bending

Figure 3. Mechanical testing of the composite samples.

Low-Velocity Impact Testing (Gardner Impact Test)

The Gardner impact test was carried out on all the composite samples to estimate the amount of impact energy absorbed by using ASTM D5420-21 [31]. The drop weight impact tester (Model: HIT230F, Zwick Roell Group, Ulm, Germany) was used to measure the energy absorbed by the samples. A hemispherical indenter at a constant rate of 5 m/s was used to carry out this test. A predetermined weight (1.04 kg) and height were determined accordingly for a 10 J impact energy. The weight was allowed to strike on composite specimens, which were supported on a horizontal platform with clamps on the edges. This test was used to measure the damage resistance of composite samples against a dropped weight impact event [32,33]. A flat rectangular sample of dimension 15 cm × 10 cm was subjected to concentrated impact using a drop weight device. The potential energy of the impactor was pre-calculated using the weight and height of the impactor. Damage was imparted through an out-of-plane, concentrated impact that was perpendicular to the plane of the sample. For each sample, 30 measurements were carried out. The peak load and total energy absorbed by the sample was obtained from the test results. The mean and standard deviation were calculated. The device for the impact test is shown in Figure 3b.

The maximum energy of 10 J was applied on the samples through the dropped weight impact tester, where the absorbed energy was calculated as follows:

Impact energy absorbed

$$(E) = M \times g \times h \tag{2}$$

where M—mass of the dropped weight, g-acceleration due to gravity and h—height of the fall of the impactor.

Thirty measurements were carried out for each sample and the mean was reported with a coefficient of variation (CV% < 5%). All results were considered for the calculation of the mean. In case some results were outside the variation (CV > 5%), the measurements were repeated.

Flexural Test

The flexural properties of composite samples were evaluated using the three-point bending test according to the standard test method ASTM-D7264 [34]. The machine speed was set to 1 mm/min and a force was continuously applied on the specimen until it fractured or the value of the force reduced to 40% of the maximum force. The Zwick/Roell universal testing device (Zwick Roell Group, Ulm, Germany) was used by changing the clamps. It measures the flexural stiffness and strength properties of polymer matrix composites. A specimen of rectangular shape having dimensions 120 mm × 13 mm was supported at the ends and deflected at the center point [35,36]. As the force was applied on the specimen and it started deflecting from the center, its deflection and force were measured and recorded until failure occurred or the maximum force was reduced to 40%. For each sample, 30 measurements were carried out. The mean and standard deviation were calculated with a coefficient of variation (CV% < 5%). The principle of three-point bending is shown in Figure 3c.

Flexural strength is defined as the maximum stress in the outermost fiber while flexural modulus is calculated from the slope of the stress vs. strain curve. A gauge length/support span of 80 mm, deformation rate of 1 mm/min and load cell of 5 kN was applied as per ASTM-D7264 [35,36]. The flexural properties were calculated using Equations (3)–(5).

Flexural stress

$$\sigma_f = \frac{3PL}{2bd^2} \tag{3}$$

Flexural strain

$$\epsilon_f = \frac{6Dd}{L^2} \tag{4}$$

Flexural modulus

$$E_f = \frac{\sigma_f}{\epsilon_f} \tag{5}$$

where

σ_f—stress in outer fibers at the midpoint (MPa);
ϵ_f—strain in the outer surface (mm/mm);
E_f—flexural modulus of elasticity (MPa);
P—load at a given point on the load deflection curve (N);
L—support span (mm);
b—width of test beam (mm);
d—depth of tested beam (mm);
D—maximum deflection of the center of the beam (mm).

3. Results and Discussion

3.1. Tensile Properties of Composites

The tensile properties of the 3D woven composite samples were evaluated and compared. The effects of the processing and curing parameters were studied.

3.1.1. Effect of Add-On Percentage on Tensile Properties of 3D Woven Composites

Add-on percentage is defined as the final weight fraction of the matrix in a composite system. It significantly affects the tensile properties of reinforced composites. Four different levels of matrix add-on percentage were selected within a reasonably practical range (45–60%). The developed 3D woven glass fabric reinforced composite samples were tested on Instron tensile tester as per the ASTM-D 3039 standard. The results are shown in Figure 4.

As the matrix add-on percentage increased, the tensile strength and tensile modulus increased due to the better interface between the resin and 3D woven reinforcement. The elongation at break of the composite samples decreased as the add-on percentage increased. This was due to better bonding between the resin and reinforcement. There was a steady increase in the tensile strength and modulus when the add-on percentage was increased from 45% to 55%. However, when the add-on percentage was increased beyond 55%, there was no further improvement in tensile properties. Thus, a 55% matrix and 45% fiber could be treated as optimal for 3D woven glass–epoxy composites.

The tensile strength, elongation at break and tensile modulus were maximum for the angle interlock structure and the other two structures showed similar tensile properties.

The geometry of the angle interlock fabric was responsible for the higher load bearing. Further, the angular disposition of the binder yarn enabled higher elongation at break compared to the other 3D woven structures. The results showed significance at the 95% confidence interval. The CV was lower than 5%.

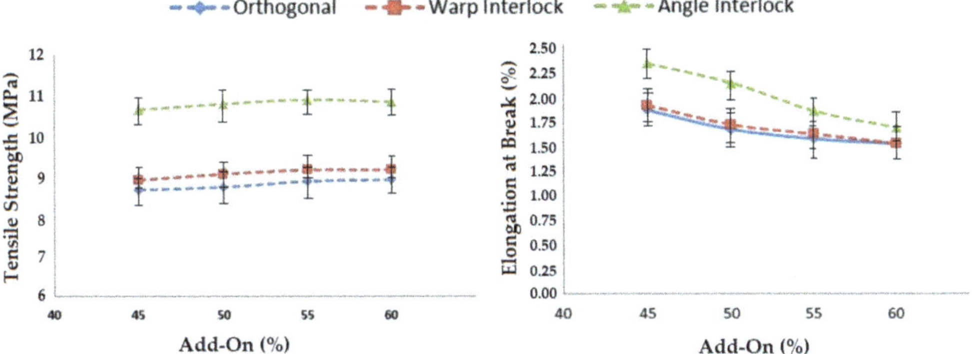

(a) Effect of add-on percentage on tensile strength   (b) Effect of add-on percentage on elongation at break (%)

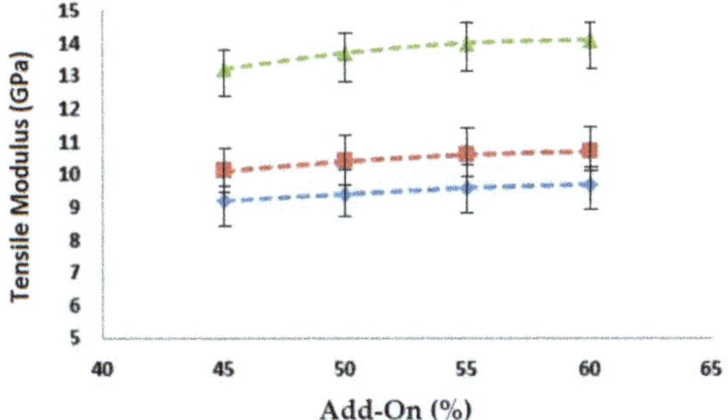

(c) Effect of add-on percentage on tensile modulus

Figure 4. Effect of add-on percentage on the tensile properties of 3D woven composites.

3.1.2. Effect of the Amount of Hardener on Tensile Properties

Tensile properties, such as tensile strength, elongation at break and tensile modulus, are also affected by changing the amount of hardener added with resin during composite preparation. In this study, four different amounts of hardener were selected as per the guidelines of the supplier. The composite samples with 3D woven fabric reinforcement were tested in order to study the influence of the hardener on the tensile properties. The results obtained for different 3D glass woven fabric reinforced epoxy composites are shown in Figure 5.

An increase in hardener percentage increased the tensile strength as well as the modulus. There was a consistent increment up to the addition of 12% hardener in the matrix. However, thereafter, the strength and modulus remained almost constant. There was no further increase in strength or modulus when the amount of hardener was increased to 14%. Therefore, 12% hardener could be considered optimal for 3D woven glass–epoxy composites. Angle interlock structure-based epoxy composites exhibited superior tensile properties with varying amounts of hardener (%) compared to orthogonal and warp interlock fabrics. The composite samples with different hardener percentages after tensile testing are shown in Figure 6.

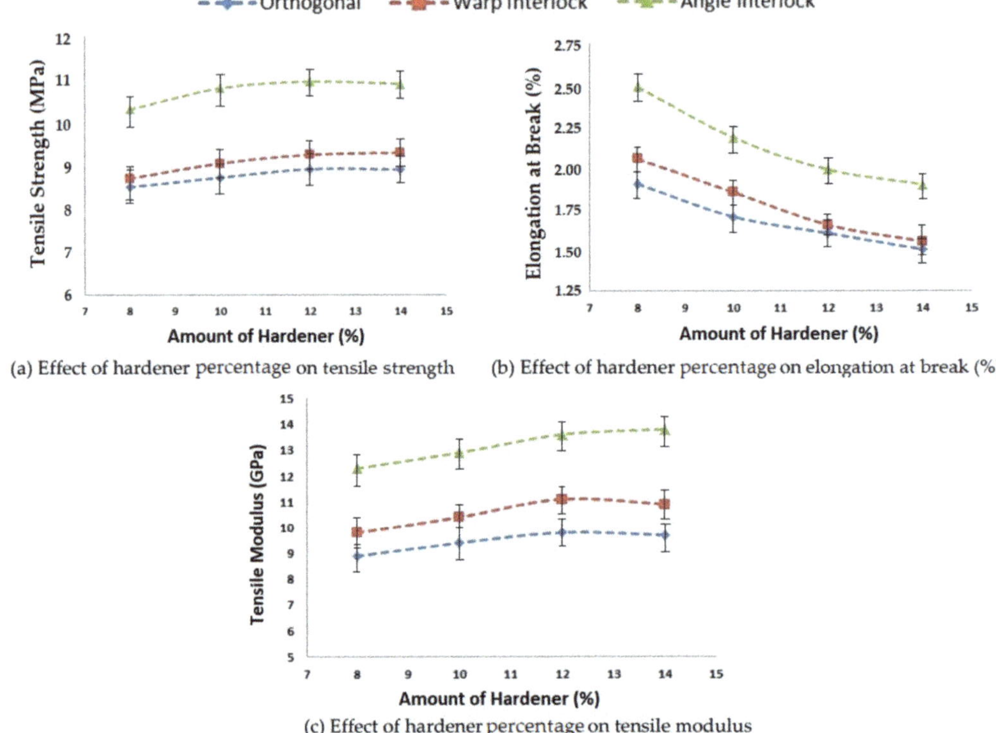

(a) Effect of hardener percentage on tensile strength   (b) Effect of hardener percentage on elongation at break (%)

(c) Effect of hardener percentage on tensile modulus

Figure 5. Effect of the amount of hardener percentage on the tensile properties of 3D woven composites.

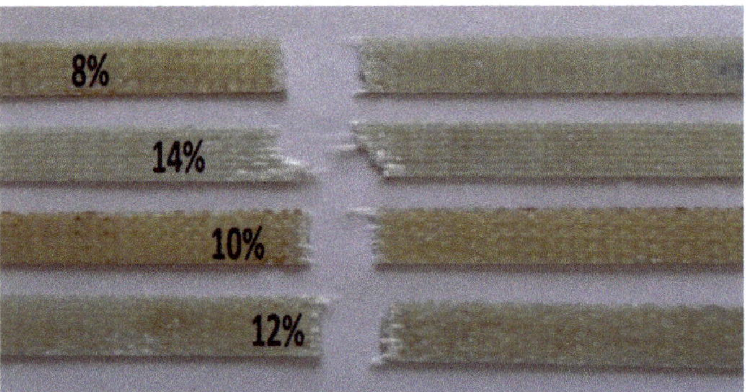

Figure 6. Nature of tensile failure in 3D woven glass–epoxy composites with different hardener percentages.

As can be observed, the samples with 8%, 10% and 12% hardener showed a straight-line failure after tensile testing. This was a good indication of uniform load bearing by all the constituent filaments/tows in the testing direction. Such behavior also indicated a homogenous impregnation of the fiber and a good interface with the matrix. However, with 14% hardener, there was an angled line of failure. This indicated that the higher hardener percentage (more than 12%) led to nonuniformity of resin impregnation and nonuniform load bearing by the constituent reinforcing elements in the 3D woven glass–epoxy composite.

3.1.3. Effect of Curing Temperature on Tensile Properties

Curing is an important process in composite preparation and the parameters, e.g., curing time and temperature, of the process affect the tensile properties of composites. Curing temperature also affects the viscosity of the resin. On increasing the temperature, the viscosity of resin reduces. Viscosity is directly associated with resin impregnation into reinforcement, which ultimately affects the tensile properties, such as the strength, elongation at break and tensile modulus. The effect of curing temperature between 100 °C and 160 °C on the tensile performance of 3D woven polymer composites is shown in Figure 7.

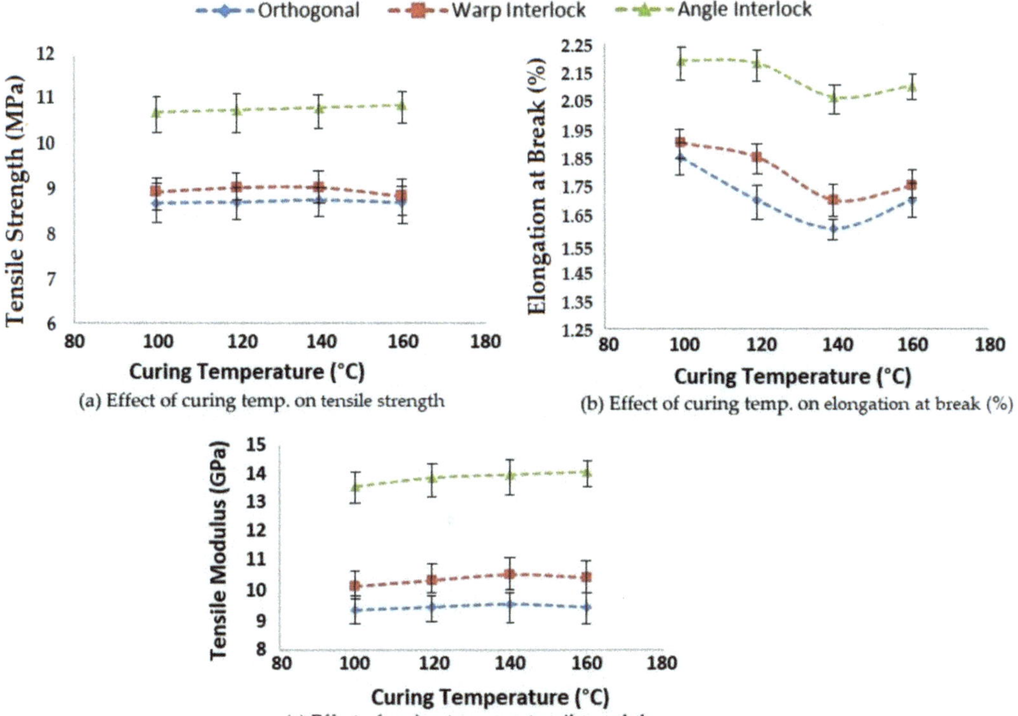

(a) Effect of curing temp. on tensile strength

(b) Effect of curing temp. on elongation at break (%)

(c) Effect of curing temp. on tensile modulus

Figure 7. Effect of curing temperature on the tensile properties of 3D woven composites.

As the curing temperature increased, the viscosity of resin decreased, which facilitated easier resin impregnation into the fibrous structure (3D woven fabric). The resin was able to bind the reinforcing 3D structure more effectively. This ultimately increased the tensile strength and modulus to an optimal value. This trend was observed to be almost similar for all three types of 3D woven fabric reinforced composites. A curing temperature of about 140 °C seemed to be optimal for glass–epoxy composites. The results showed significance at the 95% confidence interval. The CV was lower than 5%.

3.1.4. Effect of Curing Time on Tensile Properties

The influence of curing time on the tensile properties of 3D woven polymer composites was studied. The results obtained for curing times between 720 s and 1260 s are shown in Figure 8. This range was selected based on the recommendation of the resin supplier.

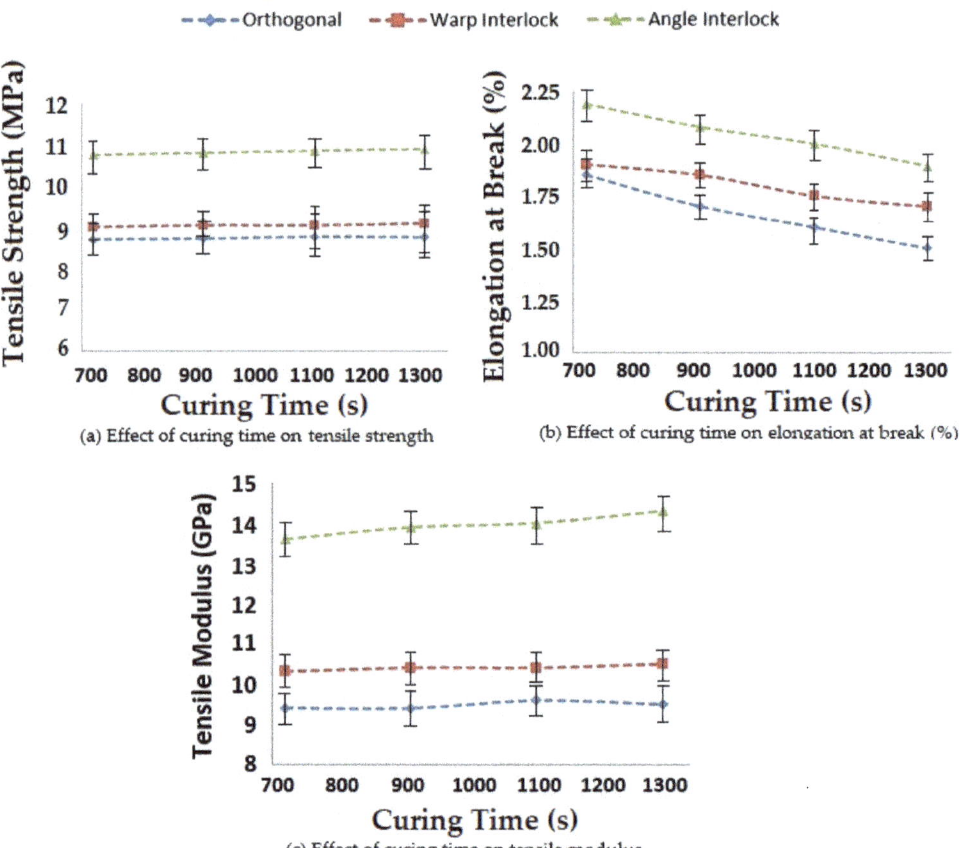

(a) Effect of curing time on tensile strength

(b) Effect of curing time on elongation at break (%)

(c) Effect of curing time on tensile modulus

Figure 8. Effect of curing time on the tensile properties of 3D woven composites.

The effect of curing time on the tensile properties of the composites seemed to follow a similar trend to the curing temperature. As the curing time increased, the tensile strength and tensile modulus increased for all three types of glass–epoxy composites using 3D woven structures. With increasing time, there was better impregnation of resin, which was responsible for the improved tensile properties. However, excessive curing time might lead to degradation of the resin. As usual, the angle interlock structure showed superior tensile properties based on the internal geometry of the fabric.

3.1.5. Effect of Molding Pressure on Tensile Properties

Pressure is applied in the mold when the curing of a composite is carried out. That pressure directly affects the resin impregnation into the reinforcement structure and, consequently, the resin impregnation affects the tensile properties, such as tensile strength, elongation at break and tensile modulus. Four levels of molding pressure, i.e., 8, 10, 12 and 14 bar, were selected, and the obtained results are shown in Figure 9.

Increasing the molding pressure helped with providing better penetration of the resin into the fabric, resulting in higher tensile strength and modulus, as is visible from Figure 9. This trend was visible up to a molding pressure of 12 bar. However, a further increase in molding pressure did not seem to increase the strength and modulus.

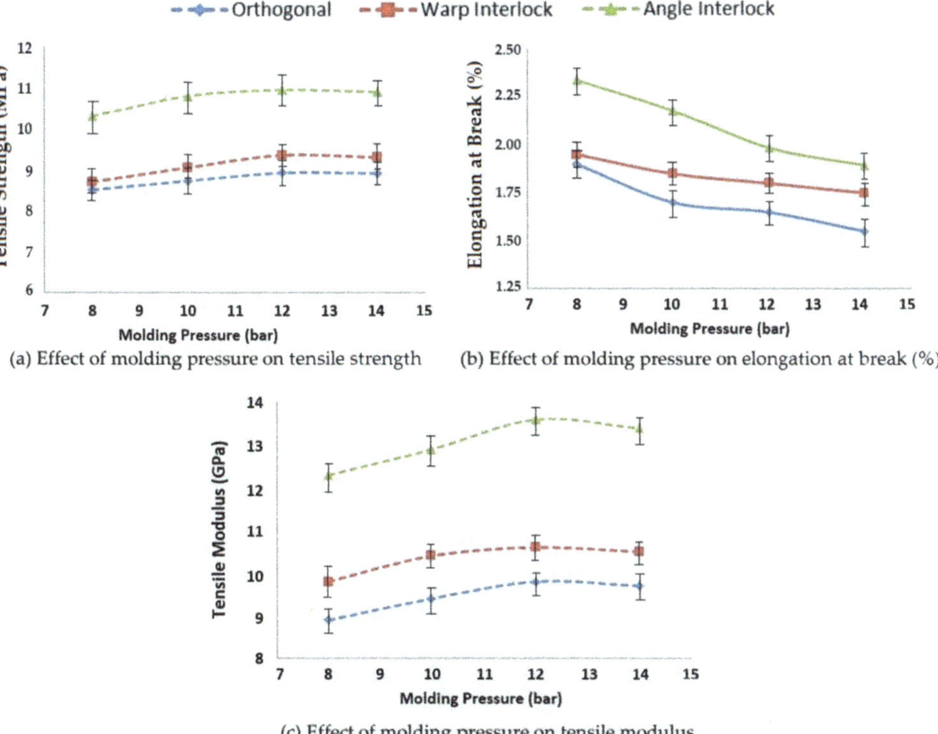

(a) Effect of molding pressure on tensile strength

(b) Effect of molding pressure on elongation at break (%)

(c) Effect of molding pressure on tensile modulus

Figure 9. Effect of molding pressure on the tensile properties of 3D woven composites.

With the increase in pressure, there was more consolidation of the fibers and resin in the composite, resulting in a stronger interface. Thus, the elongation at break decreased. This trend was observed for all three types of 3D woven reinforcement of glass fibers. The results showed significance at the 95% confidence interval. The CV was lower than 5%.

The 3D woven composite samples developed with different molding pressures and that fractured after the tensile test are shown in Figure 10.

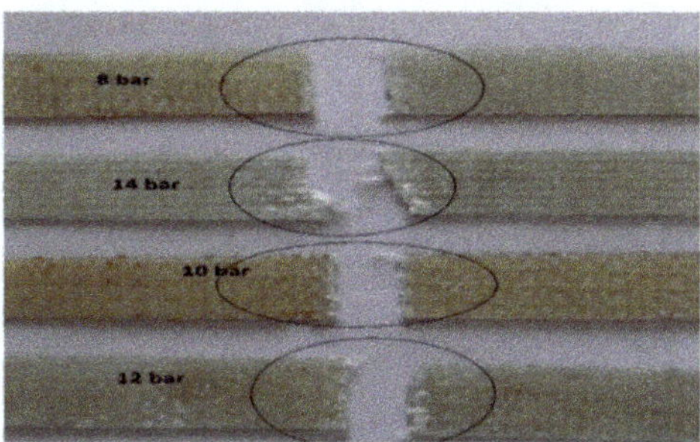

Figure 10. Nature of tensile failure in the 3D woven glass–epoxy composites manufactured with different molding pressures.

The nature of the failure was uniform till a pressure of 12 bar, as is visible from Figure 10. Therefore, an optimal pressure of 12 bar can be used during molding for 3D woven glass–epoxy composites. Some SEM images of tensile tested samples prepared at

12 bar molding pressure are shown in Figure 11. The fractured surface showed sharp edges and catastrophic failure of the fibers. The fiber rupture (in red circles) indicated that they uniformly bore the load from the resin.

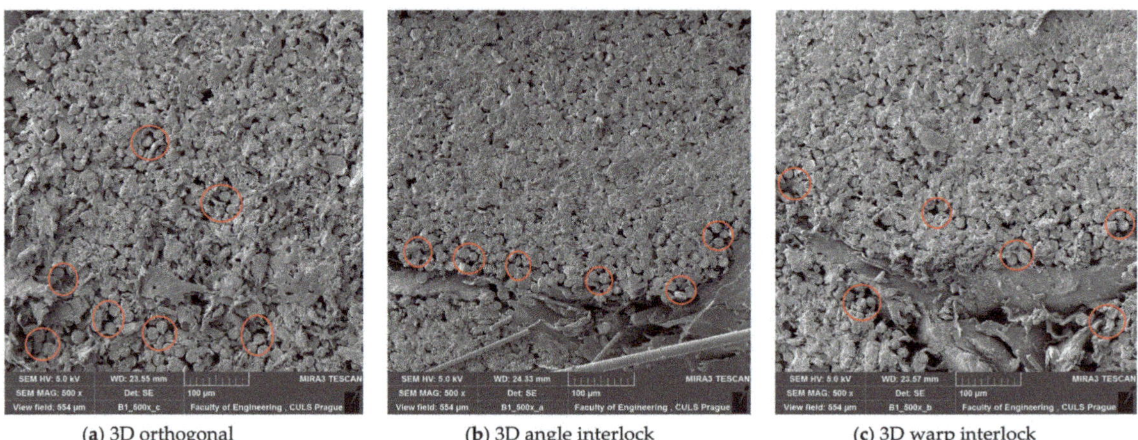

(**a**) 3D orthogonal   (**b**) 3D angle interlock   (**c**) 3D warp interlock

Figure 11. SEM images of tensile tested 3D woven composite samples prepared at 12 bar molding pressure.

3.2. Impact Properties of Composites

The impact properties of the different 3D woven composite samples were evaluated and compared. The effects of resin processing and curing parameters were studied in detail.

3.2.1. Effect of Matrix Add-On Percentage on Impact Properties

Matrix add-on percentage affects the impact properties (impact energy, peak force and deformation). Impact testing was carried out using ASTM D5420-21 for the developed composite samples with different 3D woven reinforcements impregnated with epoxy resin at various levels of matrix add-on percentage, i.e., 45, 50, 55 and 60%. The obtained results are compared and shown in Figure 12.

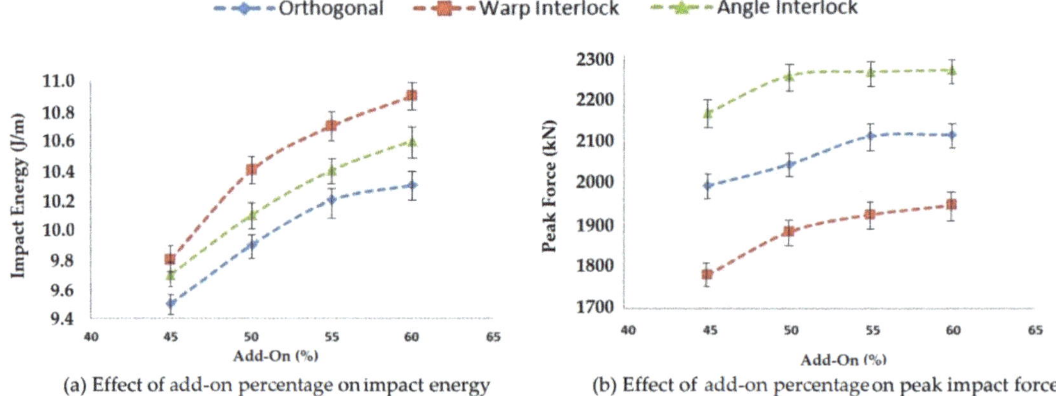

(a) Effect of add-on percentage on impact energy   (b) Effect of add-on percentage on peak impact force

Figure 12. Effect of add-on percentage on the impact properties of 3D woven composites.

With an increasing level of matrix add-on percentage, all the composite samples with different reinforcement structures followed a similar trend. The impact energy and the peak impact force increased to an optimal value at 55% add-on and after that, there was no significant increase.

The highest impact energy was observed for warp interlock fabric composite samples and lowest for orthogonal structure reinforced samples. This may have been due to the interlaminar integrated structure of the warp interlock fabric, where the Z-direction yarn showed maximum total deflection, leading to the maximum impact energy (area under the impact force and deflection curve). In the case of the orthogonal structure, the un-crimped orthogonal yarns/tows did not facilitate much high energy absorption.

The peak impact force increased with the increase of matrix add-on percentage. The trend was similar for all three structures (orthogonal, angle interlock and warp interlock) to a certain level. On the further increase of matrix add-on, there was no penetration inside the reinforcement structure. This optimal level was attained at about 55% of resin add-on.

The highest impact force was observed for the angle interlock structure and the lowest value was for the warp interlock structure. In the angle interlock fabric, the consolidation of the different layers was possible to the maximum extent. The angular disposition of the Z yarn, which enabled maximum tensile strength and modulus, also helped in providing the maximum impact strength. The warp interlock structure did not integrate all layers into a single interlacement. It bound the layers step by step. Thus, the load bearing was not the maximum, but energy absorption was facilitated. That is why the warp interlock fabric-based composite showed maximum impact energy absorption. The orthogonal fabric reinforced composite showed an intermediate peak impact force due to the lowest deformation pertaining to non-crimped constituent yarns. The results showed significance at the 95% confidence interval. The CV was lower than 5%.

3.2.2. Effect of the Amount of Hardener on Impact Properties

A hardener is used for the hardening of a matrix system. The hardening rate of a matrix is affected by the amount of hardener used in the composite preparation. This ultimately affects the impact properties of composites, such as impact energy, peak impact force and deformation. For analyzing the effect of the hardener percentage, four different levels, i.e., 8%, 10%, 12% and 14%, were used, and the samples were tested as per standard ASTM D5420-21. The results are shown in Figure 13.

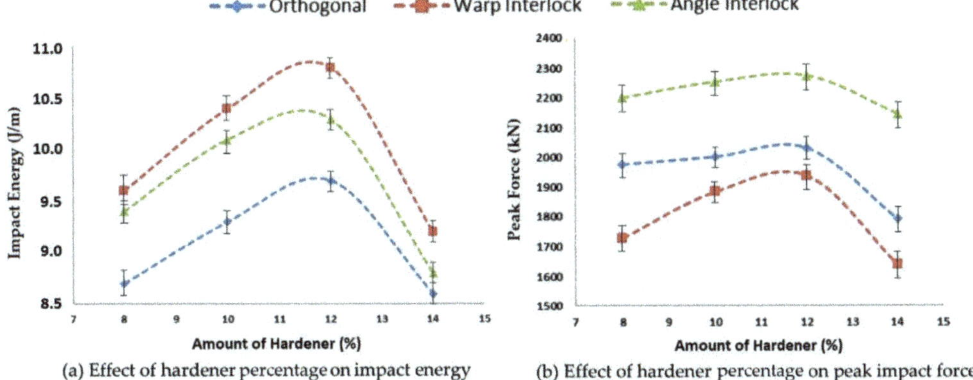

(a) Effect of hardener percentage on impact energy (b) Effect of hardener percentage on peak impact force

Figure 13. Effect of the amount of hardener percentage on the impact properties of 3D woven composites.

On increasing the amount of hardener, the impact energy and peak impact force increased for all three types of 3D woven composite structures up to the optimal level. This was due to the better hardening rate and strengthening of the interface with resin. However, beyond a 12% hardener amount, the impact energy, as well as peak impact force, decreased, which may have been due to excessive hardening of the matrix. The reinforcement structure could not be impregnated uniformly and thus was not able to bind the structure effectively.

The highest impact energy was observed for the warp interlock composite and the lowest for orthogonal fabric-based samples. This could be attributed to the energy absorption capacity, which depended on the disposition of binder yarns/tows (Z yarns). The highest deformation occurred in the warp interlock structure and the lowest deformation occurred in the orthogonal structure. The angle interlock fabric composites exhibited the highest impact strength pertaining to load-bearing capacity, resulting from the angular interlacement and maximum consolidation of the reinforcement fabric. In contrast, the warp interlock fabric composite shows the lowest peak force level due to the layer-by-layer binding, which does not allow for maximum consolidation. However, a maximum displacement/post-impact deformation helped with the highest impact energy absorption in this case.

3.2.3. Effect of Curing Temperature on Impact Properties

Curing is an important process of composite preparation and the variables, such as curing time and temperature, of the process affect the impact properties of a composite. Curing temperature influences the resin impregnation into reinforcement structure, which will affect the impact properties, such as impact energy, peak impact force and deformation. The curing temperature was varied between 100 °C and 160 °C in four steps. The effect of curing temperature on impact energy and peak impact force is shown in Figure 14.

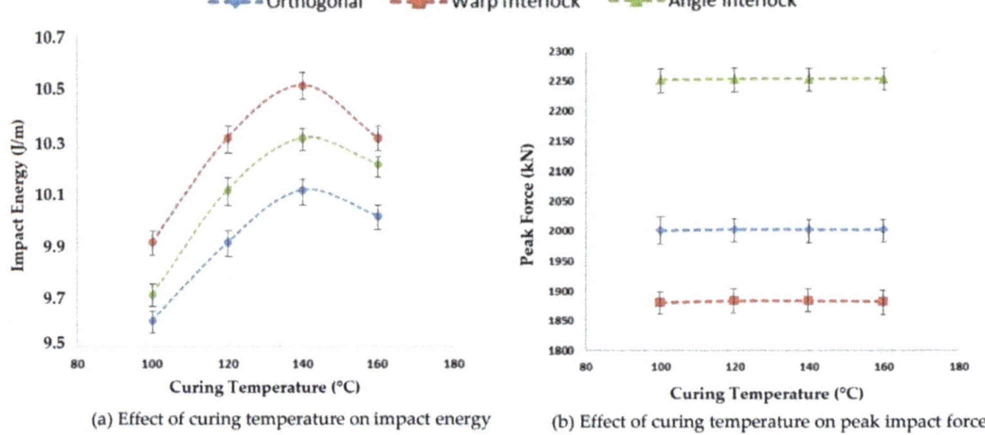

(a) Effect of curing temperature on impact energy

(b) Effect of curing temperature on peak impact force

Figure 14. Effect of curing temperature on the impact properties of 3D woven composites.

As the curing temperature was increased, the impact energy also increased for all 3D woven composites. The increase was within a very narrow range (9.6 J/m–10.5 J/m). The increase in impact energy was due to better impregnation of the resin as the viscosity of the resin reduced with increased temperature. Beyond a curing temperature of 140 °C, the resin started degrading, which reduced the adhesion properties of the resin. Thus, deterioration of impact energy was observed.

The highest impact energy was observed for warp interlock fabric composites and was the lowest for orthogonal structure composites. The peak impact force was observed to be at a maximum for angle interlock composites and at a minimum for warp interlock fabric composites. The peak impact force was not very dependent on the curing temperature. A sufficient temperature was required, at which, the resin would bind the reinforcement assembly. The optimal temperature for curing 3D woven fabric glass–epoxy composites was found to be 140 °C. The results showed significance at the 95% confidence interval. The CV was lower than 5%.

3.2.4. Effect of Curing Time on Impact Properties

The curing time is also an important parameter that influences the impact performance of polymer composites. The variation of curing time between 720–1260 s was studied and its influence on the impact properties is shown in Figure 15.

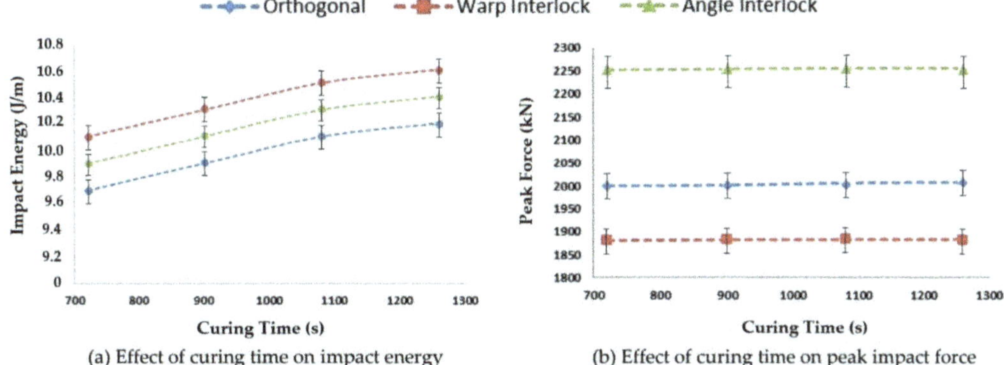

(a) Effect of curing time on impact energy

(b) Effect of curing time on peak impact force

Figure 15. Effect of curing time on the impact properties of 3D woven composites.

The curing time showed a similar trend to curing temperature. However, the impact energy kept increasing beyond the curing time of 1260 s. The peak impact force remained unchanged with the variation in curing time. A higher curing time might cause degradation of the resin.

3.2.5. Effect of Molding Pressure on Impact Properties

When pressure is applied into the mold during curing, it directly affects the resin impregnation into the reinforcement, where the resin impregnation affects the impact properties, such as impact energy, peak impact force and deformation. Four levels of molding pressure, i.e., 8, 10, 12 and 14 bar, were applied, and the results are shown in Figure 16.

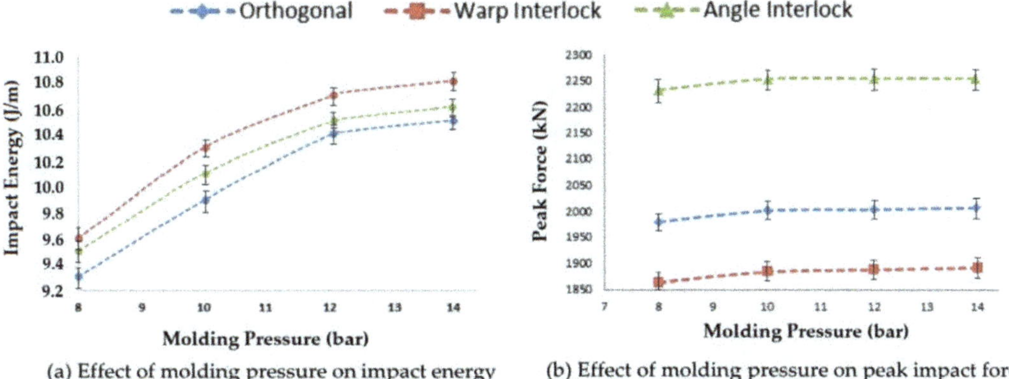

(a) Effect of molding pressure on impact energy

(b) Effect of molding pressure on peak impact force

Figure 16. Effect of molding pressure on the impact properties of 3D woven composites.

On increasing the molding pressure, the impact energy and peak impact force increased for all three types of 3D woven composite structures. The molding pressure increased the resin impregnation in the reinforcement, which improved the binding between the resin and reinforcement and ultimately improved the impact performance. After the optimal pressure of 12 bar, there was not much increase in impact energy and peak impact force. The impact energy was found to be highest for the warp interlock structure composite and at a minimum for the orthogonal 3D woven fabric glass–epoxy composite. This was due to the interlacement pattern. Further, the peak impact force was highest for the angle interlock structure reinforced composite and at a minimum for the warp interlock structure composite, as observed in previous results in this study. SEM images of the impact-tested composite samples prepared at 12 bar molding pressure are shown in Figure 17. The images show sharp edges, which indicate catastrophic fiber breakage. The fiber rupture (in red circles) indicated that they bore a uniform load from the resin.

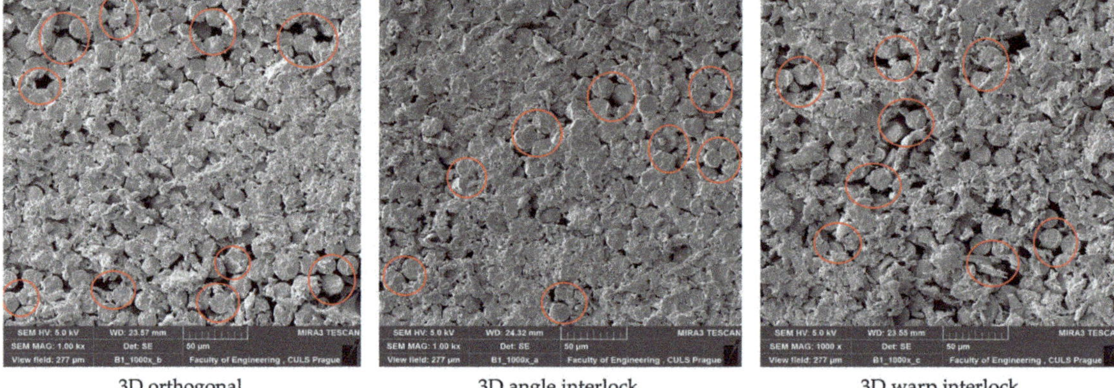

3D orthogonal   3D angle interlock   3D warp interlock

Figure 17. SEM images of impact-tested 3D woven composite samples prepared at 12 bar molding pressure.

3.3. Flexural Properties of Composites

3.3.1. Effect of Matrix Add-On Percentage on Flexural Properties

The add-on percentage of the matrix affects the bending properties of composites. Four different levels of matrix add-on percentage were studied and the prepared composite samples with different 3D woven fabric reinforcement were tested in terms of three-point bending properties as per ASTM-D7264. The results are shown in Figure 18.

As the matrix add-on percentage increased, the flexural strength increased for all three types of 3D woven reinforcement structures due to the increase of the binding element (resin) in the composite. However, after an optimum value of add-on (about 50%), the strength decreased, which may have been due to an insufficient fraction of fiber element in the composite. The flexural strain decreased with an increase of matrix add-on percentage for all three fabric structures. This was due to the higher restriction to deformation of fiber components when an excessive amount of resin was applied.

The flexural modulus followed a similar trend to that of flexural strength for all three types of 3D woven composite structures. There was an increase in flexural modulus while using up to 50% matrix add-on. Beyond this limit, a steady decrease in flexural modulus was observed.

Flexural strength was observed to be at a maximum for the angle interlock structure due to better resin impregnation. The angular arrangement of Z yarns consolidated the structure and increased the strength in all deformation modes. The lowest flexural strength was observed for orthogonal structure due to the perpendicular/uncrimped Z yarns, which provide no interlacement.

The flexural strain was at a maximum for the orthogonal structure, as the Z-direction yarn was at 90° and resulted in high strain.

The flexural modulus was at a maximum for the warp interlock fabric composites due to the optimal flexural stress/strength and strain. The layer-by-layer bindings in the warp interlock fabric helped resist the bending deformation. A 50% matrix add-on was optimal for flexural strength, as well as modulus. The results showed significance at the 95% confidence interval. The CV was lower than 5%.

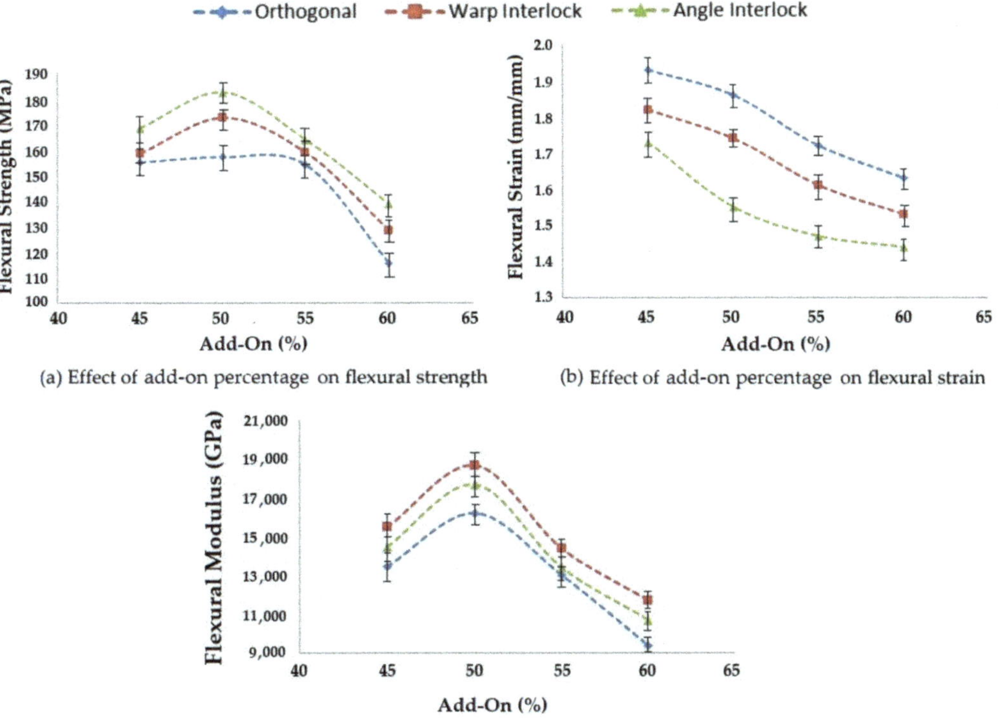

(a) Effect of add-on percentage on flexural strength

(b) Effect of add-on percentage on flexural strain

(c) Effect of add-on percentage on flexural modulus

Figure 18. Effect of add-on percentage on the flexural properties of 3D woven composites.

3.3.2. Effect of the Amount of Hardener on Flexural Properties

Flexural properties, such as flexural strength, flexural strain and flexural modulus, are also affected by changing the amount of hardener added with resin during composite preparation. In this study, four levels of hardener percentage were selected and the effects on bending properties were observed. The results are shown in Figure 19.

As the amount of hardener increased, the flexural strength and modulus also increased due to the increase in the adhesion properties of the resin. An optimal value of flexural strength and modulus was achieved at 12% hardener. With the further increase of the hardener amount, there was a decrease in flexural properties due to excessive hardening of the resin before curing. This trend was observed for all three types of 3D woven reinforcements. The flexural strain decreased until an optimum amount of hardener (12%) and then increased. This could be attributed to the hardening of resin before curing and improper adhesion with the reinforcement. As usual, the angle interlock fabric composite showed the highest flexural strength, followed by the warp interlock fabric composite. The orthogonal glass fabric epoxy composite showed the minimum flexural strength among all the 3D woven structures. The trends of flexural strain were opposite to that of flexural strength. The orthogonal fabric composites exhibited the maximum strain due to non-interlaced Z yarns. The results showed significance at the 95% confidence interval. The CV was lower than 5%.

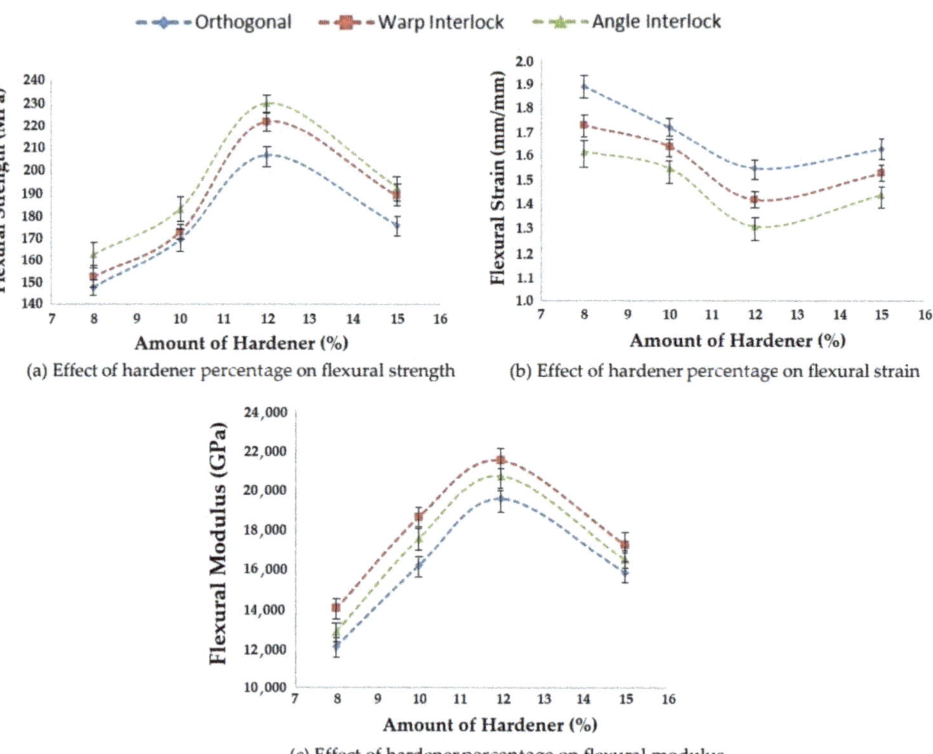

(a) Effect of hardener percentage on flexural strength

(b) Effect of hardener percentage on flexural strain

(c) Effect of hardener percentage on flexural modulus

Figure 19. Effect of the amount of hardener on the flexural properties of 3D woven composites.

3.3.3. Effect of Curing Temperature on Flexural Properties

Curing process parameters, e.g., temperature and time, significantly affect the bending properties of composites. By increasing the curing temperature, the viscosity of the resin reduces and changes the impregnation efficiency. This consequently affects the flexural strength, flexural strain and flexural modulus. The influence of different curing temperatures on flexural properties of 3D woven fabric glass–epoxy reinforced composites is shown in Figure 20.

As the curing temperature increased, the viscosity of resin decreased, which increased the resin impregnation into the structure and helped with binding the reinforcement structure more effectively. This increased the flexural strength and modulus. Such a trend was observed for all the three 3D woven reinforcements up to a curing temperature of 140 °C. However, after the optimal level, the flexural properties started decreasing, probably due to the degradation of the resin.

The flexural strain consistently decreased when the curing temperature was increased from 100 °C to 140 °C. However, beyond this temperature, there was an increase in flexural strain. This trend was similar for all three types of 3D woven fabric structures. The increase in flexural strain, along with the decrease of flexural stress, resulted in an overall reduction in flexural modulus. The results showed significance at the 95% confidence interval. The CV was lower than 5%.

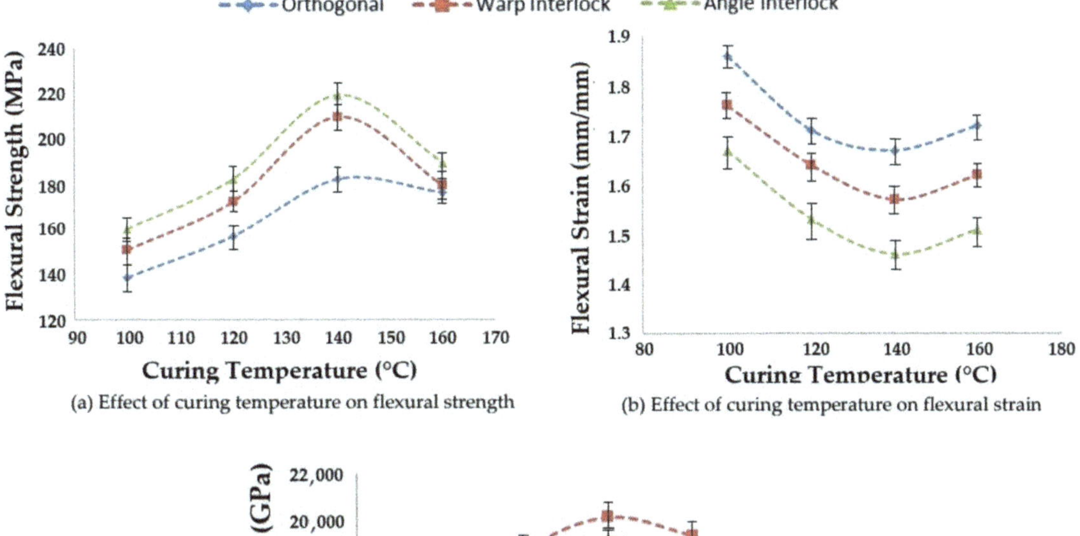

(a) Effect of curing temperature on flexural strength

(b) Effect of curing temperature on flexural strain

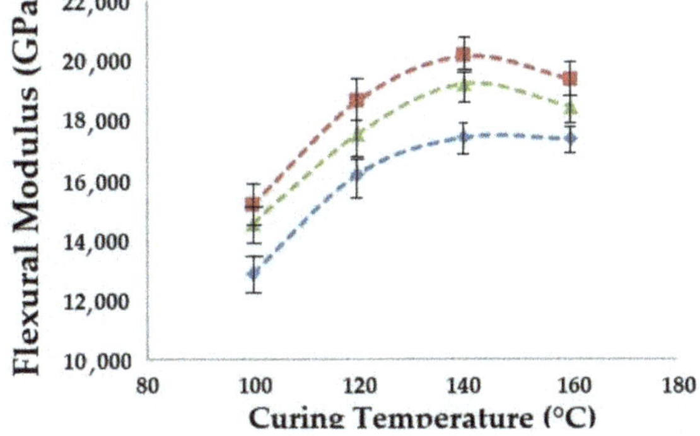

(c) Effect of curing temperature on flexural modulus

Figure 20. Effect of curing temperature on the flexural properties of 3D woven composites.

3.3.4. Effect of Curing Time on Flexural Properties

The duration of curing was changed in order to study its influence on the bending properties of polymer composites reinforced with 3D woven fabrics. The results are shown in Figure 21.

A curing time of 900 s was found to be optimal for all types of 3D woven glass fabric reinforced epoxy composites. The maximum flexural strength and modulus were obtained with this curing duration. When the curing time was increased further, there was no improvement in flexural properties. The flexural strain systematically decreased with increasing curing time. Angle interlock fabric-based composites showed maximum flexural strength and minimum flexural strain. However, the warp interlock fabric composites resulted in maximum flexural modulus due to the optimal values of flexural stress and strain. This was again attributed to the interlacement pattern and layer-by-layer binding in the reinforcement structure.

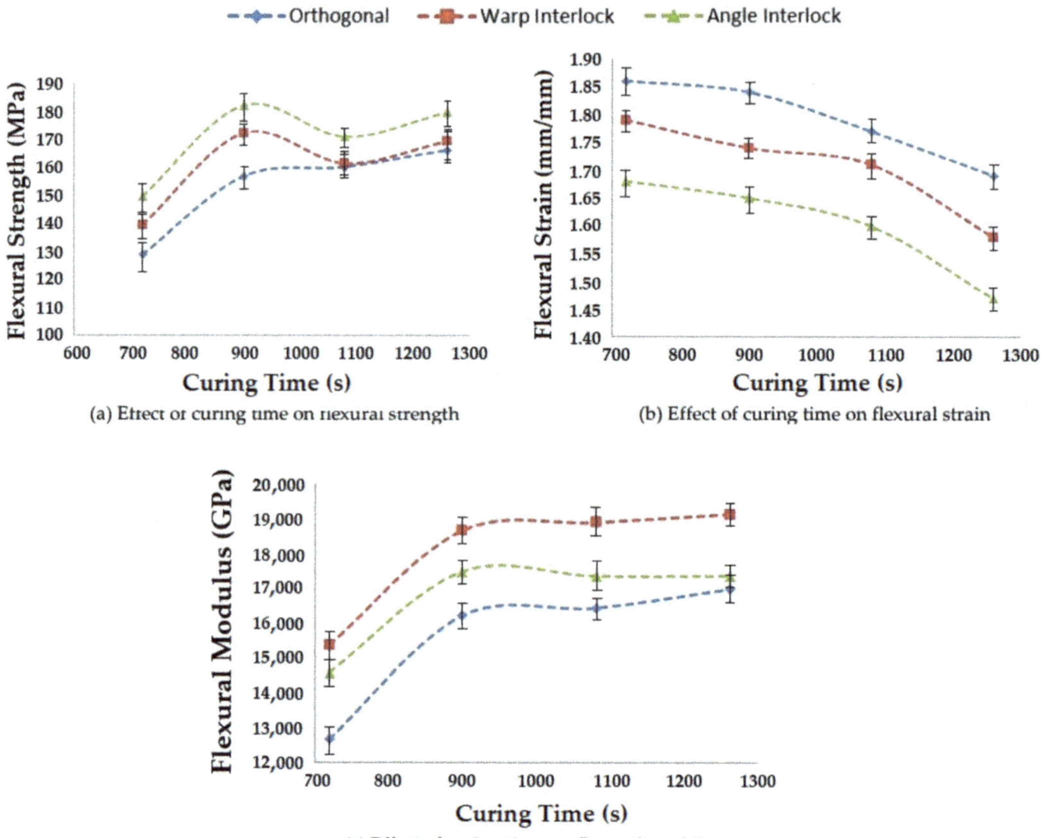

Figure 21. Effect of curing time on the flexural properties of 3D woven composites.

3.3.5. Effect of Molding Pressure on Flexural Properties

The molding pressure directly affected the resin impregnation into the reinforcement and thereby the flexural properties, such as flexural strength, flexural strain and flexural modulus. The influence of molding pressure (at 8, 10, 12 and 14 bar) on the flexural properties of 3D woven glass–epoxy composites was studied. The results are shown in Figure 22.

On increasing the molding pressure, the flexural strength and flexural modulus increased for all three types of 3D woven composite structures. This was due to the improved resin impregnation through the reinforcement, which increased the binding between resin and reinforcing fibers. A molding pressure of 12 bar resulted in the highest flexural strength and flexural modulus. The results showed significance at the 95% confidence interval. The CV was lower than 5%.

The flexural modulus was observed to be the highest for the warp interlock structure epoxy composite and minimum for orthogonal 3D structure composite. In all three types of 3D woven fabric glass–epoxy reinforced composites, the flexural strength and modulus decreased when the molding pressure was increased beyond 12 bar.

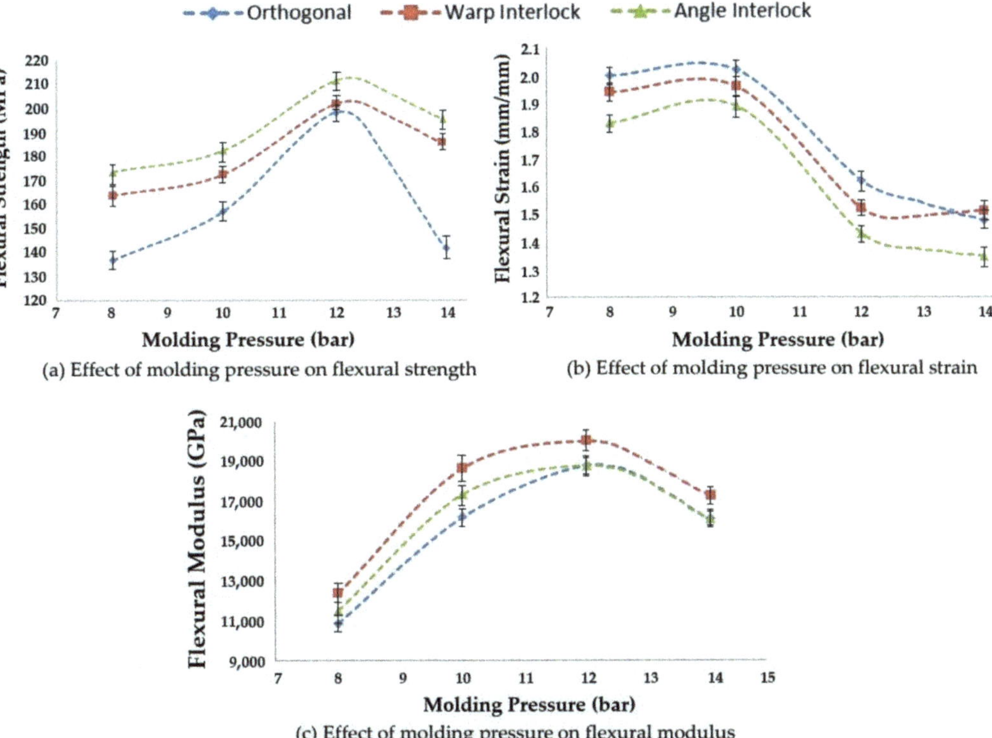

(a) Effect of molding pressure on flexural strength

(b) Effect of molding pressure on flexural strain

(c) Effect of molding pressure on flexural modulus

Figure 22. Effect of molding pressure on the flexural properties of 3D woven composites.

4. Conclusions

Three-dimensional woven glass fabrics with three different architectures, i.e., orthogonal, angle interlocked and warp interlocked structures were developed on a CCI sample weaving machine with an arrangement with an extra beam and a negative let-off motion. The matrix add-on (%), amount of hardener (%), curing temperature, curing time and molding pressure were selected as important parameters that affect resin infusion in the 3D woven structure and, thus, the ultimate mechanical properties of polymer composites.

The tensile strength, elongation at break and tensile modulus were found to be at a maximum for the angle interlock structure among the various 3D woven composites. As the matrix add-on percentage increased, the tensile strength and tensile modulus increased due to the better interface between the resin and 3D woven reinforcement. The elongation at break decreased as the add-on percentage increased. This was due to better bonding between the resin and reinforcement. A structure with 55% matrix and 45% fiber can be treated as optimal for 3D woven glass–epoxy polymer composites. A 12% addition of hardener (based on weight of the matrix) was found to be optimal for 3D glass woven fabric epoxy composites. A curing temperature of about 140 °C seemed to be optimal for glass–epoxy composites. As the curing time increased, the tensile strength and tensile modulus increased for all three types of glass–epoxy composites. Increasing the molding pressure up to 12 bar helped provide better penetration of the resin, resulting in higher tensile strength and modulus.

The highest impact energy was observed for the warp interlock fabric composites and the lowest for the orthogonal structure reinforced samples. With an increasing level of matrix add-on percentage, all composite samples with different reinforcement structures showed improvement in impact performance up to an optimal add-on value of 55%.

Beyond this level, there was no significant improvement in impact properties. A 12% addition of hardener, curing temperature of 140 °C, maximum curing time and 12 bar molding pressure resulted in the highest impact energy and peak impact force.

Flexural strength was observed to be at a maximum for the angle interlock structure composites, whereas the maximum flexural modulus was observed for warp interlock fabric composites due to optimal flexural stress/strength and strain. A 50% add-on of the epoxy matrix resulted in the highest flexural strength and modulus. The optimal conditions for the best flexural performance in 3D woven glass–epoxy polymer composites were 12% hardener, 140 °C curing temperature, 900 s curing time and 12 bar molding pressure.

The most significant parameters were identified as the add-on percentage of the matrix, amount of hardener and molding pressure. These optimized processing parameters can serve as a major guideline for further research and development in glass multifilament-based 3D woven epoxy composites.

Author Contributions: R.K.M., M.P., B.K.B. and P.K.B.: experiment design; R.K.M., M.P., B.K.B. and P.K.B.: testing of mechanical properties and data analysis; R.K.M., M.P. and B.K.B.: wrote and edited the paper; R.K.M., M.P. and B.K.B.: language correction; R.K.M.: communication with the editors; R.K.M., M.P. and B.K.B.: revision; R.K.M., M.P. and B.K.B.: administration and funding; R.K.M., M.P. and B.K.B.: supervision. All authors have read and agreed to the published version of the manuscript.

Funding: Supported by the Ministry of Education, Youth and Sports of the Czech Republic, the European Union (European Structural and Investment Funds—Operational Program Research, Development and Education) in the frames of the project "Modular platform for autonomous chassis of specialized electric vehicles for freight and equipment transportation", Reg. No. CZ.02.1.01/0.0/0.0/16_025/0007293, and the Internal grant agency of Faculty of Engineering, Czech University of Life Sciences Prague.

Institutional Review Board Statement: Not applicable.

Informed Consent Statement: Not applicable.

Data Availability Statement: The data presented in this study are available on request from the corresponding author.

Conflicts of Interest: The authors declare no conflict of interest.

References

1. Walter, T.; Subhash, G.; Sankar, B.; Yen, C. Damage modes in 3D glass fiber epoxy woven composites under high rate of impact loading. *Compos. Part B Eng.* **2009**, *40*, 584–589. [CrossRef]
2. Kendall, K.; Rudd, C.; Owen, M.; Midleton, V. Characterization of the resin transfer molding process. *Compos. Manuf.* **1992**, *3*, 235–249. [CrossRef]
3. Potter, K. In-plane and out-of-plane deformation properties of unidirectional preimpregnated reinforcement. *Compos. Part A Appl. Sci. Manuf.* **2002**, *33*, 1469–1477. [CrossRef]
4. Karbhari, V.M.; Chin, J.W.; Hunston, D.; Benmokrane, B.; Juska, T.; Morgan, R.; Lesko, J.J.; Sorathia, U.; Reynaud, D. Durability gap analysis for fiber-reinforced polymer composites in civil infrastructure. *J. Compos. Constr.* **2003**, *7*, 238–247. [CrossRef]
5. Corrado, A.; Polini, W. Measurement of high flexibility components in composite material by touch probe and force sensing resistors. *J. Manuf. Process.* **2019**, *45*, 520–531. [CrossRef]
6. Wang, F.; Xie, Z.; Liang, J.; Fang, B.; Piao, Y.; Hao, M.; Wang, Z. Tourmaline-Modified FeMnTiOx Catalysts for Improved Low-Temperature NH 3-SCR Performance. *Environ. Sci. Technol.* **2019**, *53*, 6989–6996. [CrossRef] [PubMed]
7. Behera, B.K.; Mishra, R. 3-Dimensional weaving. *Ind. J. Fiber Text. Res.* **2008**, *33*, 274–287.
8. Parnas, R.; Fly, D.; Dal-Favero, M. A permeability database for composites manufacturing. *Polym. Compos.* **1997**, *18*, 623–633. [CrossRef]
9. Hu, H.; Zhang, M.; Fangueiro, R.; Araujo, M. Mechanical properties of composite materials made of 3D stitched woven-knitted preforms. *J. Compos. Mater.* **2010**, *44*, 1753–1767. [CrossRef]
10. Qiang, L.; Montgomery, T.S.; Richard, S.P.; Anne-Marie, M. Investigation of basalt fiber composite aging behavior for applications in transportation. *Polym. Compos.* **2006**, *27*, 475–483. [CrossRef]
11. Ouyang, J.; Zhao, Z.; Yang, H.; Zhang, Y.; Tang, A. Large-scale synthesis of sub-micro sized halloysite-composed CZA with enhanced catalysis performances. *Appl. Clay Sci.* **2018**, *152*, 221–229. [CrossRef]
12. Piao, Y.; Jiang, Q.; Li, H.; Matsumoto, H.; Liang, J.; Liu, W.; Pham-Huu, C.; Liu, Y.; Wang, F. Identify Zr Promotion Effects in Atomic Scale for Co-Based Catalysts in Fischer–Tropsch Synthesis. *ACS Catal.* **2020**, *10*, 7894–7906. [CrossRef]

13. Bocci, E.; Prosperi, E.; Mair, V.; Bocci, M. Ageing and Cooling of Hot-Mix-Asphalt during Hauling and Paving—A Laboratory and Site Study. *Sustainability* **2020**, *12*, 8612. [CrossRef]
14. Mishra, R.; Behera, B.K. Impact simulation of three-dimensional woven kevlar-epoxy composites. *J. Ind. Text.* **2016**, *45*, 978–994. [CrossRef]
15. Shubhra, Q.T.H.; Alam, A.K.; Quaiyyum, M.A. Mechanical properties of polypropylene composites: A review. *J. Thermoplast. Compos. Mater.* **2013**, *26*, 362–391. [CrossRef]
16. Tsai, K.H.; Chiu, C.H.; Wu, T.H. Fatigue behavior of 3D multi-layer angle interlock woven composite plates. *Compos. Sci. Technol.* **2000**, *60*, 241–248. [CrossRef]
17. Yang, T.; Hu, L.; Xiong, X.; Mishra, R. Sound absorption properties of natural fibers: A review. *Sustainability* **2020**, *12*, 8477. [CrossRef]
18. Thwe, M.M.; Liao, K. Durability of bamboo-glass fiber reinforced polymer matrix hybrid composites. *Compos. Sci. Technol.* **2003**, *63*, 375–387. [CrossRef]
19. Jacob, M.; Joseph, S.; Pothan, L.A.; Thomas, S. A study of advances in characterization of interfaces and fiber surfaces in lignocellulosic fiber-reinforced composites. *Compos. Interface* **2005**, *12*, 95–124. [CrossRef]
20. Mishra, R. Drape behavior of 3D woven glass-epoxy composites. *Polym. Compos.* **2016**, *37*, 472–480. [CrossRef]
21. Rudov-Clark, S.; Mouritz, A.P. Tensile fatigue properties of a 3D orthogonal woven composite. *Compos. Part A Appl. Sci. Manuf.* **2008**, *39*, 1018–1024. [CrossRef]
22. Kalaprasad, G.; Thomas, S.; Pavithran, C.; Neelakantan, N.R.; Balakrishnan, S. Hybrid effect in the mechanical properties of short sisal/glass hybrid fiber reinforced low density polyethylene composites. *J. Reinf. Plast. Comp.* **1996**, *15*, 48–73. [CrossRef]
23. Kalaprasad, G.; Joseph, K.; Thomas, S. Influence of short glass fiber addition on the mechanical properties of sisal reinforced low density polyethylene composites. *J. Compos. Mater.* **1997**, *31*, 509–527. [CrossRef]
24. Mishra, R.; Gupta, N.; Pachauri, R.; Behera, B.K. Modelling and simulation of earthquake resistant 3D woven textile structural concrete composites. *Compos. Part B Eng.* **2015**, *81*, 91–97. [CrossRef]
25. Gude, M.; Hufenbach, W.; Koch, I. Damage evolution of novel 3D textile-reinforced composites under fatigue loading conditions. *Compos. Sci. Technol.* **2010**, *70*, 186–192. [CrossRef]
26. Mishra, R. Impact tolerance of 3D woven nanocomposites: A simulation approach. *J. Text. Inst.* **2013**, *104*, 562–570. [CrossRef]
27. Kumar, L.R.; Datta, P.K.; Prabhakara, D.L. Dynamic instability characteristics of laminated composite doubly curved panels subjected to partially distributed follower edge loading. *Int. J. Solids Struct.* **2005**, *42*, 2243–2264. [CrossRef]
28. *ASTM D3039/D3039M-17*; Standard Test Method for Tensile Properties of Polymer Matrix Composite Materials. American Society for Testing and Materials: West Conshohocken, PA, USA, 2017.
29. Hassan, T.; Jamshaid, H.; Mishra, R.; Khan, M.Q.; Petru, M.; Novak, J.; Choteborsky, R.; Hromasova, M. Acoustic, Mechanical and Thermal Properties of Green Composites Reinforced with Natural Fibers Waste. *Polymers* **2020**, *12*, 654. [CrossRef]
30. Friedrich, K.; Jacobs, O. On wear synergism in hybrid composites. *Compos. Sci. Technol.* **1992**, *43*, 71–84. [CrossRef]
31. *ASTM D5420-21*; Standard Test Method for Impact Resistance of Flat, Rigid Plastic Specimen by Means of a Striker Impacted by a Falling Weight (Gardner Impact). American Society for Testing and Materials: West Conshohocken, PA, USA, 2021.
32. Mishra, R. Specific functional properties of 3D woven glass nanocomposites. *J. Compos. Mater.* **2014**, *48*, 1745–1754. [CrossRef]
33. Ishikawa, T.; Chou, T.W. Stiffness and strength behaviour of woven fabric composites. *J. Mater. Sci.* **1982**, *17*, 3211–3220. [CrossRef]
34. *ASTMD 7264M-07*; Standard Test Method for Flexural Properties of Polymer Matrix Composite Materials. American Society for Testing and Materials: West Conshohocken, PA, USA, 2007.
35. Abot, J.L.; Daniel, I.M. Through-thickness mechanical characterization of woven fabric composites. *J. Compos. Mater.* **2004**, *38*, 543–553. [CrossRef]
36. Geethamma, V.G.; Kalaprasad, G.; Groeninckx, G.; Thomas, S. Dynamic mechanical behavior of short coir fiber reinforced natural rubber composites. *Compos. Part A Appl. Sci. Manuf.* **2005**, *36*, 1499–1506. [CrossRef]

Article

Thermoplastic Composites for Integrally Woven Pressure Actuated Cellular Structures: Design Approach and Material Investigation

Michael Vorhof [1,*], Cornelia Sennewald [1], Philipp Schegner [1], Patrick Meyer [2], Christian Hühne [3], Chokri Cherif [1] and Michael Sinapius [2]

1. Institute of Textile Machinery and High Performance Material Technology, Technische Universität Dresden, 01062 Dresden, Germany; cornelia.sennewald@tu-dresden.de (C.S.); philipp.schegner@tu-dresden.de (P.S.); chokri.cherif@tu-dresden.de (C.C.)
2. Institute of Mechanics and Adaptronics, Technische Universität Braunschweig, Langer Kamp 6, 38106 Braunschweig, Germany; pat.meyer@tu-braunschweig.de (P.M.); m.sinapius@tu-braunschweig.de (M.S.)
3. Institute of Composite Structures and Adaptive Systems, German Aerospace Center, Lilienthalplatz 7, 38108 Braunschweig, Germany; christian.huehne@dlr.de
* Correspondence: michael.vorhof@tu-dresden.de

Abstract: The use of pressure-actuated cellular structures (PACS) is an effective approach for the application of compliant mechanisms. Analogous to the model in nature, the Venus flytrap, they are made of discrete pressure-activated rows and can be deformed with high stiffness at a high deformation rate. In previous work, a new innovative approach in their integral textile-based manufacturing has been demonstrated based on the weaving technique. In this work, the theoretical and experimental work on the further development of PACS from simple single-row to double-row PACS with antagonistic deformation capability is presented. Supported by experimental investigations, the necessary adaptations in the design of the textile preform and the polymer composite design are presented and concretized. Based on the results of pre-simulations of the deformation capacity of the new PACS, their performance was evaluated, the results of which are presented.

Keywords: pressure-actuated cellular structure; shape morphing; compliant mechanism; anisotropic flexure hinges; textile-reinforced polymer composite; integrally woven structure

Citation: Vorhof, M.; Sennewald, C.; Schegner, P.; Meyer, P.; Hühne, C.; Cherif, C.; Sinapius, M. Thermoplastic Composites for Integrally Woven Pressure Actuated Cellular Structures: Design Approach and Material Investigation. *Polymers* **2021**, *13*, 3128. https://doi.org/10.3390/polym13183128

Academic Editor: Rajesh Mishra

Received: 28 August 2021
Accepted: 14 September 2021
Published: 16 September 2021

Publisher's Note: MDPI stays neutral with regard to jurisdictional claims in published maps and institutional affiliations.

1. Introduction

Compliant mechanisms offer immense advantages in their use. In contrast to conventional mechanisms, complex assemblies can be combined into a single part with the same functionality, and the mass and the number of necessary parts are significantly reduced. Components for coupling and guiding the individual parts are eliminated, as is the need for their maintenance and lubrication [1]. In general, the purpose of most mechanisms or mechanical systems is indirectly the transmission of energy or information between various interfaces. In addition to distribute the necessary material and energy flows, they make this possible directly by converting and transmitting forces and comparable mechanical quantities. In addition to the necessary energy sources and to an application-specific periphery, conventional mechanisms are made up of several links and coupling elements (joints). While the links are considered as idealized stiff structures, the joints have degrees of freedom (e.g., rotatory, translatory) so that kinematically determined mechanisms can be represented from their specific combination [2]. The advantage of this approach is that the kinematics and the stability (e.g., against mechanical and other loads) can be evaluated separately. This contrasts with the approach of using mechanisms that consist of only a single element and whose kinematic characteristics, e.g., their operational behavior, are represented by a locally differentiated compliance. The prerequisite for this is that the kinematics and the mechanical response (stresses, deformations) are considered together. The effort in the development of such compliant mechanisms is, therefore, depending on the application, significantly higher than for conventional mechanisms [3].

However, the advantages in the use of compliant mechanisms make them particularly interesting for aerospace applications. Shape-flexible structures and components made of fiber-reinforced composites enable mass savings but can also achieve significantly improved aerodynamic efficiency, for example, as a component of morphing wing technologies. Examples of this are the work on the 'FlexFoil system' (Kota et al. [4]), the 'vertebrate structure' (Elzey et al. [5]), or the 'droop nose' (Vasista et al. [6]). However, the problem with these concepts is the system-inherent contradiction between the required compliance and the necessary structural stiffness and strength. The stiffness is of particular importance for the creation of an accurately controlled aerodynamic shape. Cellular compliant structures with integrated pneumatic or hydraulic actuation have been identified as a promising solution, e.g., in [7–9]. Pagitz et al. and, later, Gramüller et al. developed a bionically inspired approach based on the leaf movement of the Venus flytrap (Figure 1). The Venus flytrap is able to rapidly change the internal pressure within discrete areas of the trap leaves. This allows effective trapping and safe containment of insects and spiders. This approach was further developed so that, on the basis of an algorithm, the cross section of such pressure-actuated cellular structures (PACS) can be optimized with respect to a desired structural deformation by adjusting the length of the cell walls [10].

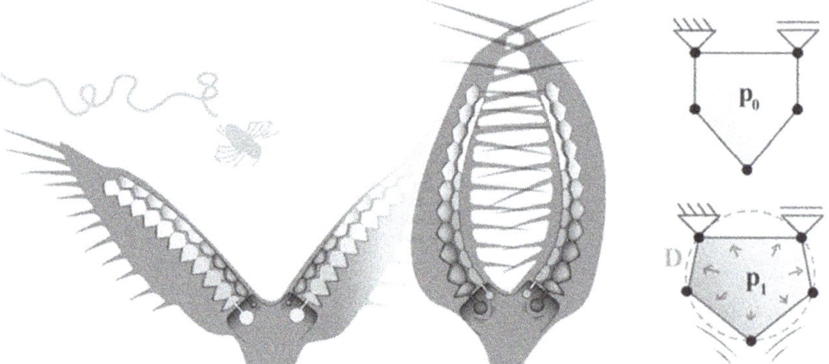

Figure 1. Schematic cross-sectional view of Venus flytrap and derived compliant mechanism of a single PACS Cell. Reprinted/adapted by permission from Springer Nature [11].

The result is the cross-sectional geometry of an optimized PACS, whose cells have a complex shape analogous to the cells of real plants and must exhibit extreme differences within the wall thickness for the necessary compliance (up to a factor of 12). For the production of the PACS, processes and materials are required with which the complex geometry specifications of the PACS cross section can be. As the first prototypes [9] have shown, the use of glass-fiber reinforced plastics (GFRP) is very promising here. However, the pre-impregnated fibers (prepregs) used for this must be manually placed in the form of the individual PACS cells. Automated deposition of the used prepregs within the cellular, non-planar geometry of the PACS is not possible. Manual positioning and draping of the prepregs prevent reproducible, precise adjustment of the wall thickness, which has a negative effect on the compliance characteristic. Furthermore, between adjacent cells is only a polymer-matrix-based connection. A continuous fiber-based connection of the cells to each other using prepregs is not feasible. As a result, PACS of this type have only limited resistance to fatigue strength. This type of production of complex PACS is, therefore, not scalable for high quantities or large components.

However, textile manufacturing technologies offer a high potential for the automation of preform operations for the production of complex fiber-reinforced polymer (FRP) structures with high reproducibility [12,13]. In comparison to different technologies such as knitting [14] and braiding [15], weaving is one of the most commonly used processes for the fabrication of textile-reinforced composite structures [16,17]. By integrating fiber reinforcement in a three-dimensional manner, which means in both in-plane and thickness directions [18], woven fabrics can be used as reinforcing structures in composites, which have a superior delamination resistance [19]. In addition, various approaches exist to provide fabrics with a cellular spacer structure in order to achieve a significant increase in stiffness and load-bearing capacity with a small increase in structural mass [20]. However, what these approaches have in common is that the compliance of the structures was not the driving factor in their development and, as structures with inherently too high stiffness, they are not suitable for implementing the basic principle of PACS. However, this does not mean that there are no actuated textile-reinforced composites [21,22]. There are so-called shape memory alloy fiber-reinforced polymers (SMA-FRPs), which can be activated via a textile-integrated SMA component (e.g., based on nickel and titanium) according to their compliance, which is also determined by their textile reinforcement structure. The mentioned SMA-FRP show great potential for the development of adaptive structures but have not yet been considered for transfer to the PACS principle in the context of the applications researched thus far.

A new approach for automated manufacturing using weaving technology was derived by Sennewald et al. [23]. This is based on the process combination of the technological approaches of terry weaving and spacer weaving [24]. For the woven textile PACS preform, an algorithm was developed to derive the simultaneously manufactured fabric layers under consideration of weaving technology restrictions. The high flexibility of the weaving technology enabled a continuous fiber-based connection of all PACS cells, as well as the representation of the required wall thickness gradient by process-integrated incorporation of solid pre-consolidated GFRP inlays. Using a subsequent process step to consolidate the woven preform in compression molding, new fabric-reinforced PACS could be developed but only in a single-row configuration [25,26] (see Figure 2).

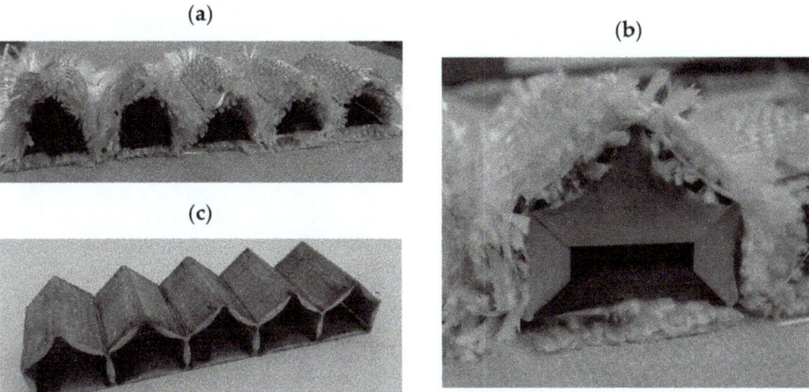

Figure 2. Textile reinforced single-row PACS: (**a**) woven fabric preform; (**b**) prepared preform with mold inlays before consolidation; (**c**) PACS composite.

This article represents the fundamental work and experiments to extend these findings regarding new double-row PACS, which enable a deflection not only in one direction but antagonistically adjustable deformation states. These are of great importance, for example, for the application in morphing wing concepts in order to specifically adapt the wing cross section for different operational conditions. This deformation behavior can be enabled analogously to the Venus flytrap by a double-row configuration of the PACS, whereby the cell rows can be separated and subjected to different internal pressures. Both for the necessary further development of the preform weave and for the technological implementation of the required geometry and wall thicknesses, the present results of single-row PACS can no longer be applied in a practicable way. The further development of the technological approach for double-row PACS requires extensive adaptation of weaving technology parameters (e.g., yarn arrangement, weave) and of the PACS cross section. These aspects are related in a complex way to the deformation behavior of the PACS and must therefore also be investigated in a complex manner.

As a starting point for further development (Figure 3), the form-finding algorithm further developed by Meyer et al. is therefore used, on the basis of which a preliminary design approach for the cross section of double-row PACS can be derived. Based on the experimental results obtained in this study, this can be concretized up with information on the characteristics of the textile preform and the most important parameters of the achievable composite properties. A finite element (FE)-based simulation enables the iterative optimization of the textile and composite process parameters to be applied on the concretized cross-sectional design. For this purpose, the simulation results were compared with calculations for a design draft for equivalent PACS made of pure polyamides, such as a design that could be produced by selective laser sintering (SLS).

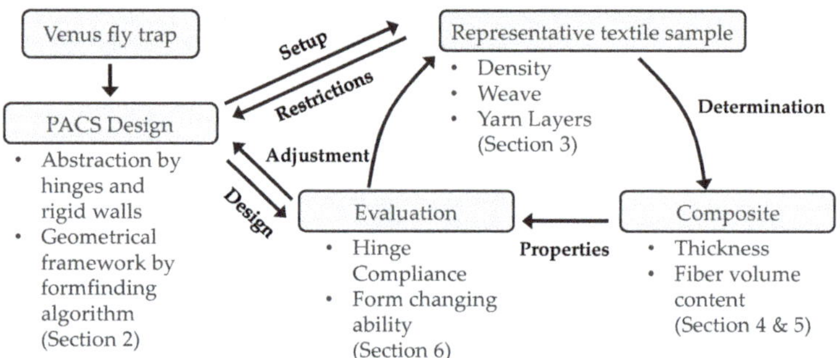

Figure 3. Process chain for the development and evaluation of textile, polymer composite, and design parameters for double-row PACS.

2. New Geometry Concept for Textile-Reinforced PACS in Antagonistic Configuration

The fundamental premise in the development of PACS is a cell cross section that remains constant along the spanwise direction. Therefore, all development steps can be performed using the 2D cross-sectional geometry. Compared to PACS in single-row configuration, the adequate determination of an optimized cell cross section is significantly more complex since the deformation behavior of the PACS is characterized by interactions between the cell rows. Furthermore, technological restrictions with regard to the weaving of the fabric preform, as well as the consolidation in compression molding, have to be taken into account. The creation of the PACS cross section is, therefore, described in detail in this section (Section 2.1), as well as its further development and modification (Section 2.2). Analogous to the implementation of single-row PACS, GFRP continues to be the material of choice due to its advantageous strength/stiffness ratio. In addition to the flexibly selectable geometrical design parameters (see Section 2.1), the mechanical parameters are the most important input variables for generating the cell cross sections. These were determined in advance of the design development using tensile specimens (DIN EN ISO 527, type 3) [27] from seven layers of orthotropic plain-woven fabric (Enka® TecTape Hybrid Roving GF-PA 6733, warp and weft density 1.5/cm, further information in Section 4.1.1) with a total composite thickness of 2.15 mm in warp and weft direction. A universal testing machine from the manufacturer Instron (type 5584) with a 150 kN load cell was used as the testing device. The strain was measured using an extensometer. The evaluation of the tensile tests showed a Young's modulus E_{11} of 19,842 $\pm$ 458 MPa, a tensile strength of 375 $\pm$ 10 MPa, and an elongation at break of 2.65 $\pm$ 0.09% (confidence interval 95%). Since no significant difference could be found between the warp and weft testing direction, the material properties for both series of specimens were combined for statistical evaluation.

2.1. PACS Cross Section and Design

The basis for the development of the multi-row PACS with an antagonistic configuration of the cell rows and high deformation capacity is the form-finding algorithm described by Gramüller et al. [11]. The implementation of an improved cell design by Meyer et al. [28] also allows a better model prediction accuracy due to an optimized load-following alignment of the flexure hinges. Furthermore, a smooth and gap-free surface contour can now be realized, which is advantageous for aerodynamic applications. The underlying form-finding algorithm adapts the cell walls of a double-row structure of arbitrary polygonal cells in such a way that the structure can move between two predefined target shapes depending on its internal pressure. Based on geometric and performance-driven constraints, optimized coordinates for flexure hinges and cell walls were obtained, which could be used to derive the cross section of the PACS. The algorithm can be divided into two steps.

In a first step, the cell wall lengths are optimized on a reduced-order truss model with rigid cell walls and discrete point joints. Figure 4a shows the truss geometry obtained from the form-finding process for a double-row structure with four and five cells in the top and bottom cell rows, respectively. The following parameters were used as input for the form-finding process. These were derived from the size of the intended test structure:

- Cell width of 52 mm;
- Cell wall thickness of 6 mm;
- Flexure hinge thickness of 0.35 mm;
- Flexure hinge length of 5 mm;
- Flexure hinge eccentricity of 5 mm.

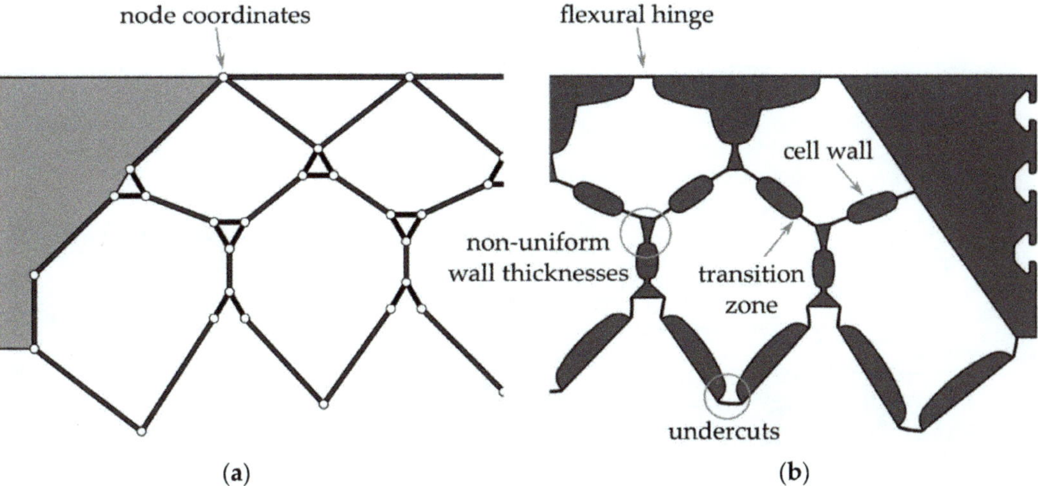

Figure 4. PACS design: (**a**) results from reduced-order truss model; (**b**) cross-sectional design optimized for additive manufacturing.

An internal pressure of 0.5 MPa and a deflection per cell of 5° were defined as performance-driven parameters. The input parameters were determined from boundary conditions of the weaving and consolidation process, as well as from previous experience in the textile–technical implementation of PACS [26].

In the second design step, the determined node coordinates are transformed into a cross section with solid flexure hinges, transition regions, and cell walls of specific thickness. Figure 4b shows the 2D cross section derived using the approach described in [11], which is optimized for manufacturing using additive manufacturing processes. However, a direct transfer of this cross-sectional geometry to FRP manufactured by the weaving process is not possible due to undercuts and non-uniform wall thicknesses. Therefore, a new approach to derive the cross-sectional geometry from the nodal coordinates has to be developed.

2.2. Cross-Sectional Adaption to FRP Process Restrictions

As a result of the algorithm, wall thicknesses and structural features of the preliminary optimized PACS cross section were distributed irregularly. In consequence, as a basis for the later weave development and consolidation tool design, the coordinates were aligned using a grid and a unified cell geometry. The efficiency of the weaving development for the woven fabric preform could thus be significantly increased, and representative weave areas could be identified for the necessary fundamental experimental work of this article. At the same time, a uniform structure of the PACS allowed a reduced effort in mold development so that the mold of the consolidation tool could later be created by assembling several identical parts and manufactured much more efficiently. Furthermore, areas of the PACS could be added that allowed for fiber-based coupling of cell rows, and potential undercuts could be eliminated. As a result, the post-processed coordinate set was used to develop a parameterized model for the PACS that includes the position of the flexure hinges with high accuracy and enables the adjustment of the wall thicknesses. Figure 4 shows a schematic depiction of the coordinate data set of the calculations and, based on it, the developed geometry of the PACS cross section.

To generate the cross section, all relevant areas and structural features could be adjusted by the internal parameters of the model. This is of essential importance for the achievable thickness of the flexure hinges, the geometric expression of the cell areas, and the additionally required contact area between the cell rows. The additional contact areas can later be used to implement an interlaced fiber structure within the PACS composite regarding the fabric structure of the woven preform. Thus, delamination-induced fatigue failure of the PACS in this area can be avoided. The following experimental work thus serves to determine the parameters for a material- and process-suitable design of the flexure hinges, the cell walls, and the required contact areas (see Figure 5). Furthermore, for future research, it is planned to subsequently fill the V-shaped cavities above the contact areas with inlays, since these are not to absorb any deformation. The concretized design will be evaluated in an FE simulation with regard to the deformation capacity, and the cross-sectional design will be iteratively adapted.

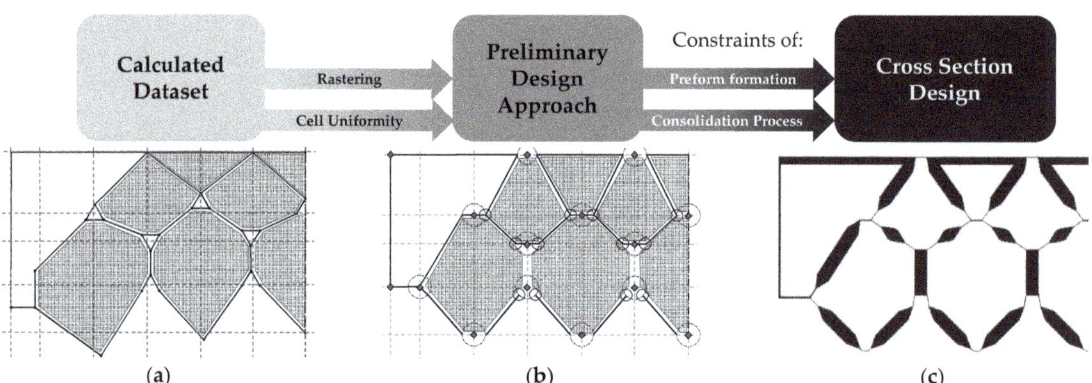

Figure 5. Steps of the geometry derivation for double row PACS (half-cross-sectional images of the symmetrical structure displayed): (**a**) calculated raw dataset and rastering; (**b**) processed dataset with aligned cell geometry; (**c**) derived cross section.

3. Mechanical Properties of the Composite and the Weave Structure

The description of composites includes information about the reinforcing fiber and matrix components used and the arrangement in relation to the FRP component. In addition to these parameters, the fiber volume content ρ_F determines the component properties. In terms of PACS, stiffness (Young's modulus E_{11}) and tensile strength (R_{11}) in the fiber direction are crucial for targeted structural design. In advance of composite fabrication and without test data, these can be estimated using the mixing rules for the intended fiber–matrix combination [29]. In the first instance, these relationships were used for the calculation of preliminary geometry datasets in preparation for the development of the textile and PACS design.

$$\rho_C = \rho_F \varphi_C + \rho_M (1 - \varphi_F) \tag{1}$$

$$E_{11} = E_F \varphi_F + E_M (1 - \varphi_F) \tag{2}$$

$$R_{11} = R_F \varphi_F + R_M (1 - \varphi_F) \tag{3}$$

The deformation behavior of the PACS is determined by both Young's modulus and material thickness t of the flexure hinge areas of the PACS. Material thickness is directly dependent on the characteristics of the fabric preform. Assuming mass equivalence of the fabric and composite, the material thickness t_C in hinge areas can be determined as follows with knowledge of the areal density of the fabric preform AD_W:

$$t_C = \frac{AD_C}{\rho_C} = \frac{AD_W}{\rho_C} = \frac{AD_W}{\rho_F \varphi_F + \rho_M (1 - \varphi_F)} \tag{4}$$

The areal density of the fabric preform corresponds to the sum of the basis weights of the fabric layers, which can be determined in a sufficient approximation on the basis of the linear thread densities $t_{T,F}$ in warp and weft direction, as well as the average distance (see Figure 6) between the threads in warp and weft.

$$AD_W = \sum_{l=1}^{l_W} \frac{t_{T,Warp,l} \, t_{T,Weft,l}}{a_{Weft,l} \, a_{Warp,l}} \tag{5}$$

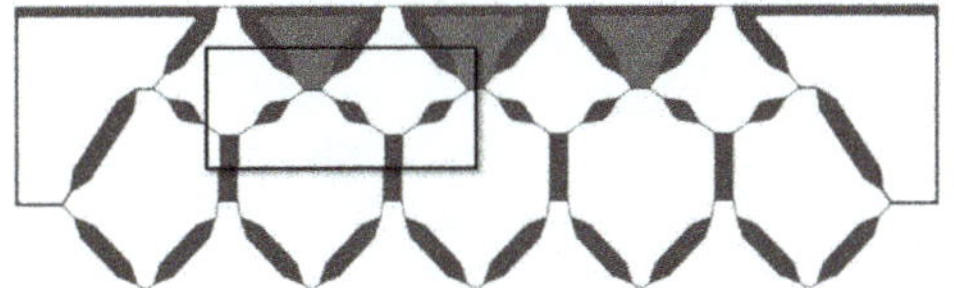

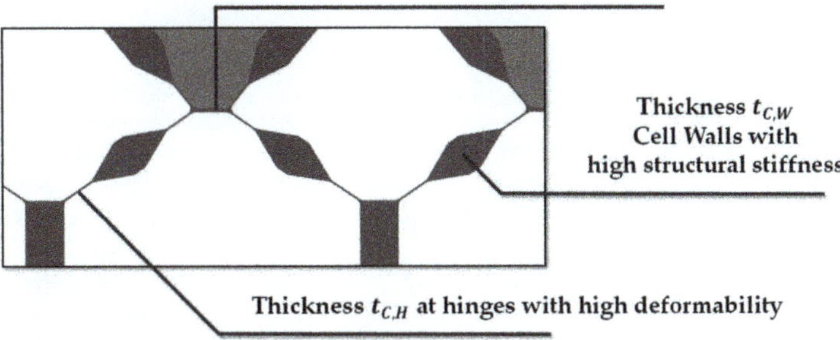

Figure 6. Cell design, geometrical characteristics, and thickness parameters.

With knowledge of the fabric weave and the average thread spacing within the preform layers, the area density can be calculated. The approach is based on the assumption that warp and weft yarns within the fabric have a quasi-circular cross section. In this way, it is possible to compare yarn positions available with actually occupied positions. Based on this assumption, it is then also possible to compare woven fabrics made of different materials and weaves [30,31]. Composites, on the other hand, are mostly composed of yarn materials with flat ribbon-like cross sections, and various modeling approaches exist to describe the yarn cross section (circular [32], racetrack [33], lenticular [34]). These allow accurate modeling and calculation of fabric properties in diverse applications (e.g., drape behavior, permeability) but are not practical for the intended evaluation of fabric properties for fabrication of the textile-generated and -reinforced PACS. Based on the material data of the reinforcement yarns, a substitute cross section was formed from several circular cross sections $d_{Warp,l,sub}/d_{Weft,l,sub}$ for the calculation of the fabric density d_W and the fabric and material parameters (Figure 6).

Based on the definition of the equivalent cross sections, the relative fabric density d_W is obtained as follows:

$$d_W = \frac{p_W \left(d_{Warp,l,sub} + d_{Weft,l,sub} \right)^2}{a_{Weft,l,sub} a_{Warp,l,sub}} \tag{6}$$

Physically, the parameters fabric density and areal density/composite thickness are closely related. However, the direct analytical coupling is not readily possible. The reason for this is that the formulas mentioned do not contain any information about the porosity of the fabric preform. This varies considerably depending on the material and processing. Therefore, one objective of this work was the experimental determination of the compaction of the fabric preform during consolidation as a measure for the preform porosity. The compaction of the fabric preform during consolidation occurred in two stages (see Section 4). The first stage began with the closing of the mold. First, a hybrid yarn was used, which had a porosity of approx. 50–70%, and was compacted to the densest packing. A residual porosity remained within the preform, which was filled during the second phase. Under the influence of pressure and temperature, the polyamide component in the hybrid yarn melts and impregnates the glass filaments as the matrix. At the same time, further densification of the glass filaments occurs. The way in which stress within the preform is reduced by the melting of the polyamide of the hybrid yarn and the reinforcing fibers can thus assume the required position in the PACS in the composite according to the geometry of the mold. However, there are limits to this compaction due to the characteristic interlacing of the warp and weft yarns, since only yarn portions running parallel can be optimally compacted. Another compaction effect is represented by nesting, in which the proportion of crossing points that are shifted relative to one another and arranged one above the other enables additional compaction [35,36]. The determination and description of the compaction characteristics is, not at least, a decisive basis for the successful implementation of a tool design for the consolidation of the PACS.

4. Materials and Methods

4.1. Materials

4.1.1. Basic Fabric Setup

In consequence, a certain main objective of this study was to determine the achievable wall thickness in the composite component as a function of the number of yarn layers and the fabric density. An *Enka*® *TecTape Hybrid Roving GF-PA 6733* (PHP Fibers, Obernburg am Main, Germany) in warp and weft direction was used for fabric production. The roving has a linear density of 1855 tex with a glass fiber content of 67% by mass (corresponds to a volume content of approx. 46% in the composite). The sample was produced on a rapier loom *P1* from Lindauer DORNIER, Lindau, Germany. In particular, for the production of fabrics with several warp and weft layers (LTL1-3, Figure 7), this machine was equipped with a special thread take-up. Here, the warp threads run in successively staggered double flat steel heddles (e.g., *TWINtec* by DERIX, Grefrath, Germany), which is very advantageous when processing coarse warp material in high warp thread densities [37].

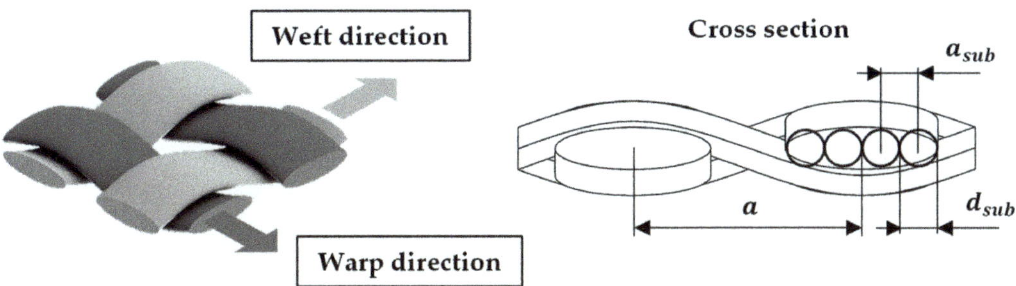

Figure 7. Yarn arrangement in woven fabric preforms (left) and cross-sectional view with substitute diameters.

On the one hand, single-layer samples of the respective fabric configuration were investigated and, on the other hand, multilayer stacked samples. In this way, the resulting composite wall thickness in each case is determined using 2 up to 10 fabric layers stacked on top of each other. The structure of the individual layers investigated and an overview of the fabric configurations are shown in Figure 5 and Table 1.

Table 1. Overview of examined woven fabric preform configurations.

Weave Setup	Number of Yarn Layers		Number of Warp Yarns per cm		Number of Weft Yarns per cm	
	Warp	Weft	Per Layer	Σ	Per Layer	Σ
PL1	1	1	1.50	1.50	1.50	1.50
PL1a	1	1	1.50	1.50	1.62	1.62
PL1b	1	1	1.50	1.50	1.74	1.74
PL1c	1	1	1.50	1.50	1.86	1.86
PL1d	1	1	1.50	1.50	2.00	2.00
LTL1	1	2	1.50	1.50	1.50	3.00
LTL2	2	3	1.50	3.00	1.50	4.50
LTL3	3	4	1.50	4.50	1.50	6.00

4.1.2. Test Series and Processing

According to the objective of analyzing the influence of weft density and the number of yarn plies, two different sample series were used. On the one hand, single-ply plain weaves with different weave densities, and on the other hand, fabric stacks from different weaves were investigated (Table 2).

Table 2. Test series numbering and assignment.

Examined Parameter	Series Identification No.	Weave Setup
	1	PL1
	2	PL1a
Relative Weave Density	3	PL1b
	4	PL1c
	5	PL1d
Number of Yarn Layers	1 *; 6–16	various (see Table 3)

* Series no. 1 has been used for both examination objectives.

Table 3. Selection of possible partitioning options in the structure of the fabric samples with 2 to 10 yarn layers. The examined variants are marked with x. The numbering is consecutive for the identification of the respective sample.

Preform Layer	Number of Yarn Layers of the Preform																					
	2	3	4	5	5	6	6	7	7	7	8	8	8	9	9	9	9	10	10	10	10	10
2/(a) **	1		2		1	3		2	1		4	1		3		2	1	5	2	1		
3/(b)		1			1		2	1				2	1	1	3				2	1		1
5/(c)				1					1				1			1				1	2	
7/(d)										1							1					1
Tested:	x	x	x	x	x	x			x	x	x			x				x				
No.:	1	6	7	8	9	10			11	12	13			14				15				

** according to Figure 1.

For the systematic investigation of the composite wall thickness as a function of the number of layers (samples No. 1, 6–16), various combinations are possible with the given weaves (Figure 8). These can be formed on the basis of the partitioning (summation decomposition) [38] of the required number of layers. To limit the testing effort, technologically equivalent variants were selected with respect to the number of plies so that the yarn plies 2 to 10 were represented at least once and, for control purposes, the number of layers 5 and 7 were represented twice (Table 3).

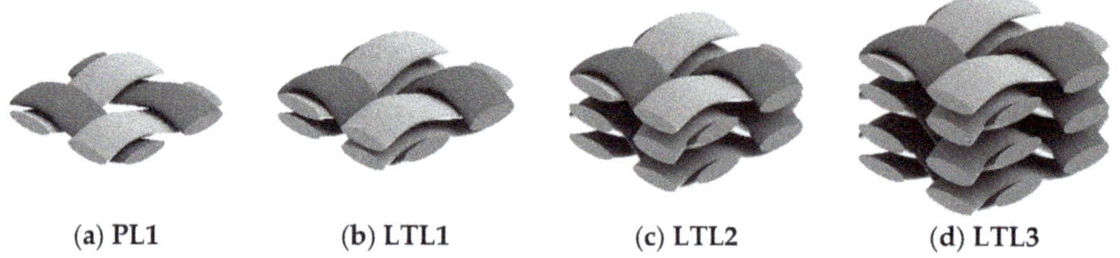

(a) PL1   (b) LTL1   (c) LTL2   (d) LTL3

Figure 8. Overview of the basic fabric weave and their structure: (**a**) plain-woven fabric; (**b–d**) layer-to-layer weaves.

The processing of all composite specimens was carried out by compression molding on a *COLLIN P300 PV* laboratory heating press (COLLIN Lab & Pilot Solutions, Maitenbeth, Germany) in a platen tool (Figure 9). The temperature–pressure–time cycle during specimen consolidation was matched to the processing of the PA6 matrix, ensuring full impregnation of the glass fibers and a minimum pore content (Figure 10, maximum temperature 280 °C, maximum pressure 65 N/cm^2, heating and cooling rate 10 K/min). The consolidation process was carried out in a residual ambient pressure of 0.01 MPa.

Figure 9. Sample consolidation: laboratory platen-press COLLIN 300 PV and plate tool.

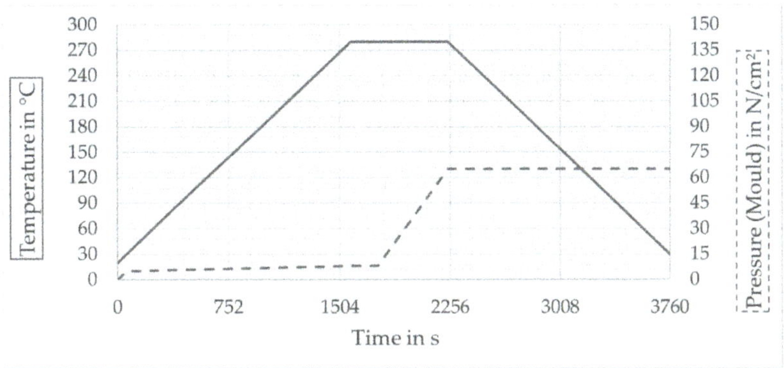

Figure 10. Temperature–pressure–time diagram of the sample consolidation.

4.2. Methods

All measurements on textile and composite samples were carried out under standard climate conditions (DIN EN ISO 139: 20 °C, 65% rel. humidity) [39] and after a minimum of 48 h sample conditioning, in order to establish an equilibrium with regard to the water absorption of polyamide from the environment.

The fabric thickness was measured with a *Rainbow* digital textile thickness tester from KARL SCHRÖDER (Weinheim, Germany) in combination with a digital measuring device (Sylvac S229, SYLVAC, Yverdon, Schwitzerland, Figure 11) according to DIN EN ISO 139 [40] (measuring plunger pressure area 25 cm^2, test pressure 2 kPa). The measurements were made 30 s after specimen gripping. Eight measured values were distributed over the specimen areas, and the average material thickness was determined from them. Paper support (measured: thickness 0.3 mm; area density 0.231 kg/m^2) was used to transfer the specimens between cutting, measurement, and consolidation, and its values were subtracted from the measured values in the evaluation. The consolidated samples were also tested according to the same methodology. Furthermore, the fiber volume content was determined on random samples according to DIN EN ISO 1172 [41]. The areal density was determined gravimetrically on all samples.

Figure 11. Thickness measurement device rainbow (KARL SCHRÖDER/SYLVAC).

5. Results

The diagrams (Figures 12 and 13) show the results of the tests. The diagrams in Figure 11 show the results of the investigation of fabric preforms with the same ply structure and fabric weave. To analyze the influence of relative fabric density on fabric and composite thickness, the weft density was increased successively (see Tables 1 and 2). In addition to the measured data, the results of the calculations (according to Equations (4) and (5)) are also entered. The comparison shows that measurement and calculation agree well. Figure 13 shows the results of varying the layer structure of the preforms by successively increasing the number of yarn layers. The resulting increase in composite wall thickness is clearly visible. On the one hand, the comparison of the measured values with the results of the calculations shows that, especially for preform structures with a low number of plies, measurement and calculation agree well with each other. On the other, with increasing preform thickness, an increasing deviation between measured and calculated values can be observed.

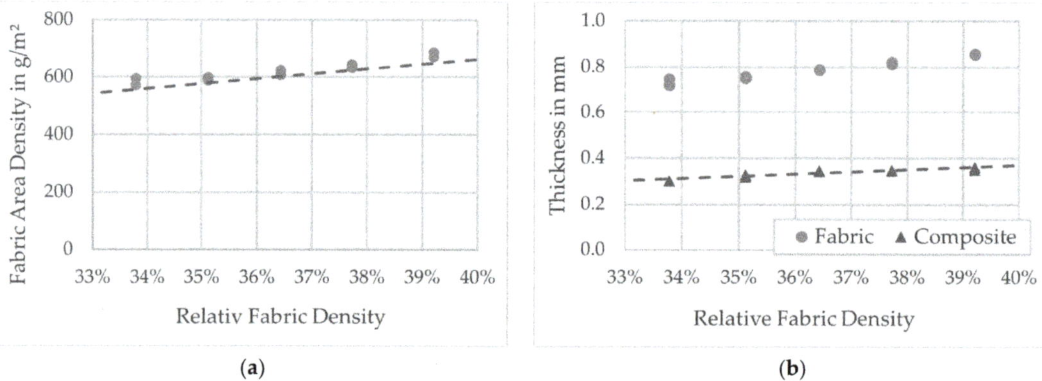

Figure 12. Results of the fabric testing for various weft yarn densities: (**a**) fabric area density and calculation result as trendline; (**b**) fabric and composite thickness with calculation result as the trendline.

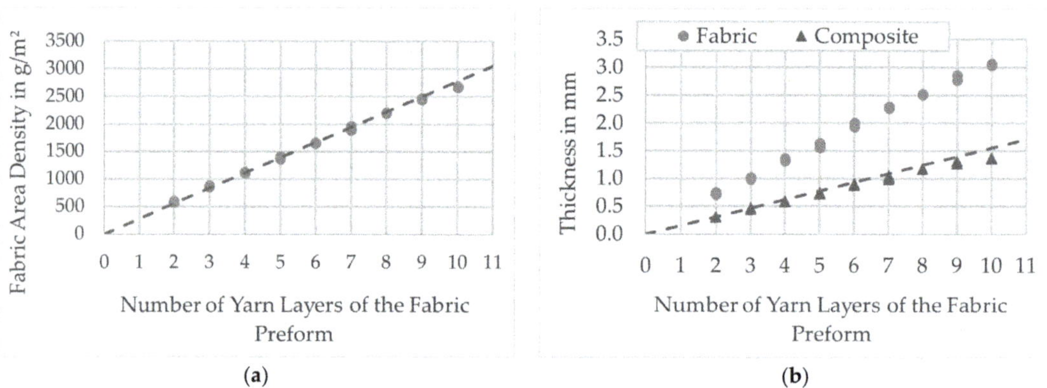

Figure 13. Results of the fabric testing for different number of yarn layers of the preform: (**a**) fabric area density and calculation result as trendline; (**b**) fabric and composite thickness and calculation result as the trendline.

As already shown in Table 3, the greater the number of layers, the greater the possibilities of forming them from the different fabric layers within the preform. However, the measured values do not show any effects of the arrangement of the fabric layers on the preform and composite thickness. Thus, the basic assumption can be confirmed that, with similar preform surface mass, it is above all the number of superimposed yarn layers that determines the composite thickness. Thus, preform thickness and composite thickness can be defined with a high degree of certainty as decisive factors for the detailed PACS consolidation tool design. Figure 13 shows the average fiber volume content of all samples.

6. Discussion

Based on the experimental results, the design for integrally fabric-reinforced PACS can now be concretized and the deformation capacity can be adequately evaluated. The determined correlations between the fabric preform and composite design are presented and discussed together with the results for the simulation of the deformation capacity.

6.1. Weave Density and Preform Layup

The deformation behavior and load capacity of the PACS are highly dependent on the wall thickness within the flexure hinge areas. The wall thickness in the hinges determines the stress that can be applied normally in the direction of the hinge by applying pressure to the cells. The capacity to change, on the other hand, is determined by the bending stiffness of the hinges (see [16]). In order to match the fundamental design framework for the flexure hinges with a length of about 5 mm within the fabric preform, it is necessary to increase the weft density and thus the fabric density from approx. 34% to approx. 39%. For this purpose, concrete values for the expected wall thicknesses within the hinges could be obtained on the basis of the tests. Thus, an increase in the weft density from 1.5/cm to 2/cm results in an equally approx. 18% higher wall thickness in the hinge from 0.30 mm to approx. 0.35 mm. Based on the measurement results and the calculations, a directly proportional relationship between fabric density and composite wall thickness for the thin-walled hinges can be demonstrated with these investigations (assumption: mass equivalence of preform and composite).

Another factor influencing the thickness of the composite is the sequencing of the layers within the fabric preform. According to Table 3, the preform structure can consist of individual layers with different bond-determined numbers of filaments. A fundamental characteristic of the compaction and consolidation behavior of the fabric preforms is that crossed filament layers "lock" each other, i.e., glass filaments of a yarn system cannot be displaced into adjacent, parallel yarns during consolidation. This relationship can be observed on the basis of the experimental analysis of the fiber volume content. The share of reinforcing fibers is at the very same level of approx. 45% for all samples, which corresponds to the fiber volume content of the hybrid polyamide/glass raw material. The test results showed that this effect is also effective when stacking different individual layers up to a certain preform thickness. Irrespective of the weave, preforms with cumulatively the same number of filament layers (e.g., test series no. 11: $1 \times 2 + 1 \times 5$ vs. test series no. 12: 1×7) exhibit similar compacting behavior (see Figure 14).

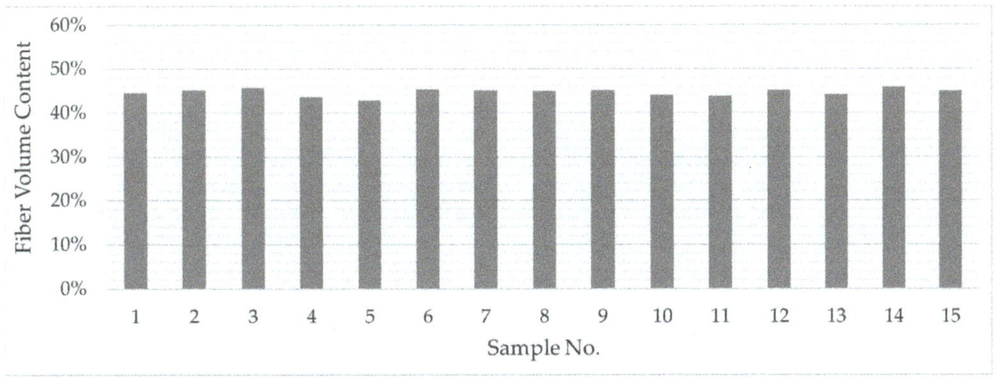

Figure 14. Overview of the measured values for the fiber volume content.

Above a yarn layer count of 9, an increasing deviation of the composite thickness between the calculated values and the measured values can be observed (e.g., test series 15 with 14% difference). This is due to increasing nesting effects between the preform layers. The decisive factor for the occurrence of the overall composite thickness is the concentration of statistically distributed areas with parallel oriented filament layers within the individual preform layers. During consolidation, these areas are locally more densified, so that the expected proportional increase in the composite wall thickness is then no longer given. This was taken into account when deriving the mold geometry of the cell wall area.

6.2. Preform and Consolidated PACS

The measurement results allow the preform design to be specified and the resulting PACS geometry to be estimated with a high degree of accuracy and certainty. Compared to other thermoplastic reinforcement materials, the used TecTape shows a significantly reduced volume shrinkage of less than 50% in the consolidation (See Figure 15). For comparable thermoplastic materials, this is 70% and more [42]. This finding significantly simplifies the development of a PACS consolidation tool, since the measures required to compensate for volume shrinkage can be much lower. Furthermore, the determination of values and basic assumptions for the estimation of the deformation of the PACS for the simulation are possible with a high degree of certainty.

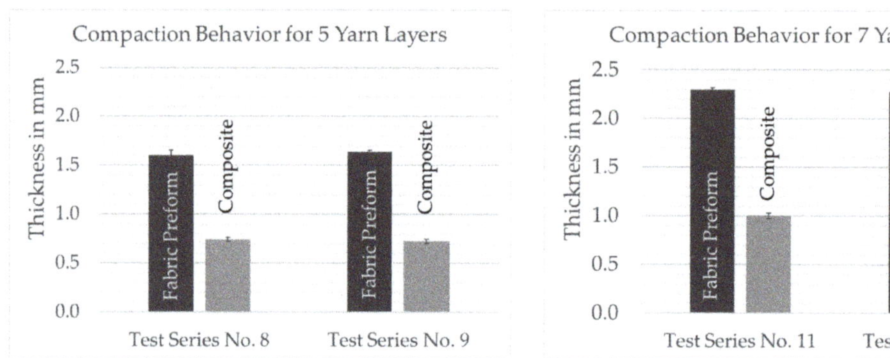

Figure 15. Comparison between different layup variants and preform compaction behavior during consolidation.

The realization of the necessary maximum length of the hinges of 5 mm requires a weave density of the preform of 39%. This results in a thickness of the woven preform of approx. 0.86 mm and the consolidated PACS of 0.35 mm in the hinge areas. This value is significantly lower than the one reported in [29] of approx. 0.5 mm. This improvement can be attributed to the modified production setup, which had to be adapted to the production of the double-row PACS. On the other hand, interlacement influences can consequently be regarded as quasi-homogeneously distributed over the cross section of the flexure hinges. The contact area between the hinges (see Figures 6 and 16) must be double layered for a stable coupling between the cell rows. This is achieved in the production of the PACS preforms by weaving the upper and middle warp layers interlocked together. This avoids shear and cross-tension-induced delamination as a result of the large structural distortion during the actuation of the PACS. The thickness of the PACS preform is 1.62 mm in the contact areas and approx. 0.74 mm after consolidation.

The overall stiffness of the PACS is adjusted via the thickness of the cell walls. For a secure closure [26], the cell walls should simultaneously be used for screwing on appropriate closure caps. These can be realized by weaving-integrated inserts or screws directly inserted into the composite. The implementation will be concretized in the context of the pending tool development for the consolidation of the new PACS. The target value for the geometry development of the PACS is therefore a composite wall thickness of 6 mm. With linear regression of the measured values, the fabric preform would have to consist of 39 filament layers and have a thickness of 11.7 mm. Since the influence of nonlinear nesting effects increases significantly with the number of yarn layers, this value will be even greater in reality. The purely weaving implementation of such thicknesses is not reasonable. Nevertheless, in order to represent such wall thicknesses, different strategies are available, including monolithic composite using composite inserts [25,26], sandwich, or hollow chamber structure. The selection was made considering the results of the numerical evaluation of the deformation behavior.

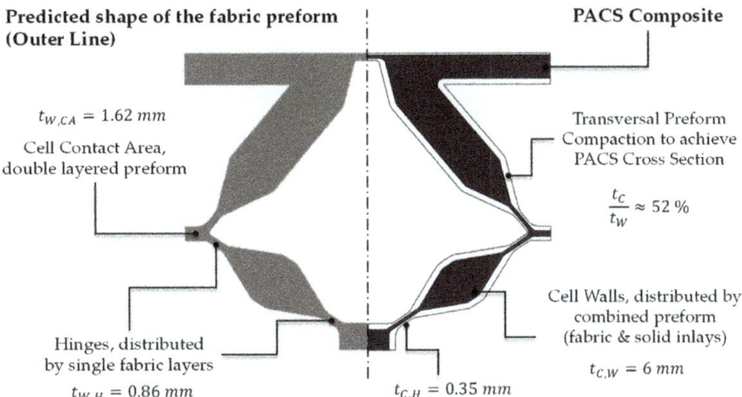

Figure 16. Cross section of the fabric preform and the consolidated PACS.

6.3. Evaluation of the PACS Deformation Behavior

A finite element analysis (FEA) was used to evaluate the deformation capacity of the PACS. Two cross-sectional designs were investigated: (i) "CS-AM", with the cross section optimized for additive manufacturing (AM) according to [11] (Figure 4b); (ii) "CS-FRP", with the design adapted to integral woven GFRP manufacturing (Figure 5c). An overlay of both designs is shown in Figure 17.

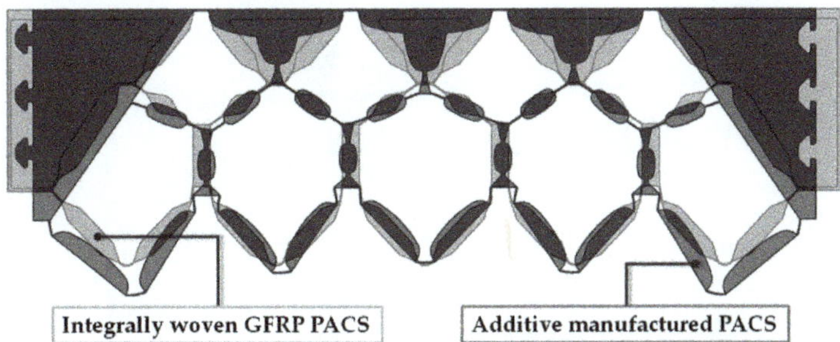

Figure 17. Comparison of two PACS cross-sectional designs: "CS-AM" optimized for additive manufacturing; "CS-FRP" adapted to restrictions from integral woven GFRP manufacturing.

For the FEA, a 2D model was used in ANSYS assuming a plane strain state. The meshing was performed with quadratic plane-183 elements. The mesh density was controlled by the number of elements across the gap. In a mesh refinement study, convergence was obtained for six elements in the thickness direction of the flexure hinges, resulting in a total of 38,053 elements for "CS-AM" and 29,975 elements for "CS-FRP". The internal pressure was applied via surf153 elements as a surface effect.

Figure 18 shows the cross-sectional design of "CS-FRP" in the deformed state for the upper and lower deflection when only one cell row was pressurized with the design pressure of 0.5 MPa in each case. For the calculations, the upper triangles and the two outer connection elements were considered as solid materials, since they do not contribute to the compliance and are later filled with inlays. The left edge of the model was fixed.

The angular displacement $\Delta\beta$ of the right outer edge was considered as a measure of the change in shape. By selectively adjusting the pressure difference between the two cell rows, any deflections between the two extreme states could be achieved.

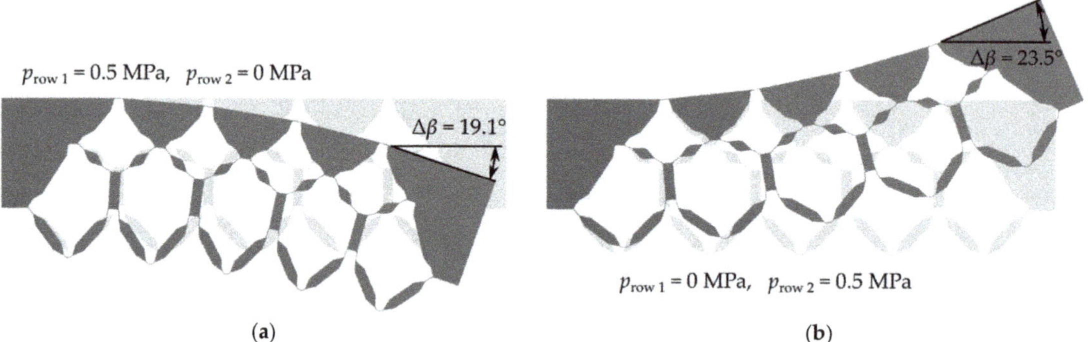

Figure 18. Deflection of "CS-FRP" calculated with FEA for both targeting states: (**a**) pressurization of the upper cell row; (**b**) pressurization of the lower cell row.

Figure 19 shows the mechanical behavior of the cellular structure as a function of the internal pressure when only one cell row was pressurized. The results from the reduced-order truss model were compared with the FEA results for the two cross-sectional designs "CS-AM" and "CS-FRP". Differences between the reduced-order truss model and the FEA on the design "CS-AM" have already been explained in [26] and arise primarily from the simplification in the reduced-order model. The hinge transition areas were not taken into account and the deformation of the cell walls was neglected, both resulting in lower deflections, compared to the FEA.

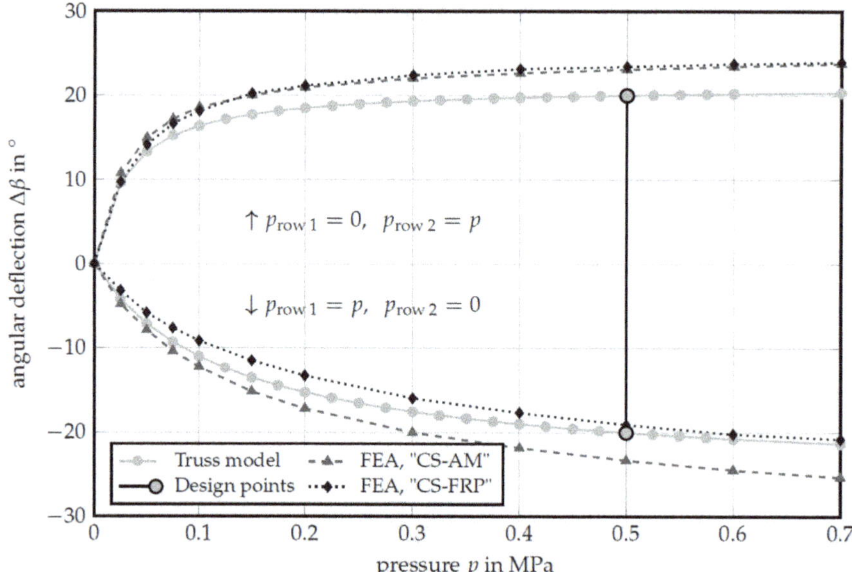

Figure 19. Angular deflection $\Delta\beta$ as a function of the cell pressure. Comparison of the integrally woven "CS-FRP" and the "CS-AM" optimized for additive manufacturing. Pressurization was applied to only one cell row at a time, while the pressure in the other was set to zero.

A comparison of the two cross-sectional designs "CS-AM" and "CS-FRP" shows that pressurizing the lower cell row results in an almost identical deformation curve for both designs. Various restrictions had to be taken into account when converting the cross section into a weave- and tool-compatible cross-sectional design (Section 2.2). As a result of the simulation, it can be shown that a simplification of the cross-sectional geometry does not have to result in any drawbacks with regard to the deformation capacity restrictions of the compliant mechanism of the PACS.

When the upper cell row is pressurized, the "CS-FRP" design appears stiffer than "CS-AM". This deviation needs further investigation. Currently, it is assumed that it results from the unification of the cells and the nodal coordinates. This leads to a reduced eccentricity of the undeformed upper cell in this row and to a reduction of the deformation capacity. In general, the upper cell row reacts more sensitively to changes in the loading condition, since the lever arm is lower relative to the bending neutral axis.

7. Conclusions

The presented work established the essential basis for the development of integrally textile-reinforced double-row PACS. In the beginning, it was possible to draw on extensive empirical data, which enabled the successful previous development of textile-reinforced single-row PACS. In the course of the consistent extension of the PACS approach to double-row structures, it became apparent that the technological and theoretical bases used thus far were not sufficient to capture and produce the significantly more complex geometry of double-row PACS. As a result, a new PACS design was derived on the basis of existing experience in the development of single-row PACS and an extended form-finding algorithm and was further developed for implementation as a textile-reinforced FRP structure. An essential criterion in deriving the highest possible deformability of the PACS is the compliance of the flexure hinges, the material-dependent stiffness, elongation limit, and, last but not least, the wall thickness as a geometry-based coefficient. On the basis of the new experimental data, it could be shown that the latter in particular can be significantly improved with the use of a new setup in fabric production, whereby the wall thickness can be reduced from the previous 0.5 mm to the current 0.35 mm. This corresponds to a reduction of 30% and, in terms of hinge stiffness, approx. 65%, which means that the potential deformation capacity of the PACS can be significantly increased in the future. The experimental investigations identified a strict relationship between hybrid yarn material, fabric weave, and the resulting composite attributes. Consequently, the knowledge about these findings enables the effective further development of fabric structure for integrally woven double-row PACS. The textile design is strongly influenced by the properties of the subsequent composite structure, in particular the wall thickness within the flexure hinges. These influences and essential correlations are now more describable. In connection with the further developed unified and adapted design, the tool design for the consolidation of the PACS preform can be derived. Numerical simulations prove that these necessary adaptations do not cause significant disadvantages for the later form-changing capability of the PACS and that a high degree of functionality can be achieved by the integrally woven fabric-reinforced PACS. These results are a significant milestone in the development of textile-reinforced PACS. Further studies will therefore be carried out to validate the mechanical model assumptions for flexure properties of flexure hinges based on plain-woven fabric reinforcement. This will be carried out experimentally using a test method close to the component and published in a near future. Based on the collected data and simulative evaluation of the current geometry design of the PACS, it is now possible to develop a concrete geometry and component-compatible design of the required fabric weaves. Last but not least, the finalized PACS geometry will be used to design a further developed tool. This concerns in particular the design of the cell rows and the necessary displacement of the mold inserts for axis-appropriate compression and consolidation of the preform.

Author Contributions: Conceptualization, M.V., P.M., C.S., C.C. and C.H.; methodology, M.V., C.S. and P.M.; software, P.S. and P.M.; validation, P.M.; formal analysis, C.S.; investigation, M.V. and

Polymers **2021**, 13, 3128

P.M.; resources, P.S.; data curation, P.M. and P.S.; writing—original draft preparation, M.V.; writing—review and editing, P.M., C.S., P.S., C.C. and M.S.; visualization, C.S. and P.S.; supervision, C.C., M.S. and C.H.; project administration, M.V., P.M. and C.H.; funding acquisition, C.C., M.S. and C.H. All authors have read and agreed to the published version of the manuscript.

Funding: This research and the APC were funded by the Deutsche Forschungsgemeinschaft (DFG, German Research Foundation) Grant Number 280656304, CH174/42/2.

Institutional Review Board Statement: Not applicable.

Informed Consent Statement: Not applicable.

Conflicts of Interest: The authors declare no conflict of interest. The funders had no role in the design of the study; in the collection, analyses, or interpretation of data; in the writing of the manuscript, or in the decision to publish the results.

Abbreviations

The following abbreviations are used in this manuscript:

AM	Additive manufactured
CS-AM	Cellular structure—additive manufactured
CS-FRP	Cellular structure—Fiber-reinforced plastic
FE	Finite elements/Finite elements analysis
GF	Glass Fiber
GFRP	Glass-fiber-reinforced plastic
LTL	Layer-to-layer woven fabric
PA	Polyamide
PACS	Pressure-actuated cellular structure
PL	Plain-woven fabric
SLS	Selective laser sintering
SMA	Shape memory alloy

References

1. Howell, L.L.; Magleby, S.P.; Olsen, B.M. *Handbook of Compliant Mechanisms*, 1st ed.; John Wiley & Sons: Chichester, UK, 2013.
2. Dresig, H.; Vul'fson, I.I. Kinematik zwangläufiger Mechanismen. In *Dynamik der Mechanismen*; Springer: Vienna, Austria, 1989; pp. 15–54.
3. Shuib, S.; Ridzwan, M.; Kadarman, A.H. Methodology of Compliant Mechanisms and its Current Developments in Applications: A Review. *Am. J. Appl. Sci.* **2007**, 4, 160–167. [CrossRef]
4. Kota, S.; Flick, P.; Collier, F.S. Flight Testing of FlexFloil^{TM} Adaptive Compliant Trailing Edge. In *54th AIAA Aerospace Sciences Meeting*; American Institute of Aeronautics and Astronautics: San Diego, CA, USA, 2016.
5. Elzey, D.M.; Sofla, A.Y.N.; Wadley, H.N.G. A bio-inspired, high-authority actuator for shape morphing structures. In Proceedings of the Smart Structures and Materials 2003: Active Materials: Behavior and Mechanics, San Diego, CA, USA, 3–6 March 2003; Lagoudas, D.C., Ed.; SPIE: Bellingham, WA, USA, 2003; pp. 92–100.
6. Vasista, S.; Titze, M.; Schäfer, M.; Bertram, O.; Riemenschneider, J.; Monner, H.P. Structural and Systems Modelling of a Fluid-driven Morphing Winglet Trailing Edge. In Proceedings of the AIAA Scitech 2020 Forum, Orlando, FL, USA, 6–10 January 2020; American Institute of Aeronautics and Astronautics: Reston, VA, USA, 2020; pp. 1–19.
7. Pagitz, M.; Lamacchia, E.; Hol, J.M. Pressure-actuated cellular structures. *Bioinspir. Biomim.* **2012**, 7, 016007. [CrossRef]
8. Luo, Q.; Tong, L. Adaptive pressure-controlled cellular structures for shape morphing I: Design and analysis. *Smart Mater. Struct.* **2013**, 22, 055014. [CrossRef]
9. Gramüller, B.; Boblenz, J.; Hühne, C. PACS—Realization of an adaptive concept using pressure actuated cellular structures. *Smart Mater. Struct.* **2014**, 23, 115006. [CrossRef]
10. Gramüller, B.; Köke, H.; Hühne, C. Holistic design and implementation of pressure actuated cellular structures. *Smart Mater. Struct.* **2015**, 24, 125027. [CrossRef]
11. Sinapius, M.; Hühne, C.; Sadri, H.; Riemenschneider, J. Active Shape Control. In *Adaptronics—Smart Structures and Materials*; Springer: Berlin/Heidelberg, Germany, 2021; pp. 155–225.
12. Cherif, C. *Textile Materials for Lightweight Constructions—Technologies, Methods, Materials, Properties*; Springer: Berlin/Heidelberg, Germany, 2016.
13. Chen, X.; Taylor, L.W.; Tsai, L.-J. An overview on fabrication of three-dimensional woven textile preforms for composites. *Text. Res. J.* **2011**, 81, 932–944. [CrossRef]
14. Chen, F.; Hu, H.; Liu, Y. Development of weft-knitted spacer fabrics with negative stiffness effect in a special range of compression displacement. *Text. Res. J.* **2015**, 85, 1720–1731. [CrossRef]

15. Rawal, A.; Saraswat, H.; Sibal, A. Tensile response of braided structures: A review. *Text. Res. J.* **2015**, *85*, 2083–2096. [CrossRef]
16. Ansar, M.; Xinwei, W.; Chouwei, Z. Modeling strategies of 3D woven composites: A review. *Compos. Struct.* **2011**, *93*, 1947–1963. [CrossRef]
17. Mountasir, A.; Hoffmann, G.; Cherif, C.; Kunadt, A.; Fischer, W.-J. Mechanical characterization of hybrid yarn thermoplastic composites from multi-layer woven fabrics with function integration. *J. Thermoplast. Compos. Mater.* **2012**, *25*, 729–746. [CrossRef]
18. Boussu, F.; Cristian, I.; Nauman, S. General definition of 3D warp interlock fabric architecture. *Compos. Part B Eng.* **2015**, *81*, 171–188. [CrossRef]
19. Tripathi, L.; Behera, B.K. Review: 3D woven honeycomb composites. *J. Mater. Sci.* **2021**, *56*, 15609–15652. [CrossRef]
20. Mountasir, A.; Löser, M.; Hoffmann, G.; Cherif, C.; Großmann, K. 3D Woven Near-Net-Shape Preforms for Composite Structures. *Adv. Eng. Mater.* **2016**, *18*, 391–396. [CrossRef]
21. Bollas, D.; Pappas, P.; Parthenios, J.; Galiotis, C. Stress generation by shape memory alloy wires embedded in polymer composites. *Acta Mater.* **2007**, *55*, 5489–5499. [CrossRef]
22. Ashir, M.; Hindahl, J.; Nocke, A.; Cherif, C. A statistical approach for the fabrication of adaptive pleated fiber reinforced plastics. *Compos. Struct.* **2019**, *207*, 537–545. [CrossRef]
23. Sennewald, C.; Vorhof, M.; Schegner, P.; Hoffmann, G.; Cherif, C.; Boblenz, J.; Sinapius, M.; Hühne, C. Development of 3D woven cellular structures for adaptive composites based on thermoplastic hybrid yarns. *IOP Conf. Ser. Mater. Sci. Eng.* **2018**, *369*, 012041. [CrossRef]
24. Mountasir, A.; Hoffmann, G.; Cherif, C.; Löser, M.; Großmann, K. Competitive manufacturing of 3D thermoplastic composite panels based on multi-layered woven structures for lightweight engineering. *Compos. Struct.* **2015**, *133*, 415–424. [CrossRef]
25. Sennewald, C.; Vorhof, M.; Hoffmann, G.; Cherif, C. Overview of Necessary Development Steps for the Realization of Woven Cellular Structures for Adaptive Composites. *J. Fash. Technol. Text. Eng.* **2018**, s5. [CrossRef]
26. Meyer, P.; Boblenz, J.; Sennewald, C.; Vorhof, M.; Hühne, C.; Cherif, C.; Sinapius, M. Development and Testing of Woven FRP Flexure Hinges for Pressure-Actuated Cellular Structures with Regard to Morphing Wing Applications. *Aerospace* **2019**, *6*, 116. [CrossRef]
27. Deutsches Institut für Normung, e.V. *DIN EN ISO 527: Kunststoffe—Bestimmung der Zugeigenschaften*; Beuth Verlag: Berlin, Germany, 2003.
28. Meyer, P.; Lück, S.; Spuhler, T.; Bode, C.; Hühne, C.; Friedrichs, J.; Sinapius, M. Transient Dynamic System Behavior of Pressure Actuated Cellular Structures in a Morphing Wing. *Aerospace* **2021**, *8*, 89. [CrossRef]
29. Schürmann, H. *Konstruieren Mit Faser-Kunststoff-Verbunden*; Springer: Berlin/Heidelberg, Germany, 2007.
30. Walz, F.; Luidbrandt, J. Die Gewebedichte I. *Textilpraxis* **1947**, *2*, 330–335.
31. Walz, F.; Luidbrandt, J. Die Gewebedichte II. *Textilpraxis* **1947**, *2*, 366–370.
32. Peirce, F.T. The Geometry of Cloth Structure. *J. Text. Inst. Trans.* **1937**, *28*, T45. [CrossRef]
33. Kemp, A. An Extension of Peirce's Cloth Geometry to the Treatment of Non-circular Threads. *J. Text. Inst. Trans.* **1958**, *49*, T44. [CrossRef]
34. Hearle, J.W.S.; Shanahan, W.J. An Energy Method for Calculations in Fabric Mechanics, Part I: Principles of the Method. *J. Text. Inst.* **2008**, *69*, 81–91. [CrossRef]
35. Lomov, S.V.; Verpoest, I.; Peeters, T.; Roose, D.; Zako, M. Nesting in textile laminates: Geometrical modelling of the laminate. *Compos. Sci. Technol.* **2003**, *63*, 993–1007. [CrossRef]
36. Potluri, P.; Sagar, T.V. Compaction modelling of textile preforms for composite structures. *Compos. Struct.* **2008**, *86*, 177–185. [CrossRef]
37. Vorhof, M.; Weise, D.; Sennewald, C.; Hoffmann, G. New method for warp yarn arrangement and algorithm for pattern conversion for three-dimensional woven multilayered fabrics. *J. Ind. Text.* **2020**, *49*, 1334–1356. [CrossRef]
38. Steger, A. *Diskrete Strukturen*; Springer: Berlin/Heidelberg, Germany, 2007.
39. Deutsches Institut für Normung e.V. *DIN EN ISO 139: Textilien—Normalklimate für die Probenvorbereitung und Prüfung*; Beuth Verlag: Berlin, Germany, 2011.
40. Deutsches Institut für Normung e.V. *DIN EN ISO 5084: Bestimmung der Dicke von Textilien und Textilen Erzeugnissen*; Beuth Verlag: Berlin, Germany, 1996.
41. Deutsches Institut für Normung e.V. *DIN EN ISO 1172: Prepregs, Formmassen und Laminate—Bestimmung des Textilglas- und Mineralfüllstoffgehalts*; Beuth Verlag: Berlin, Germany, 1998.
42. FOREL—3DProCar Bericht. Available online: https://plattform-forel.de/3dprocar-bericht/ (accessed on 8 September 2021).

Article

Investigation of Impacts on Printed Circuit Board Laminated Composites Caused by Surface Finish Application

Denis Froš *, Karel Dušek and Petr Veselý

Department of Electrotechnology, Faculty of Electrical Engineering, Czech Technical University in Prague, Technická 2, 166 27 Prague 6, Czech Republic; dusekk1@fel.cvut.cz (K.D.); veselp13@fel.cvut.cz (P.V.)
* Correspondence: frosdeni@fel.cvut.cz; Tel.: +420-224-355-851

Abstract: The purpose of this study was to compare the strength of the bond between resin and glass cloth for various composites (laminates) and its dependence on utilized soldering pad surface finishes. Moreover, the impact of surface finish application on the thermomechanical properties of the composites was evaluated. Three different laminates with various thermal endurances were included in the study. Soldering pads were covered with OSP and HASL surface finishes. The strength of the cohesion of the resin upper layer was examined utilizing a newly established method designed for pulling tests. Experiments studying the bond strength were performed at a selection of laminate temperatures. Changes in thermomechanical behavior were observed by thermomechanical and dynamic mechanical analyses. The results confirmed the influence of the type of laminate and used surface finish on bond strength. In particular, permanent polymer degradation caused by thermal shock during HASL application was observed in the least thermally resistant laminate. A response to thermal shock was detected in thermomechanical properties of other laminates as well, but it does not seem to be permanent.

Keywords: resin; glass cloth; substrate; glass transition temperature; surface finish; pad cratering; thermal resistance

Citation: Froš, D.; Dušek, K.; Veselý, P. Investigation of Impacts on Printed Circuit Board Laminated Composites Caused by Surface Finish Application. *Polymers* **2021**, *13*, 3203. https://doi.org/10.3390/polym13193203

Academic Editor: Rajesh Mishra

Received: 29 August 2021
Accepted: 17 September 2021
Published: 22 September 2021

Publisher's Note: MDPI stays neutral with regard to jurisdictional claims in published maps and institutional affiliations.

1. Introduction

The interconnection of electrical components through printed circuit boards (PCBs) has been conducted for several decades. In terms of continual development in the electronics industry that allows placing electronic devices in areas with demanding conditions, the properties of the materials used for PCB production must correlate with the requirements of those devices. Therefore, the evaluation of the properties of commonly utilized materials is important for confirming the compatibility for operation in a supposed environment. If ordinary materials are not suitable, materials with enhanced parameters must be selected for manufacturing. However, testing of materials with better properties should be carried out to assess deployment with existing technology and other materials involved in PCBs. Laminates for PCB manufacturing are composites, i.e., it is a system of filler (reinforcement) and resin. A significant property to be tested is the temperature endurance of the composite (usually called a substrate in the PCB industry) that is dominantly determined by the resin. PCBs can be affected by high temperature during the soldering process and, subsequently, by operating conditions that the device is subjected to. Unresolved thermal compatibility may lead to a failure directly on the PCB and to a subsequent dysfunction of the complete device.

Resins that are used for impregnating the filler consist of amorphous polymers. For these polymers, glass transition temperature (Tg) is defined. Under Tg, the polymer is in solid-state, but short-range interconnections break up by heating the material closer to Tg temperature. Molecules of polymer with temperatures over Tg receive enough energy to move freely around. It results in a rubbery state of the material including a change in the physical properties such as the coefficient of thermal expansion (CTE), heat capacity, and mechanical properties [1,2]. CTE changes in the direction of the z-axis caused by crossing Tg must be carefully monitored, because multiple soldering or thermal cycles with an upper-temperature limit above the Tg raises the possibility of damaging the plated barrels and vias [3]. Cracking is caused by considerable differences in the CTEs of copper and the substrate in the vertical direction.

Epoxy resins are compounds comprising two or more epoxy groups. These groups react with a wide range of curing agents [4]. Another aspect governing the Tg is the number of epoxide groups in a polymer structure. In addition, additives, the rate of cross-linking determined by the curing agent, and the curing conversion impact on the Tg value as well as other properties and the resin's performance [5–7]. Higher cross-link density is responsible for an increase in temperature, chemical, and moisture resistance. On the other hand, it also results in a less flexible and more brittle material. A typical range of Tg that can be achieved for epoxy resin is 135–185 °C. Ehrler [8] compared the properties of two curing agents used in FR4 laminates. The laminate with a phenol novolac curing system is more suitable for higher thermal requirements than the dicyandiamide (DICY) hardener. Another resin must be chosen in case of specialized higher temperature applications. For example, bismaleimide triazine or polyimide resin produces high-Tg laminates. The possibility of a fall in Tg as a consequence of thermal degradation or moisture absorption must be taken into account, too [9,10]. Applicable techniques for experimental Tg determination are thermomechanical analysis (TMA), dynamic mechanical analysis (DMA), or differential scanning calorimetry (DSC). The two former methods were also used in this study.

Warpage is an effect occurring when the PCB is exposed to thermal stress. Current PCBs in mobile devices with complicated designs are prone to this adverse behavior. Xia et al. [11] experimentally proved that twisting or bowing is more significant for thinner PCBs. Warpage may be further aggravated by the release of residual stress induced in the substrate during the lamination process [12]. Several other studies [13,14] focused on modeling this issue in order to predict the warpage of PCBs. It helps to avoid connection failure following re-design of a product. Moreover, in the case of a relatively larger component with more soldered connections (e.g., ball grid array) that undergoes warpage itself due to the fact of CTE mismatch of the component, mechanically weak solder joints (called head-in-pillow) may occur [15]. Tearing off of the soldering pad from the substrate may appear during soldering while the joint has already solidified and warping of the assembly is still ongoing or at field conditions when CTE differences in the substrate and mounted components raise the stress concentration [16,17]. This failure is known under the term pad cratering.

Pad cratering evaluation has been a subject of many studies. A consequence of stress concentration is crack initiation. Its propagation continues at the glass cloth and resin interface, resulting in a separation of the copper pad out of the substrate. It is a critical failure that cannot be reworked. Roggeman et al. [18] investigated the dependency of pad cratering on filled and unfilled resin systems. The filled resin contains particles that reduce the CTE in the z-direction, and according to the results in this study, they also inhibit crack propagation and failure look. Compared to the number of loading cycles until the failure occurs, the filled resin is better, but the average pull strength is higher for those unfilled. In a filled system, the glass cloth is not visible after cratering because the crack does not tend to propagate deep into the material. The same study further deals with the impact of the pull rate, pull angle, amount of reflow cycles, and degradation mechanisms on pad cratering. Godbole et al. [19] investigated the connection between the pad cratering and the pad placement within the PCB together with the effect of reflow cycles and moisture exposure. For pad cratering determination, three methods have been established, and they are described in the IPC-9708 standard. These methods are ball shear, ball pull, and pin pull testing depicted in Figure 1. They are presented in more detail in [20], including the description of their benefits and drawbacks. Cia et al. [21] took advantage of the pin pull method to evaluate pad cratering after multiple reflows and accelerated thermal cycling. Susceptibility to pad cratering is influenced by the reflow profile peak temperature, and thermal aging plays an important role if the temperature is above Tg [22]. The possibility of pad cratering is enhanced by inducing stress during the release of latent heat when the solder solidifies, as Dušek and Rudajevová mentioned in their study [23]. Latent heat locally raises the temperature under the pad and keeps the resin in a viscoelastic state, while the surrounding PCB is under Tg and is already rigid.

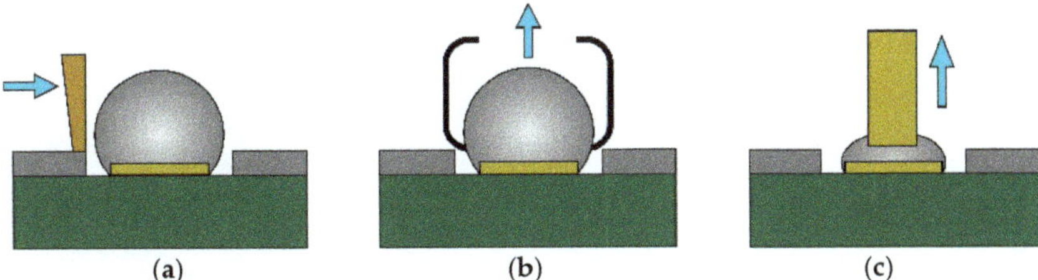

Figure 1. Methods for pad cratering evaluation: (**a**) ball shear; (**b**) ball pull; (**c**) pin pull.

Copper pads intended for soldering tend to be covered by the combination of copper oxides [24]. Surface finish utilization is practically unavoidable for preventing the reaction of copper with oxygen in the air. Moreover, its utilization is supported by the fact that current soldering pastes containing no-clean fluxes with low activity. Oxidized surface causes insufficient solderability. The solder wetting without using very aggressive flux is impossible in the case of exceeding oxide thickness threshold [25]. Surface finish influences the reliability of solder joints [26–28]. Therefore, the selection must comply with the following device operation. Surface finishes are deposited in various ways according to the requirements of the material, ensuring protection.

Two surface finishes, HASL (hot air solder leveling) and OSP (organic surface preservative), are presented in this study. Hot air solder leveling finish involves soaking the PCB into the molten alloy. In an area of lead-free soldering, the PCB is exposed to 250 °C. In addition, after taking out of the bath, the PCB undergoes a hot air knife in order to remove excess solder and make the thickness more uniform [29]. An organic solder preservative is deposited by immersing the PCB into liquid, or in horizontal conveyorized processing, the board is sprayed. The properly cleaned PCB and micro-etched copper pads are exposed to liquid consisting of an organic compound. The organic component is dissolved first in water and organic acid. Then the PCB with OSP coating is left to dry under conditions not exceeding 50 °C. The whole process is quite simple, and it does not affect the PCB in terms of thermal shock, unlike HASL. The compatibility of OSP with lead-free soldering is being solved by developing new substances, providing higher heat stability [30]. OSP coating is cheaper and more planar than HASL. On the contrary, HASL is more resistant to mechanical damages, moisture, and temperature and reduces storage demands.

A distinction between OSP and HASL in the field of soldering and PCBs was determined by several studies. Dušek et al. [31] detected stronger mechanical resistance of soldered joints for HASL. In the work of Vasko et al. [32], HASL provided better wettability than OSP. The results showed stronger joints made on pads covered by HASL, too. Reliability tests done by Zhou et al. [33] proved comparable results after thermal cycling for both finishes and better endurance for OSP in the drop tests.

Within PCB testing, another strength of the interface is studied. Peel strength is commonly performed to assess the bond between the resin and pressed conductive foil. Peel strength is mainly given by the foil roughness [34,35]. Interesting research in terms of our study was conducted by Liu et al. [36], who investigated the peel strength after thermal shock inflicted on the copper-clad laminate. The study showed the negative repercussions of the thermal shock on the adhesion of copper patterns to the substrate.

Our investigation focused on the evaluation of the thin resin layer adhesion beneath the copper pattern to the reinforcement. The strength of this adhesion (in the article called bond strength) is a crucial indicator for the formation of presented failure—pad cratering. As it was mentioned, there are no possibilities for repair. Therefore, there must be a wide range of evaluating studies dealing with this failure. Then, during the PCB's design and material selection, the experience gained by the studies helps to avoid or decrease the risk of tearing the pad out of the substrate. Resins used for the production of the laminated composites have various thermal properties, and it is expected to have a different response to thermal loads. To cover this concern, three laminates characterized by diverse Tg values were chosen. Even though the two variants of epoxy resin were subjected to similar evaluations, a comparison with polyimide resin and in relation to other testing parameters has not been performed. Another concern appears as the resins do not have the same adhesion to reinforcement. In addition, the bond might change after thermal or mechanical stress. The bond strength of the soldering pad to the substrate has not been deeply evaluated considering the surface finish. Further, an indicated issue was tested directly under an elevated temperature, which gives importance to this study. A new method was developed for the purposes of strength testing, which reduces the problems related to failure mode recognition. The used method ensures detachment at the interface of the resin and filler.

According to one method of surface finish application, the consequences left on the laminate caused by this part of PCB fabrication should not be neglected. Adverse effects in the form of deviations in thermal expansion, drops in Tg value, or changes in material reaction thermomechanical loading ought to be checked. Changes in material behavior can cause some of the negative difficulties described in the paragraphs above in the course of soldering or consequent device operation. This relevant issue is not addressed in previous studies. Hence, it was observed in more detail within this study.

2. Materials and Methods

Three types of laminates were selected based on their Tg value. A set of samples included the basic variant of DICY-cured epoxy FR4 laminate (Tg1), which is still widely utilized in consumer electronics. Phenolic-cured epoxy FR4 (Tg2) and polyimide G30 (Tg3) resins in combination with glass cloth were further laminates involved in the evaluation. This laminate was intended for high-temperature usage to ensure the required reliability. The list of used materials is shown in Table 1. The testing boards made of listed laminates were designed to contain the spacious circular soldering pad with a diameter of 5.5 mm. The overall size (12.5 × 12.5 mm^2) of one specimen (see Figure 2) was adjusted to fit the size available in the tool. For each sample version, 15 pieces were assessed.

Table 1. Laminates used for the investigation.

Producer	Type	Grade	Glass Transition Temperature	Samples Marking
JIANGSU RODA ELECTRON MATERIAL, Rudong, China	RD140	FR4	135 °C	Tg1
TECHNOLAM, Troisdorf, Germany	NP-175F	FR4	170 °C	Tg2
Göttle Leiterplattentechnik, Königsbrunn, Germany	VT-901	G30	250 °C	Tg3

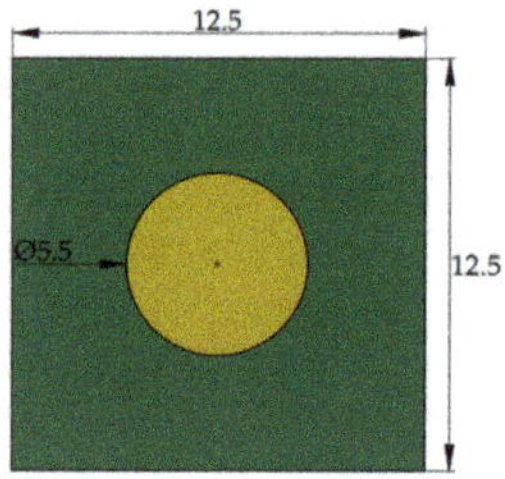

Figure 2. Design of the testing board.

The tool for placing the sample as a part of the employed tensile testing machine X250-3 (Testometric, Rochdale, Great Britain) is visible in Figure 3. As the previous section suggests, HASL (H) and OSP (O) were chosen to assess if some finishes can make the bond between the glass cloth and the thin layer of the resin under the soldering pad weaker, thereby contributing to the pad cratering phenomenon.

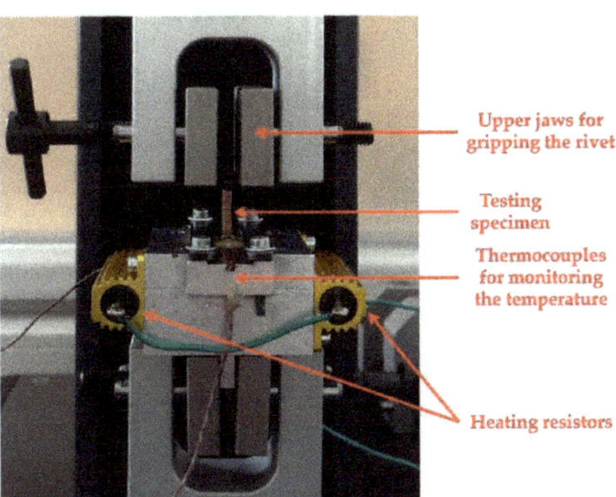

Figure 3. Sample fixed in the tool.

A novel approach for analysis was established to eliminate a majority of the aspects influencing the strength of the bond between the resin and the glass cloth during the mechanical test. The newly utilized method allows for focusing on the desired bond and thereby material and technology comparisons. There are more possibilities of resin and reinforcement suitable for substrate production and many other types of material inputs (e.g., surface finish and solder mask) involved in PCB fabrication. These materials must be applied on PCBs by a technological process that may be incompatible with the selected base materials and erode the original properties. Further, this method can be adopted for various mass soldering techniques used for surface mount technology (e.g., hot air, vapor phase, or infrared soldering). It should be noted that utilizing soldering pads with a relatively large area does not influence the mutual comparison of technology and material combinations.

A copper countersunk head rivet was mounted to transfer the tensile force to the soldering pad (see Figure 4). Rivets were mounted to the soldering pads using the conventional reflow method. At first, lead-free soldering paste SAC305 (Sn–3%Ag–0.5%Cu) was applied on the soldering pads using stencil printing. After manual assembling of the rivets, the PCBs were reflowed in a forced air convection oven Mistral 260 (Technoprint, Ermelo, The Netherlands) with three temperature zones. The temperatures of the zones were adjusted to create the temperature profile suitable for the chosen lead-free solder paste (see Figure 5). Before every measuring test, the sample was attached to the tool in that way to allow the pad to detach from the substrate freely. A rivet with a 3 mm diameter was firmly fastened into the upper jaws of the deformation device before the test. The speed of the upper jaw was set at 1 mm/min.

Tests were performed at ambient temperature (AT) and an elevated temperature (ET) of 100 °C. The preheating of samples for purposes of testing at an elevated temperature took place in an oven UN55 (Memmert, Schwabach, Germany). Preheating was used to achieve a more uniform temperature at the samples. The preheating process lasted for approximately 30 min. After, the sample was placed and fastened into the heated tool. The heat of the tool was supplied by energy dissipation in two resistors connected in a series to the voltage source. The temperature in the tool and on the sample was monitored by thermocouples (type K) located in the tool directly under the sample.

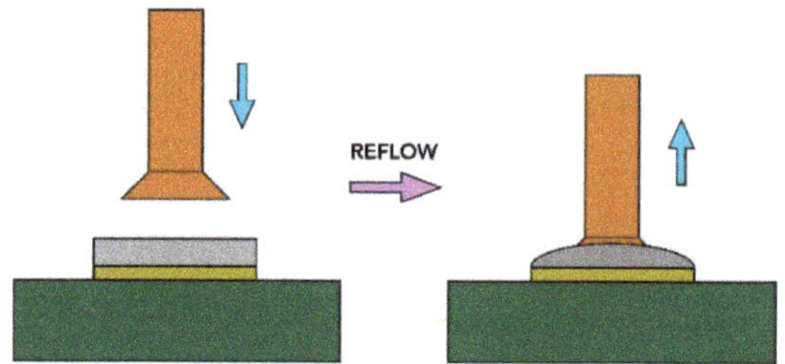

Figure 4. Testing method—rivet mounting.

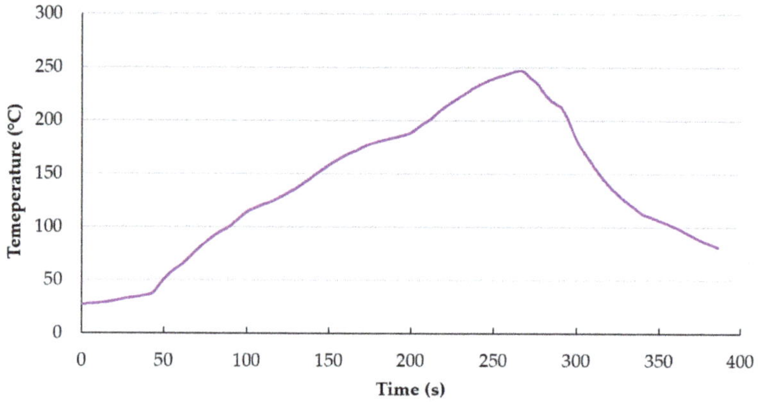

Figure 5. Temperature profile achieved for reflow in the oven Mistral 260.

Measurements to determine the Tg and possible influence of the surface finish on the behavior under thermal action were conducted on TMA Q400EM (TA Instruments, New Castle, DE, USA) equipment. This apparatus served for both the TMA and DMA procedures. During thermomechanical analysis, square-shaped samples with an edge length of 8 mm and thickness of 1.5 mm were placed on the stage and heated to a temperature that was approximately 40 °C above the transition temperature given by the datasheet. The heating ramp was established at 5 °C/min. A nitrogen atmosphere was available in the heating chamber. The Tg estimation was performed as Yong et al. [37] proposed. Accordingly, analysis of the curve (temperature dependence of dimensional change) was obtained by a sensitive probe while having plotted the first derivation of that dependency. The first derivative of dimension change with respect to temperature helped to define the onset and the end of glass transition. The sharpest dimension change identifies the onset, thereby the lower temperature boundary. The peak or stabilization of the first derivation points to the upper limit. The temperature closest to the midpoint of the stated temperature interval was taken as the Tg.

The DMA measurements were performed using an aluminum fixture with supporting rollers that formed a three-point bending system in conjunction with a flexural probe. The rollers on which the samples were put were at a 10.16 mm distance from each other. For DMA purposes, the delivered composite substrates with a thickness of 1.5 mm were cut to obtain rectangular-shaped specimens with a width of 3 mm and length of 15 mm. Temperature conditions during DMA were similar to the TMA measurements. The only maximal temperature was adjusted to capture complete significant transitions of monitored quantities. Modulated force amplitude was set to ±0.2 N that was applied to the sample at a frequency of 1 Hz. The static force was 0.25 N. Each type of sample was analyzed three times. One exemplary diagram is comprised in the results.

The methodology of the study is presented in Figure 6. The procedure diagram items are labeled to describe the motivation of the work steps. Details of several of the motivations are explained in Table 2 which support the diagram.

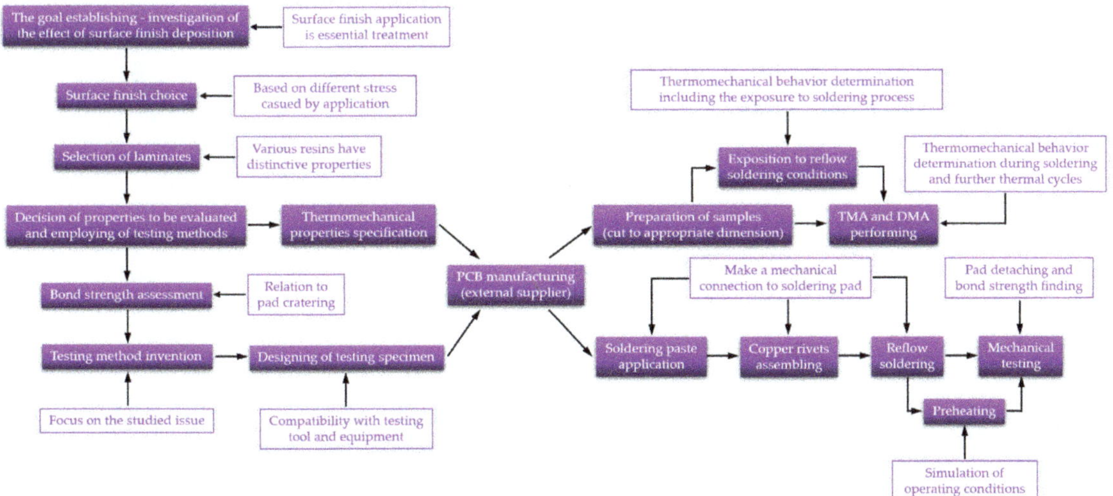

Figure 6. Research methodology.

Table 2. Explanation of the main motivations.

Operation	Motivation
Investigation of the effect of surface finish	The impact of the technological process (surface finish application) on the laminate evaluation is crucial.
Surface finish choice	Two surface finishes were chosen regarding thermal circumstances during application. HASL application is accompanied by thermal stress, whereas OSP is not.
Selection of laminates	Various resins or their modifications have different thermal properties and adhesion to filler.
Bond strength assessment	The strength of the adhesion of the soldering pad, specifically resin to filler, is significant in relation to failure–pad cratering occurring on the PCBs.
Reflow soldering	Except establishing the mechanical connection, the bond strength results respect the effect of this treatment, which is an essential step in electronic assembly.
Preheating	Specimens tested at an elevated temperature were preheated to achieve an equal temperature throughout the sample. Consequent mechanical tests performed at 100 °C were realized in order to simulate field conditions.
Exposition to reflow soldering conditions	It was included to verify the effect of surface application, i.e., comparison of slow and rapid heating.
Thermomechanical analysis (TMA) and dynamic mechanical analysis (DMA)	Observation of material behavior in the surrounding of Tg and detection of Tg value displacement. Assessment of material response during mechanical loading in conjunction with temperature rise. More measurements cycles were conducted to determine the response during soldering and, consequently, the effect of the thermal loading.

3. Results

3.1. Bond Strength Evaluation

The whole pulling process was recorded, and the maximum force for further evaluation was consequently determined from the sampled data. The course of force allows for definite detection of the highest force. Examples of pulling course are depicted in Figure 7.

This course was typical for every sample. The detachment of the pad was accomplished at one moment without any gradual tearing off.

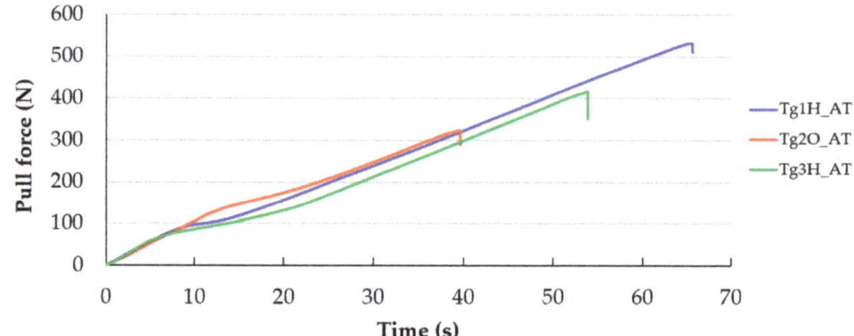

Figure 7. Dependency of pull force on the time.

The failure mode of the pulling test for each laminate is visible in the photo of tested samples shown in Figure 8. The pads are completely torn from the substrate, and the glass cloth is visible. An advantage of the used method was confirmed, because no part of a specimen was destroyed in another area than was required. Thus, any destruction at the interfaces of solder and pad, rivet and solder, nor breaking the rivet did not happen. Obviously, thanks to its design and larger cross-sectional dimensions.

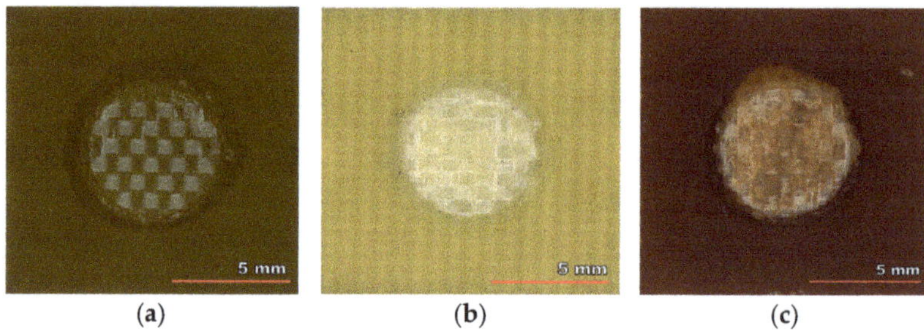

(a)   (b)   (c)

Figure 8. Exposed glass cloth after tearing the pad: (**a**) FR4 low Tg; (**b**) FR4 high Tg; (**c**) G30.

The sample marking is explained in Figure 9. It corresponds with the bond strength's evaluation. The beginning of the marking without the abbreviation that follows the underscore was used in the whole text.

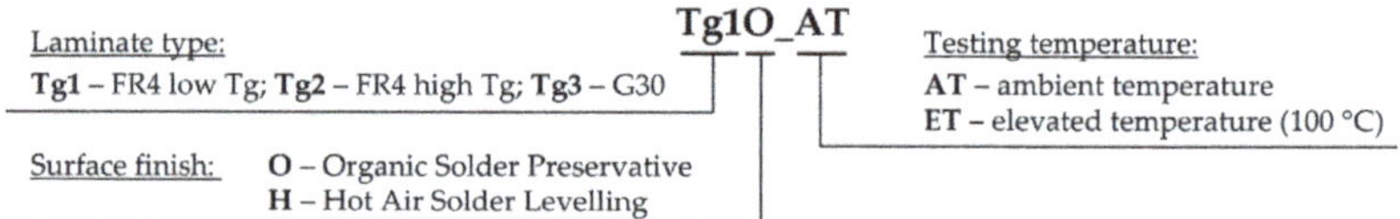

Figure 9. Explanation of the sample marking.

The pulling test results summarized in the boxplots (see Figure 10) and in Table 3 show the strong dependency of the bond strength on the resin type. Testing at ambient temperature, as reported in Figure 10a, points to a slight impact of surface finish on the studied issue for both versions of FR4 laminates. Despite the presumptions, pads with an HASL surface finish and the resin under them did not lose the rate of adhesion to glass reinforcement. Our tests showed a certain improvement in adhesion that could result from the softening of the resin during the surface finish application, because these resins have a Tg far below the temperature of solder bath. The thermal treatment provides the possibility for the resin to adhere better to the reinforcement.

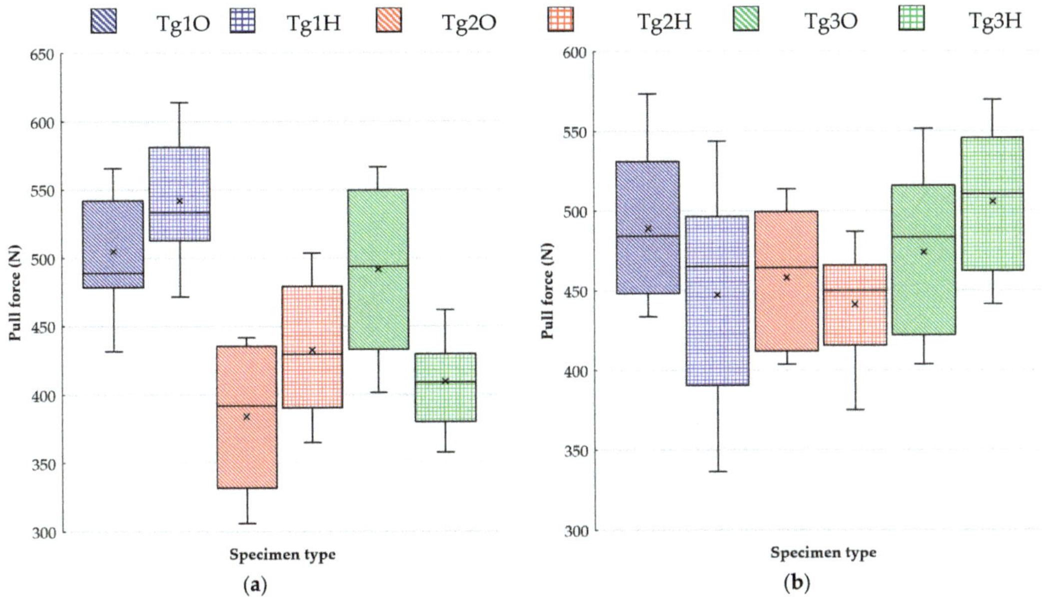

Figure 10. Pulling test results: **(a)** at ambient temperature; **(b)** at elevated temperature (100 °C).

Table 3. Summary of the measured and calculated values.

	Tg1H_AT	Tg2H_AT	Tg3H_AT	Tg1O_AT	Tg2O_AT	Tg3O_AT	Tg1H_ET	Tg2H_ET	Tg3H_ET	Tg1O_ET	Tg2O_ET	Tg3O_ET
Mean (N)	542.0	432.5	409.6	505.0	384.2	491.9	447.2	441.2	473.9	488.8	458.2	505.6
Minimum (N)	471.7	365.1	357.8	431.5	305.9	401.5	336.5	375.1	403.6	433.5	403.7	441.3
Maximum (N)	614.0	503.5	461.8	565.8	441.8	566.5	543.9	486.9	551.5	573.6	513.7	569.5
SD (N)	44.4	47.1	32.6	39.9	49.7	60.0	66.7	31.6	51.4	46.3	41.4	44.6
Bond strength (N/mm^2)	22.8	18.2	17.2	21.3	16.2	20.7	18.8	18.6	20.0	20.6	19.3	21.3

The ability of the surface finish to influence the bond between the resin and glass cloth was detected to a greater extent for the laminate with the highest Tg. It was found that the resin where soldering pads had an HASL surface finish had a smaller bond strength by 3.2 N/mm^2 on average than those with OSP.

As to testing at an elevated temperature, it can be stated that the distinctions in bond strength among the used laminates diminished. The effect of the surface finish persisted, but the test proved worse resin cohesion to glass cloth for HASL in the case of FR4 laminates. In particular, FR4 with a lower Tg indicated a noticeable difference. In contrast to the testing at 25 °C, the G30 substrate with OSP became prone to pad cratering than pads covered by an HASL surface finish, but the difference in average force values was not so evident.

The bar chart in Figure 11 offers a clearer comparison of average bond strength values between testing at room (blue bars) and elevated temperature (red bars). These values were calculated by dividing the mean value by the surface of the soldering pad. Error bars were derived from standard deviation. The higher testing temperature significantly affected the bond strength in specimens made of low-Tg FR4, G30 with HASL, and high-Tg FR4 covered by OSP. It was confirmed that the decrease in bond strength during testing at 100 °C for the least thermally resistant material, but Tg1O was affected very little. Laminates for thermal application behaved the opposite way when all the specimens showed higher strength even though the differences were almost negligible for Tg1O, Tg2H, and Tg3O. The PCB supplier declared the manufacturing (temperature and pressure during copper cladding) in accordance with the recommendations of the laminate producer. In connection with this fact, another question arose as to whether the optimizing procedure parameters of the pressing process may reveal whether the studied strength of adhesion can be improved.

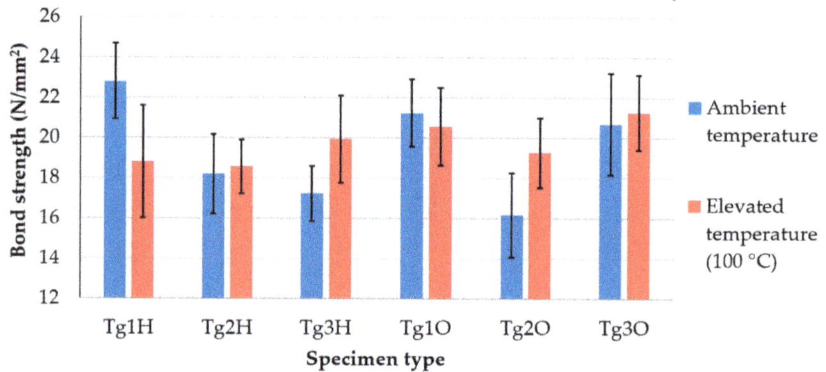

Figure 11. Comparison of the results according to the testing temperature.

The lower bond strength attained for the phenolic-cured substrate was consistent with the results obtained by Ahmad et al. [38]. Their study dealt with pull strength associated with pad cratering contained DICY and phenolic-cured laminates, too. Tests within the study revealed an approximately 50% smaller pull strength for the phenolic-cured material. The results relating to epoxy resins and surface finish may be roughly compared with [39], which dealt with a similar issue. Interlaminar strength in the epoxy resin and glass fiber system was assessed but differed in thermal shock caused not by fast heating but rapid cooling down of the heated samples. Nevertheless, the insignificant influence of this thermal load was likewise discovered.

The outcome of testing at an elevated temperature can be compared with the research conducted by Roggeman and Jones [40]. Laminates based on epoxy resin reported a moderate drop of pad strength when tested at 65 °C. Here, a discrepancy with our results obtained for high-Tg epoxy laminate might be registered, because we measured higher values when testing at 100 °C.

3.2. Analysis of Thermomechanical Properties

Observation of material mechanical properties during the heating cycle was the second part of this paper. It must be noted that at first, all thermomechanical analyses were conducted using the substrate that had not passed the reflow soldering or had not been exposed to higher temperatures after delivering the produced PCB samples. After performing these analyses and based on the results, both the TMA and DMA of the Tg1 specimens were carried out after exposure to the reflow profile shown in Figure 5. Each sample was analyzed in three cycles, but sometimes, only two cycles are included in graphs when the third cycle did not differ significantly from the second one.

The curves that are depicted in Figure 12 were obtained for the FR4 substrate with lower Tg. Shape undulation of laminate with an OSP finish is observable in the glass transition surrounding, especially during the first heating cycle. The gradual curing process and stress release of the laminate were detected by the following measurements of the same sample. This observation is in accordance with the findings in the literature conducted by Rudajevová and Dušek [41]. A new outcome can be found for the impact that causes the application of an HASL surface finish from a thermomechanical point of view. The thermal shock caused by dipping the board into the molten solder had an adverse effect on the non-fully cured substrate. Even though a specific curing process can be observed for Tg1H, it did not lead to possible improvements in the Tg value, and the magnitude remained diminished. A method of Tg derivation described previously was observable in the TMA figures, and Tg values derived from those curves are summed up in Table 4. The values in the table are stated as an average of the given measurement cycles (second or third).

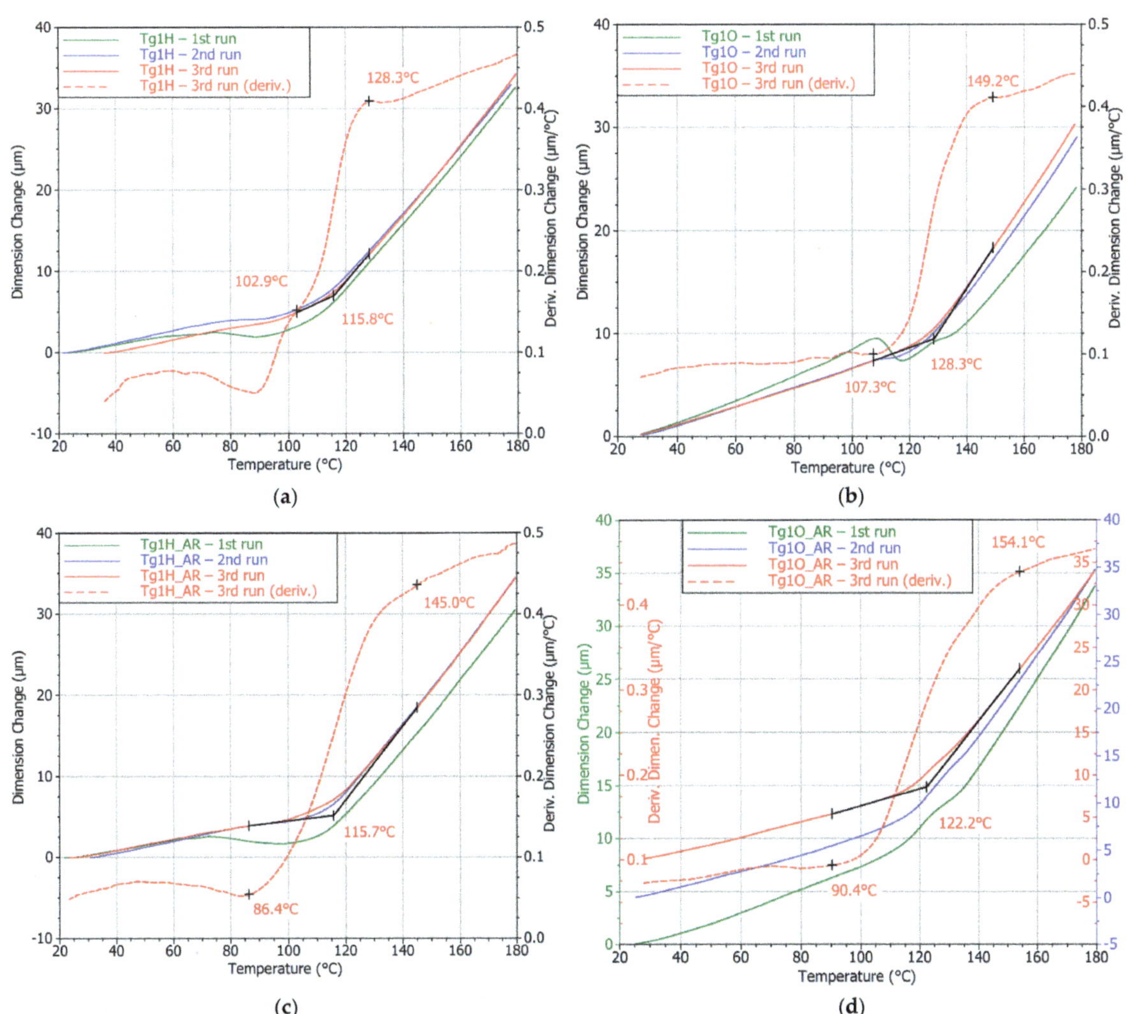

Figure 12. TMA diagrams of the Tg1 substrates and with surface finish: (**a**) HASL; (**b**) OSP; (**c**) HASL after reflow (AR); (**d**) OSP after reflow (AR).

Table 4. Glass transition temperatures (expressed in Celsius) determined by TMA.

Tg1		Tg1_After Reflow		Tg2		Tg3	
HASL	OSP	HASL	OSP	HASL	OSP	HASL	OSP
116.2	127.9	115.5	123.4	177.1	176.7	228.8	221.4

Samples that underwent the reflow process showed this treatment to serve for their curing. It can especially be seen in Figure 12d. A throb of the first cycle curve was not so enormous as it is in Figure 12b. The curves must have shifted in the y-axis; hence, three y-axes were utilized to make them visible and compare their shapes. It can be said that the changes in the shape by performing more cycles were negligible. Using TMA in the case of samples that had been subjected to HASL application and consequent reflow cycle led to the detection of an irreversible decrease in Tg.

Detections obtained by TMA were confirmed by conducting DMA measurements. Considering the monitored dynamic properties of Tg1O, subsequent curing of the resin was noticed. It resulted in the delay of the storage modulus decrease during the second cycle. Storage modulus fall takes place at a higher temperature, while Tg given by the loss modulus peaks remained unchanged. The results shown in Figure 13a verify the premature softening of the Tg1H samples at approximately 85 °C during the first heating. In the second run, a slight shift of the softening point at approximately 5 °C was observed. In addition, the decline in the point of storage modulus was related to the ability of the material to withstand mechanical stress without deformation. The residual of the proposed Tg can be recognized in recorded storage or loss modulus. It may indicate that the degradation of the polymer resin had not influenced the whole volume of the substrate. The board was dipped in the solder bath for few seconds; therefore, the warm-up was not passed into the resin located further away from the surface.

Disruption of cross-linking is a probable reason for the detected dissimilar and wavy peaks of loss modulus. Although Margem et al. [42] tested epoxy matrix using different curing agents, insufficient cross-linking rates lead to the occurrence of the same phenomena in DMA results. Polanský et al. [9] studied the effect of thermal influence with a longer duration and achieved double peaks explained by degradation of inner structure, changes in material surface profile, and the appearance of delamination.

Diagrams of DMA (eventually TMA measurements) for samples that went through the soldering oven evidently testify that a slow increase in temperature to 250 °C did not have a lasting impact on the thermomechanical properties. Laminate with OSP did not report any breaking changes in material behavior, only some slight post-curing consequences influencing the laminate were observed. Mostly, the second heating cycles of both finishes were connected with the lower absolute values of storage modulus in the glassy area. This means the lower stiffness of the laminate is associated with the higher energy stored by the system. Another reason for the varying storage modulus was the strength of the adhesion between the fiber and the matrix within the composite as Keusch and Haessler [43] researched in their work.

The effect of HASL application on the Tg2H (high Tg FR4 substrate) was demonstrated because it reduces internal stress and improves the curing rate (see Figure 14a). Deviation in the Tg region was less noticeable, and the curve of the first measurement was closer to the course achieved during the second heating run. A fall in Tg by 2 °C was found, but it was not statistically significant as in the case of Tg1H (low Tg FR4 substrate). On the other hand, DMA measurement brought the ascertainment that the laminate temporarily reported a higher storage modulus, and then its decline followed as is visible in Figure 14b. However, by heating the sample, this property disappeared. In the second cycle, the storage modulus met a typical shape comparable to the substrate with soldering pads covered by OSP, but the storage modulus remained high. The change in interface bonding can explain this finding. The storage modulus of the second analysis testified that there was no impact on the Tg value, too. The bend in the storage modulus (second cycles) suggests a Tg value of approximately 175 °C, which supports the TMA results.

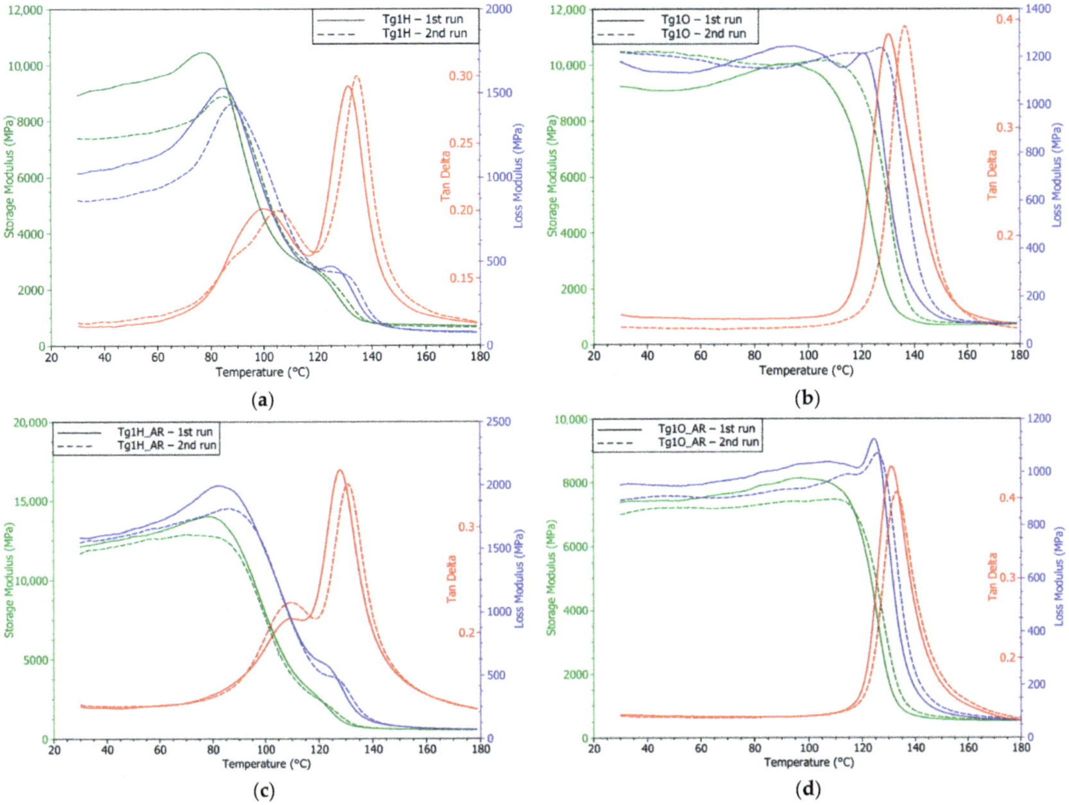

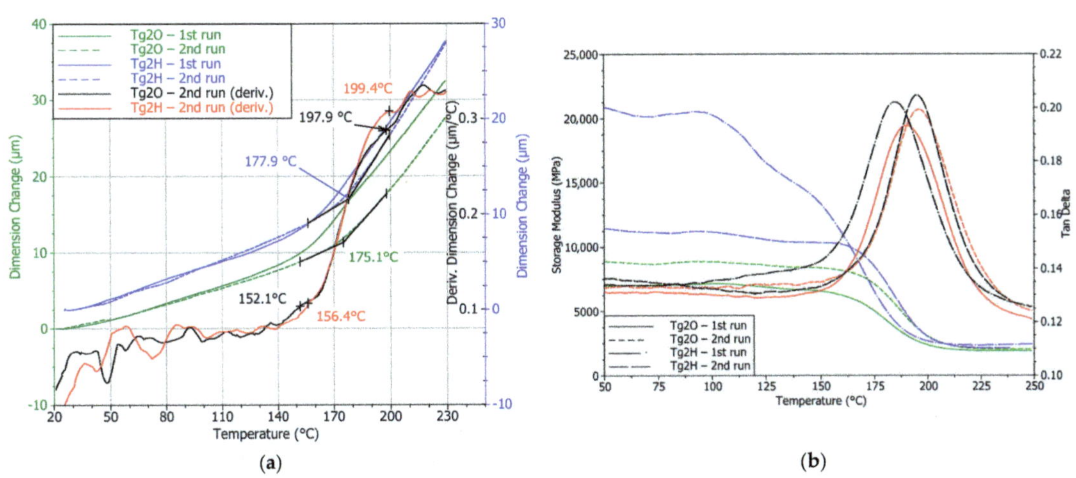

Figure 13. DMA diagrams of the Tg1 substrates and with surface finish: (**a**) HASL; (**b**) OSP; (**c**) HASL_AR; (**d**) OSP_AR.

Figure 14. Diagrams of the Tg2 substrate including both surface finishes: (**a**) TMA; (**b**) DMA.

TMA analysis revealed distinctive expansion behavior in the first heating cycle compared to the second one. Except for the undulation of the curves during the first cycles, particularly the HASL specimens exhibited higher CTEs in the glassy state. The thermomechanical performance of both specimen types became nearly the same in the second measurement run. Further, the curve bend was not as striking as it was typical for the previous laminates. Hence, inaccuracy in Tg determination may occur. Regardless of this fact, the Tg was almost independent on the surface finish, and the values increased by undergoing the thermal treatment under the conditions of the TMA measurements as can be seen in Figure 15. The Tg for Tg3H was detected to be higher by 3.5% on average, which can be found to be relatively insignificant. However, noticeably lower values than those declared in the datasheet must be considered for a potential application.

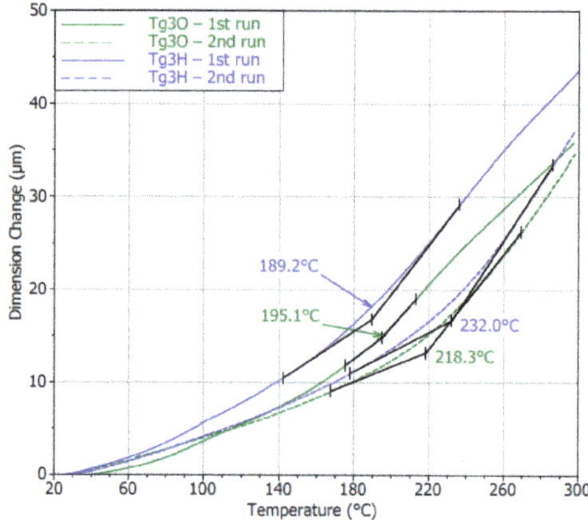

Figure 15. Comparison of the results according to the testing temperature.

Both G30 laminates using HASL and OSP showed changes in the storage modulus (see Figure 16), thus the mechanical properties comparing the first and second cycles. An increase in the storage modulus and extension of its glassy state in the OSP specimen may denote an improvement in the interface between the filler and the reinforcement. A low diminishing of the peak magnitude of tan δ correlated with this statement. Generally, it was found that the Tg3 laminate had the lowest tan δ magnitudes, which means smaller energy dissipation options. Storage modulus curves or optionally the peaks of the loss modulus suggest the Tg values were under 250 °C as the TMA results also indicated.

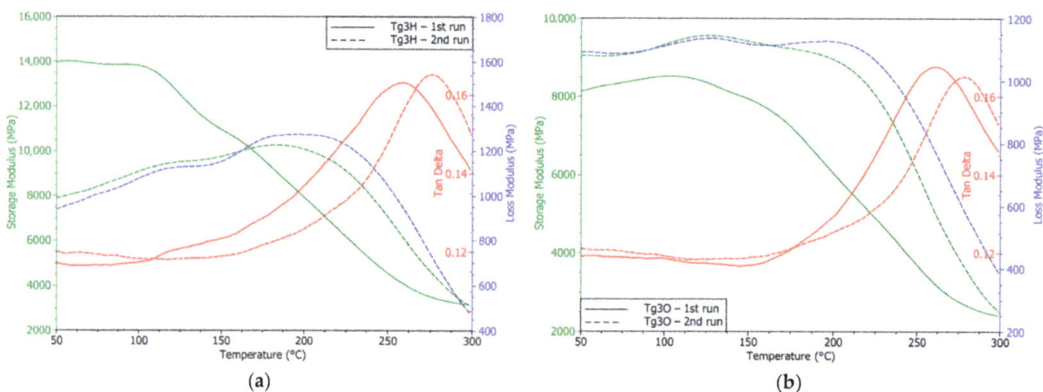

Figure 16. DMA diagrams of the Tg3 substrate with the following surface finish: (**a**) HASL; (**b**) OSP.

4. Conclusions

We evaluated the strength of the bond between the glass cloth and selected resins. A new testing approach was used considering reflow soldering. Our results revealed the dependence of adhesion on the resin type. Furthermore, the role of surface finish on bond strength was demonstrated. However, the bond strengthening or weakening rate depended on the resin. The same material combinations were tested at an elevated temperature of 100 °C. It was proved that the resin with the lowest Tg tended to lose adhesion to glass cloth. The bond strength in other assessed laminates did not show a higher tendency to become weaker when tested at 100 °C. Actually, in all cases, a higher force necessary for detaching the testing pad was measured. Simultaneously, the differences among the sample types were significantly smaller for testing at elevated temperatures. The best resistance to tearing the pad from the substrate was detected for the sample marked as Tg1H—laminate containing the resin with the glass transition temperature of 135 °C (according to the datasheet) and HASL finish. The evaluation at 100 °C revealed the strongest bond for the sample marked as Tg3O—G30 laminate on which the pads were protected by OSP.

The second part of the study utilized a thermomechanical analyzer to establish material behavior changes in consideration of the surface finish. The TMA and DMA results were included. In the matter of used surface finish, conventional substrate (marked as Tg1) was highly impacted by HASL application resulting in permanent degradation of the laminate. Both methods also indicated the decline of Tg. Specifically, the Tg shift to a lower magnitude derived by TMA was almost 10 °C. A direct effect of thermal shock caused by the application of HASL finish was confirmed. As the testing of the same laminates after simulating the reflow process (i.e., the gradual heating of the laminate to temperature reaching 250 °C and subsequent measuring of the laminate with OSP surface finish) showed observable Tg movement to a lower magnitude but not as significant as the difference between sample Tg1O (with OPS surface finish) and Tg1H (with HASL surface finish) before reflow.

Composites with higher temperature resistance did not undergo permanent deviations in the thermomechanical performance regarding the border between the glassy and rubbery region. The post-curing internal process leading to stabilization of the material in the vicinity of Tg as well as stress relief were detected for these laminated composites, too. Moreover, laminates covered by HASL tended to a transient increase in stiffness. That effect vanished, because second cycles reported common diagrams of storage modulus.

Author Contributions: D.F. and K.D. designed the experiments; D.F. conducted the laboratory experiments; K.D. and P.V. did the data analysis and interpretation; D.F. created the graphs and wrote the manuscript; K.D. and P.V. edited the draft; K.D. supervised the whole study. All authors have read and agreed to the published version of the manuscript.

Funding: This work was supported by the Grant Agency of the Czech Technical University in Prague, grant No. SGS20/123/OHK3/2T/13.

Institutional Review Board Statement: Not applicable.

Informed Consent Statement: Not applicable.

Data Availability Statement: The data presented in this study are available on request from the corresponding author.

Conflicts of Interest: The authors declare no conflict of interest.

References

1. Sirk, T.W.; Khare, K.S.; Karim, M.; Lenhart, J.L.; Andzelm, J.W.; McKenna, G.B.; Khare, R. High Strain Rate Mechanical Properties of a Cross-Linked Epoxy across the Glass Transition. *Polymer* **2013**, *54*, 7048–7057. [CrossRef]
2. Hutapea, P.; Grenestedt, J.L. Effect of Temperature on Elastic Properties of Woven-Glass Epoxy Composites for Printed Circuit Board Applications. *J. Electron. Mater.* **2003**, *32*, 221–227. [CrossRef]

3.	Knadle, K.T.; Jadhav, V.R. Proof Is in the PTH—Assuring via Reliability from Chip Carriers to Thick Printed Wiring Boards. In Proceedings of the Proceedings Electronic Components and Technology, ECTC '05, Lake Buena Vista, FL, USA, 31 May–3 June 2005; Volume 1, pp. 406–414.

4.	Wu, C.; Xu, W. Atomistic Molecular Modelling of Crosslinked Epoxy Resin. *Polymer* **2006**, *47*, 6004–6009. [CrossRef]

5.	Lesser, A.J.; Crawford, E. The Role of Network Architecture on the Glass Transition Temperature of Epoxy Resins. *J. Appl. Polym. Sci.* **1997**, *66*, 387–395. [CrossRef]

6.	Zhang, J.; Li, T.; Wang, H.; Liu, Y.; Yu, Y. Monitoring Extent of Curing and Thermal–Mechanical Property Study of Printed Circuit Board Substrates. *Microelectron. Reliab.* **2014**, *54*, 619–628. [CrossRef]

7.	Jain, R.; Kukreja, P.; Narula, A.K.; Chaudhary, V. Studies of the Curing Kinetics and Thermal Stability of Epoxy Resins Using a Mixture of Amines and Anhydrides. *J. Appl. Polym. Sci.* **2006**, *100*, 3919–3925. [CrossRef]

8.	Ehrler, S. The Compatibility of Epoxy–based Printed Circuit Boards with Lead–free Assembly. *Circuit World* **2005**, *31*, 3–13. [CrossRef]

9.	Polanský, R.; Mentlík, V.; Prosr, P.; Sušír, J. Influence of Thermal Treatment on the Glass Transition Temperature of Thermosetting Epoxy Laminate. *Polym. Test.* **2009**, *28*, 428–436. [CrossRef]

10.	Ma, L.; Sood, B.; Pecht, M. Effect of Moisture on Thermal Properties of Halogen-Free and Halogenated Printed-Circuit-Board Laminates. *IEEE Trans. Device Mater. Reliab.* **2011**, *11*, 66–75. [CrossRef]

11.	Xia, W.; Xiao, M.; Chen, Y.; Wu, F.; Liu, Z.; Fu, H. Thermal Warpage Analysis of PBGA Mounted on PCB during Reflow Process by FEM and Experimental Measurement. *Solder. Surf. Mt. Technol.* **2014**, *26*, 162–171. [CrossRef]

12.	Kim, S.-W.; Lee, S.-H.; Kim, D.-J.; Kim, S.K. Simulation of Warpage During Fabrication of Printed Circuit Boards. *IEEE Trans. Compon. Packag. Manuf. Technol.* **2011**, *1*, 884–892. [CrossRef]

13.	Rayasam, M.; Thompson, T.B.; Subbarayan, G.; Gurumurthy, C.; Wilcox, J.R. A Model for Assessing the Shape of Solder Joints in the Presence of PCB and Package Warpage. *J. Electron. Packag.* **2006**, *128*, 184–191. [CrossRef]

14.	Straubinger, D.; Bozsóki, I.; Bušek, D.; Illés, B.; Géczy, A. Modelling of Temperature Distribution along PCB Thickness in Different Substrates during Reflow. *Circuit World* **2019**, *46*, 85–92. [CrossRef]

15.	Zhao, Z.; Chen, C.; Park, C.Y.; Wang, Y.; Liu, L.; Zou, G.; Cai, J.; Wang, Q. Effects of Package Warpage on Head-in-Pillow Defect. *Mater. Trans.* **2015**, *56*, 1037–1042. [CrossRef]

16.	Albrecht, O.; Wohlrabe, H.; Meier, K. Impact of Warpage Effects on Quality and Reliability of Solder Joints. In Proceedings of the 2019 42nd International Spring Seminar on Electronics Technology (ISSE), Wroclaw, Poland, 15–19 May 2019; pp. 1–6.

17.	Zhang, Q.; Lo, J.C.C.; Lee, S.W.R. Correlation of Board and Joint Level Test Methods with Strain Dominant Failure Criteria for Improving the Resistance to Pad Cratering. In Proceedings of the 2016 17th International Conference on Electronic Packaging Technology (ICEPT), Wuhan, China, 16–19 August 2016; pp. 1–6.

18.	Roggeman, B.; Borgesen, P.; Li, J.; Godbole, G.; Tumne, P.; Srihari, K.; Levo, T.; Pitarresi, J. Assessment of PCB Pad Cratering Resistance by Joint Level Testing. In Proceedings of the 2008 58th Electronic Components and Technology Conference, Lake Buena Vista, FL, USA, 27–30 May 2008; pp. 884–892.

19.	Godbole, G.; Roggeman, B.; Borgesen, P.; Srihari, K. On the Nature of Pad Cratering. In Proceedings of the 2009 59th Electronic Components and Technology Conference, San Diego, CA, USA, 26–29 May 2009; pp. 100–108.

20.	Xie, D. Pad Cratering and Role of Thermal Stresses. In *Encyclopedia of Thermal Stresses*; Hetnarski, R.B., Ed.; Springer: Dordrecht, The Netherlands, 2014; pp. 3593–3608. ISBN 978-94-007-2739-7.

21.	Cai, M.; Xie, D.J.; Chen, W.B.; Wu, B.Y.; Yang, D.G.; Zhang, G.Q. A Novel Soldering Method to Evaluate PCB Pad Cratering for Pin-Pull Testing. *Microelectron. Reliab.* **2013**, *53*, 1568–1574. [CrossRef]

22.	Yang, C.; Song, F.; Lee, S.W.R.; Newman, K. Comparative Study of PWB Pad Cratering Subject to Reflow Soldering and Thermal Impact. In Proceedings of the 2010 Proceedings 60th Electronic Components and Technology Conference (ECTC), Las Vegas, NV, USA, 1–4 June 2010; pp. 464–470.

23.	Dušek, K.; Rudajevová, A.; Plaček, M. Influence of Latent Heat Released from Solder Joints on the Reflow Temperature Profile. *J. Mater. Sci. Mater. Electron.* **2016**, *27*, 543–549. [CrossRef]

24.	Cheat, L.W.; Sing, L.S. Copper Discoloration: Correlation Between Copper Oxidation States and Their Colors. In Proceedings of the 2018 IEEE International Symposium on the Physical and Failure Analysis of Integrated Circuits (IPFA), Singapore, 16–19 July 2018; pp. 1–3.

25.	Ramirez, M.; Henneken, L.; Virtanen, S. Oxidation Kinetics of Thin Copper Films and Wetting Behaviour of Copper and Organic Solderability Preservatives (OSP) with Lead-Free Solder. *Appl. Surf. Sci.* **2011**, *257*, 6481–6488. [CrossRef]

26.	Su, S.; Jian, M.; Hamasha, S. Effects of Surface Finish on the Shear Fatigue of SAC-Based Solder Alloys. *IEEE Trans. Compon. Packag. Manuf. Technol.* **2020**, *10*, 457–466. [CrossRef]

27.	Hamasha, S.; Akkara, F.; Abueed, M.; Rababah, M.; Zhao, C.; Su, S.; Suhling, J.; Evans, J. Effect of Surface Finish and High Bi Solder Alloy on Component Reliability in Thermal Cycling. In Proceedings of the 2018 IEEE 68th Electronic Components and Technology Conference (ECTC), San Diego, CA, USA, 29 May–1 June 2018; pp. 2032–2040.

28.	Collins, M.N.; Punch, J.; Coyle, R. Surface Finish Effect on Reliability of SAC 305 Soldered Chip Resistors. *Solder. Surf. Mt. Technol.* **2012**, *24*, 240–248. [CrossRef]

29.	Yang, Y. Failure Analysis for Bad Wetting on HASL PCB. In Proceedings of the 2017 18th International Conference on Electronic Packaging Technology (ICEPT), Harbin, China, 16–19 August 2017; pp. 126–129.

30. Tong, K.H.; Ku, M.T.; Hsu, K.L.; Tang, Q.; Chan, C.Y.; Yee, K.W. The Evolution of Organic Solderability Preservative (OSP) Process in PCB Application. In Proceedings of the 2013 8th International Microsystems, Packaging, Assembly and Circuits Technology Conference (IMPACT), Taipei, Taiwan, 22–25 October 2013; pp. 43–46.
31. Dušek, K.; Bušek, D.; Petráč, A.; Plaček, M.; Kozák, M.; Molhanec, M.; Beshajová Pelikánová, I. Comparison of Mechanical Resistance of SnCu and SnBi of Solder Joints. In Proceedings of the 2017 40th International Spring Seminar on Electronics Technology (ISSE), Sofia, Bulgaria, 10–14 May 2017; pp. 1–4.
32. Vasko, C.; Novotny, M.; Szendiuch, I. Impact of Solder Pad Shape on Lead-Free Solder Joint Reliability. In Proceedings of the 2008 31st International Spring Seminar on Electronics Technology, Budapest, Hungary, 7–11 May 2008; pp. 433–445.
33. Zhou, P.; Jiang, T.; Yang, D.; You, Z.; Ma, Y.; Ren, R. The Effect of Interface Reaction with Different Finishes and SnAgCu on the Reliability of Solder Joints. In Proceedings of the 2010 11th International Conference on Electronic Packaging Technology & High Density Packaging, Xi'an, China, 16–19 August 2010; pp. 998–1002.
34. Wiechmann, R. Experiences with Peel Strength. *Circuit World* **2006**, *32*, 30–39. [CrossRef]
35. Ikegawa, N.; Hamada, H.; Yamanouchi, M. Evaluation of Adhesion between Glass/Epoxy Composite Substrate and Copper Foil for Printed Circuit Board Applications. *Compos. Interfaces* **2002**, *9*, 235–245. [CrossRef]
36. Liu, F.; Wan, B.; Wang, F.; Chen, W. Effect of Thermal Shock Process on the Microstructure and Peel Resistance of Single–Sided Copper Clad Laminates Used in Waste Printed Circuit Boards. *J. Air Waste Manag. Assoc.* **2019**, *69*, 1490–1502. [CrossRef] [PubMed]
37. Yong, A.X.H.; Sims, G.D.; Gnaniah, S.J.P.; Ogin, S.L.; Smith, P.A. Heating Rate Effects on Thermal Analysis Measurement of Tg in Composite Materials. *Adv. Manuf. Polym. Compos. Sci.* **2017**, *3*, 43–51. [CrossRef]
38. Ahmad, M.; Burlingame, J.; Guirguis, C. Validated Test Method to Characterize and Quantify Pad Cratering under BGA Pads on Printed Circuit Boards. In Proceedings of the IPC APEX Expo, San Jose, CA, USA, 25–27 January 2022; pp. 1–13.
39. Ray, B.C. Effect of Thermal Shock on Interlaminar Strength of Thermally Aged Glass Fiber-Reinforced Epoxy Composites. *J. Appl. Polym. Sci.* **2006**, *100*, 2062–2066. [CrossRef]
40. Roggeman, B.; Jones, W. Characterizing the Lead-Free Impact on PCB Pad Craters. In Proceedings of the IPC APEX Expo, San Diego, CA, USA, 20–21 July 2010.
41. Rudajevová, A.; Dušek, K. Influence of Manufacturing Mechanical and Thermal Histories on Dimensional Stabilities of FR4 Laminate and FR4/Cu-Plated Holes. *Materials* **2018**, *11*, 2114. [CrossRef]
42. Margem, F.M.; Monteiro, S.N.; Bravo Neto, J.; Rodriguez, R.J.S.; Soares, B.G. The Dynamic-Mechanical Behavior of Epoxy Matrix Composites Reinforced with Ramie Fibers. *Matér. Rio Jan.* **2010**, *15*, 164–171. [CrossRef]
43. Keusch, S.; Haessler, R. Influence of Surface Treatment of Glass Fibres on the Dynamic Mechanical Properties of Epoxy Resin Composites. *Compos. Part Appl. Sci. Manuf.* **1999**, *30*, 997–1002. [CrossRef]

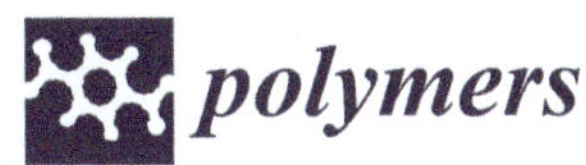

Article

Analysis of the Mechanical and Preforming Behaviors of Carbon-Kevlar Hybrid Woven Reinforcement

Zhengtao Qu [1], Sasa Gao [1,2,*], Yunjie Zhang [1] and Junhong Jia [1]

1   College of Mechanical & Electrical Engineering, Shaanxi University of Science & Technology, Xi'an 710021, China; 1905014@sust.edu.cn (Z.Q.); 200511032@sust.edu.cn (Y.Z.); jhjia@sust.edu.cn (J.J.)
2   State Key Laboratory of Applied Optics, Changchun Institute of Optics, Fine Mechanics and Physics, Chinese Academy of Sciences, Changchun 130033, China
*   Correspondence: gaosasa@sust.edu.cn

Abstract: Carbon-Kevlar hybrid reinforcement is increasingly used in the domains that have both strength and anti-impact requirements. However, the research on the preforming behaviors of hybrid reinforcement is very limited. This paper aims to investigate the mechanical and preforming behaviors of carbon-Kevlar hybrid reinforcement. The results show that carbon-Kevlar hybrid woven reinforcement presents a unique "double-peak" tensile behavior, which is significantly different from that of single fiber type reinforcement, and the in-plane shear deformation demonstrates its large in-plane shear deformability. Both the tensile and in-plane shear behaviors present insensitivity to loading rate. In the preforming process, yarn slippage and out-of-plane yarn buckling are the two primary types of defects. Locations of these defects are closely related to the punch shape and the initial yarn direction. These defects cannot be alleviated or removed by just increasing the blank holder pressure. In the multi-layer preforming, the compaction between the plies and the friction between yarns simultaneously affect the quality of final preforms. The defect location of multi-layer preforms is the same as that of single-layer, while its defect range is much wider. The results found in this paper could provide useful guidance for the engineering application and preforming modeling of hybrid woven reinforcement.

Keywords: hybrid woven reinforcement; preforming; mechanical properties; carbon-Kevlar

Citation: Qu, Z.; Gao, S.; Zhang, Y.; Jia, J. Analysis of the Mechanical and Preforming Behaviors of Carbon-Kevlar Hybrid Woven Reinforcement. *Polymers* **2021**, *13*, 4088. https://doi.org/10.3390/polym13234088

Academic Editor: Rajesh Mishra

Received: 22 October 2021
Accepted: 22 November 2021
Published: 24 November 2021

Publisher's Note: MDPI stays neutral with regard to jurisdictional claims in published maps and institutional affiliations.

1. Introduction

In recent years, the demand for excellent performance and lightweight materials has prompted the application of fiber-reinforced composite materials by replacing traditional metal materials [1]. The properties of reinforcements play an important role in composites. By controlling the orientation and volume fraction of fibers, composites can achieve the desired dimensional stability and mechanical strength while being formed into complex geometrical components [2–6]. Another approach to obtain the desired or improved performance is through hybridization among different types of reinforcements. Compared with traditional composites (single reinforcement type), hybrid composites have special properties and can meet a variety of design requirements in a more cost-effective manner. Some research has reported that there are many situations in which a high modulus material is needed, for example, carbon fiber reinforcements, but it is usually associated with a catastrophic brittle failure, which is not desirable [7]. The solution to such weakness is that the combination of carbon fibers and other types of fiber that have suitable ductility. For instance, carbon-Kevlar hybrid composite, which has excellent impact resistance, is a typical material used in military ballistic protection [8–10].

In the manufacturing of small and medium-sized composite parts, resin transfer molding (RTM) is the most common process. RTM requires the preforming of dry woven reinforcement and subsequent injection of resin. In this process, the reinforcement undergoes a change process from a 2D plane to a 3D geometric shape. The variations in the reinforcement, such as fiber volume fraction, deformation state, and other parameters, will have a great effect on the resin flow impregnation. In addition, the preforming defects observed at this stage, such as yarn slippage, buckling, and wrinkling, are closely related to the preforming behaviors [11–14]. The deformation behavior of reinforcement to form a specific shape primarily includes the deformation modes such as tension, in-plane shear, bending, and transverse compression [15–17], even the tension-shear coupling [18,19]. The deformation state of reinforcements strongly affects the mechanical properties of the final composite components [20–22]. Moreover, the reinforcement type (weaving type, number of layers, fiber orientation, etc.) and process parameters (such as punch shape, forming speed, pressure of the blank holder, etc.) can have a significant influence on the preform quality [23]. Therefore, in order to manufacture complex geometric composite parts without defects, it is important to understand the deformation behaviors during the preforming process. Numerical simulation allows researchers to adopt more effective and cost-effective design methods to study the feasibility of fabric forming. The simulation code describes the evolution of the behavior of the reinforcement through a mechanical method [14,19,24–27]. However, the accuracy of these numerical works must be verified by experimental results. It is important to study these defects and their mechanism so as to reduce their appearance. Labanieh et al. [22] studied the yarn slippage mechanism in the hemispheric preforming with woven carbon fabric. Gatouillat et al. [28] researched the yarn slippage defect and proposed an analytical model to predict the slippage. Capelle et al. [29] investigated the out-of-plane buckling of flax woven fabric during preforming and proposed a new special blank holder, which could reduce and sometimes even eliminate the occurrence of buckling. Furthermore, Shanwan et al. [30] developed a new strategy to understand the mechanism of mesoscopic defects in the preforming of woven fabrics and analyzed the effects of these defects on the mechanical properties of composites.

In practice, composite components are usually composed of multi-layer reinforcements, which can be arranged in different ply orientations to optimize the structural performance of the components. However, the deformation properties of multi-layer reinforcements have a close relationship with the relative orientation of plies. In this way, the occurrence possibility of defects is higher for multi-layer structures, especially for those with different ply orientations [31–33]. Moreover, the inter-ply friction prevents the relative sliding of plies, resulting in more severe wrinkles [31,34,35]. Guzman-Maldonado et al. [32] performed a numerical study on multi-layer preforming reinforcements. They investigated the interaction between adjacent plies during the preforming process and emphasized that the distortion of contact plies in different directions increased the severity of wrinkles. In addition, Thompson et al. [35] have proved that the stacking orientations can result in their own wrinkle pattern in the multi-layer preforming. Each ply became wrinkled due to the combined effect of its own intra-ply mechanical deformations and interactions with adjacent plies. Huang et al. [36] studied the effect of ply orientations on wrinkles of multi-layer reinforcement during bending. Severe wrinkling and interlaminar separation were observed in the alternate $0°/45°$ plies, while no wrinkling was observed in the pure $45°$ plies. Furthermore, Allaoui et al. [31] compared the preforming results of single-layer and multi-layer reinforcements by quantifying the shear angle and defects. Preforming defects of multi-layer reinforcements occurred in the same location as those of single-layer reinforcements but in a wider range.

From the above literature, it can be found that much research has been conducted on the preforming behavior of woven reinforcement fabrics. However, most of them are for the single fiber type reinforcement, while few reports are for the hybrid reinforcement. As a promising composite material, the performance of carbon-Kevlar hybrid woven reinforcement may be different from that of pure carbon fiber reinforcement or pure Kevlar fiber reinforcement due to the interaction of the two kinds of fibers. There exists some research on the carbon-Kevlar hybrid composites [8,37,38], indicating that appropriate fiber hybridization could improve the mechanical performance of composites. However, these studies are primarily focused on the cured composites composed of carbon-Kevlar hybrid reinforcements, and the knowledge of the mechanical properties of dry hybrid reinforcements is far not sufficient. Due to the brittleness of carbon fibers, the fracture strain of carbon fibers (about 1%) is much lower than that of Kevlar fibers (about 3.5%), which makes them more vulnerable to damage during stretching and even preforming processes.

The objective of this work is to investigate the mechanical properties of carbon-Kevlar hybrid woven reinforcements and their possibilities to form complex shapes. A main concern is the defects that occur during the preforming process. Several different preforming process parameters are designed to investigate the mode of defects under specific conditions. The results and defects of single-layer and multi-layer preforming are compared quantitatively.

2. Materials and Methods

2.1. Tested Materials

The carbon-Kevlar hybrid woven reinforcement (Shanghai Banglin Composite Material Technology Co., Ltd., Shanghai, China) used in this study is shown in Figure 1, which consists of two yarn networks in two directions (warp and weft), with carbon fiber yarn and Kevlar fiber yarn in each direction. The reinforcement possesses a 1/3 twill woven structure. Compared with plain woven structure, it has a reduced crimp level and better drapability, while the interlocking between warp and weft yarns is not that tight. These factors result in high permeability and better suitability for the RTM process. The main material parameters of carbon-Kevlar hybrid woven reinforcement are shown in Table 1.

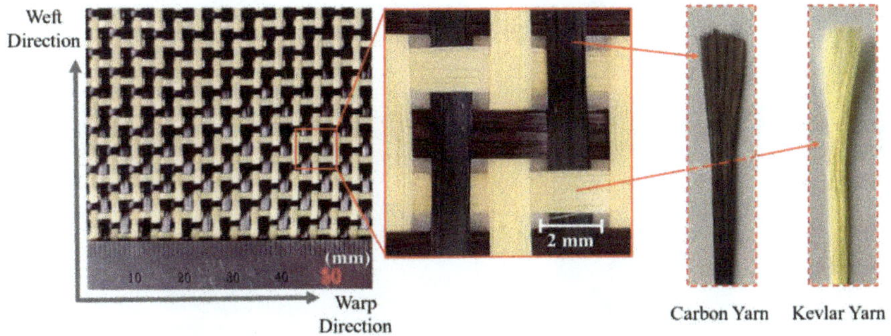

Figure 1. Tested carbon-Kevlar hybrid woven reinforcement.

Table 1. Material parameters of carbon-Kevlar hybrid woven reinforcement.

Material	Yarn Count (Yarn/cm²)	Woven Structure	Areal Density (g/m²)	Thickness (mm)	Yarn Type	Linear Density (g/m)	Bulk Density (g/cm³)
Carbon-Kevlar hybrid reinforcement	5 × 5	1/3 twill	220	0.3			
Carbon yarn					T300 3K	0.198	1.76
Kevlar yarn					1500D	0.1679	1.414

2.2. Mechanical Behavior Characterizations

There are complex relationships among mechanical properties of reinforcements, such as preforming process parameters and punch shapes, which affect the quality of final preforms. Firstly, the mechanical characterization of the hybrid woven reinforcement was investigated. The interested deformation behaviors of hybrid woven reinforcements are the tensile deformation along yarn direction and in-plane shear between two yarn directions, which are the primary deformation mode in preforming. In the tensile tests, it can be noted that the deformation behavior of hybrid woven reinforcement was significantly different from that of a single fiber type of reinforcement. Although yarns will not be broken in the preforming process, the tensile fracture behavior of carbon-Kevlar hybrid woven reinforcements was still analyzed considering their actual applications (usually used in the field of protection). The phenomenon was interesting because it visualizes the advantages of hybridization. The tensile test conducted in this study included two aspects: one was for the hybrid woven reinforcement, and the other was for single carbon yarn and single Kevlar yarn in order to better explain the failure mechanism of hybrid woven reinforcement. The selected geometry of specimens in the tensile test is sketched in Figure 2. The effective area of specimens (200×40 mm^2) contains 10 carbon yarns and 10 Kevlar yarns in the longitudinal direction (Figure 2a). Different loading speeds (2, 10, 50, 10, 200, and 400 mm/min) were applied to investigate the sensitivity of loading speed on the hybrid woven reinforcement. Moreover, in order to analyze the contribution of different kinds of yarns to the global mechanical behaviors of hybrid woven reinforcement, the tensile behavior of single carbon yarn and single Kevlar yarn extracted randomly from raw material were tested. For yarn samples, the gauge length was 200 mm (Figure 2b,c), and the loading speed was set as 10 mm/min. In addition, the deformation and failure behaviors of the specimens were recorded by a high-resolution camera at the sampling rate of 2 Hz. Each of the specimen configurations was tested at least three times to assess repeatability.

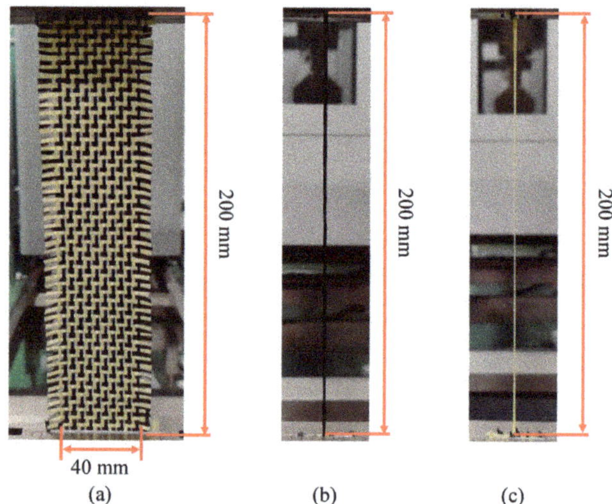

Figure 2. Tested specimen in the tensile test: (**a**) hybrid woven reinforcement; (**b**) carbon yarn; (**c**) Kevlar yarn.

The in-plane shear is the primary deformation mode in the preforming of double-curved shape. Physical phenomena such as contact and friction between yarns are related to shear properties. During the RTM process, the in-plane shear deformation could cause a significant change in the permeability of reinforcement, especially when the shear angle reaches the "locking angle". Although there is no direct relationship between shear angle and wrinkling, an excessive shear angle is more likely to cause wrinkling [16]. In this study, the in-plane shear properties of hybrid woven reinforcement were investigated by the bias-extension test. Compared with the picture frame test, the bias-extension test is relatively simple and can be performed on a tension machine without other special devices. The specimen was a rectangular piece with an angle of $\pm 45°$ between the yarn orientations and the loading direction (Figure 3), whose length (L = 160 mm) is at least twice as long as the width (W = 80 mm). Red marker lines have been drawn on the specimens, which would be used for tracking shear angle evolution. To rule out the influence of the yarn slippage on the shear angle that is directly determined from the displacement, a high-resolution camera was used to take the images. The shear angle can then be accurately measured with ImageJ software to process images taken by the camera. In the current work, the bias-extension tests were conducted under different loading speeds of 2, 10, 50, and 200 mm/min.

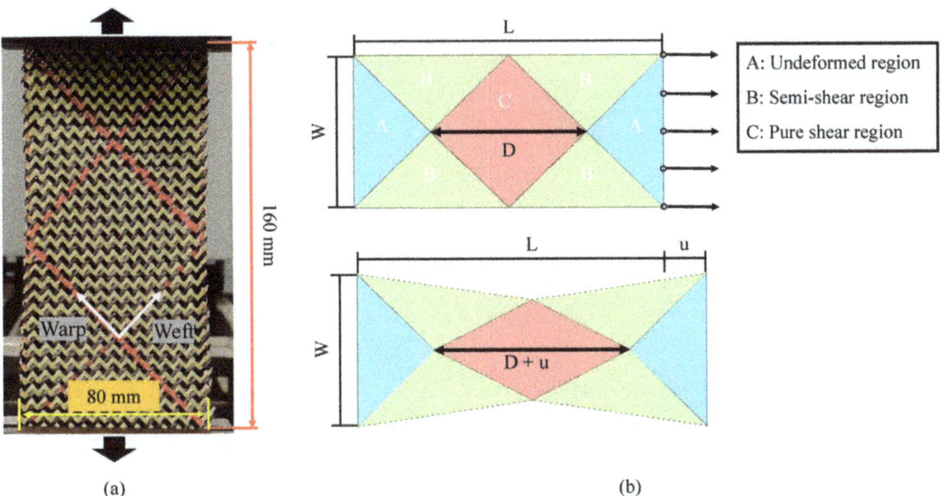

Figure 3. Tested specimen in the bias-extension test: (**a**) tested specimen; (**b**) illustration of the specimen before and after deformation.

2.3. Preforming Tests

The carbon-Kevlar hybrid woven reinforcement is usually used in cases with both strength and anti-impact requirements, such as a protective helmet or car bumper, which usually includes complex double-curved shapes. Considering its special applications, classic hemispherical and tetrahedral punches were used to investigate the feasibility of hybrid woven reinforcements to form specific shapes. The punch was installed on the universal testing machine (Shenzhen Wance Testing Machine Co., Ltd., Shenzhen, China) to realize the preforming test. The movement of the punch was controlled and measured by an electric jack equipped with a sensor, which also can measure the position and the loading speed of the punch. The first punch was a hemisphere with a diameter of 150 mm, and the punch stroke was 75 mm (Figure 4a). The gap between the punch and die was 3 mm. The other open tetrahedral die is shown in Figure 4b. The shape of the tetrahedral punch is the part cut on a cube with an edge length of 103.9 mm and a radius of 7.5 mm at its edge corner. The gap between the punch and die was 1.5 mm, and the punch stroke was 60 mm. The adjustment was installed in the four corners of the blank holder in both dies, which can achieve variation blank holder pressures by adjusting the spring compression. Additionally, the spring stiffness related to the holder pressure should be measured by independent experiments before preforming tests.

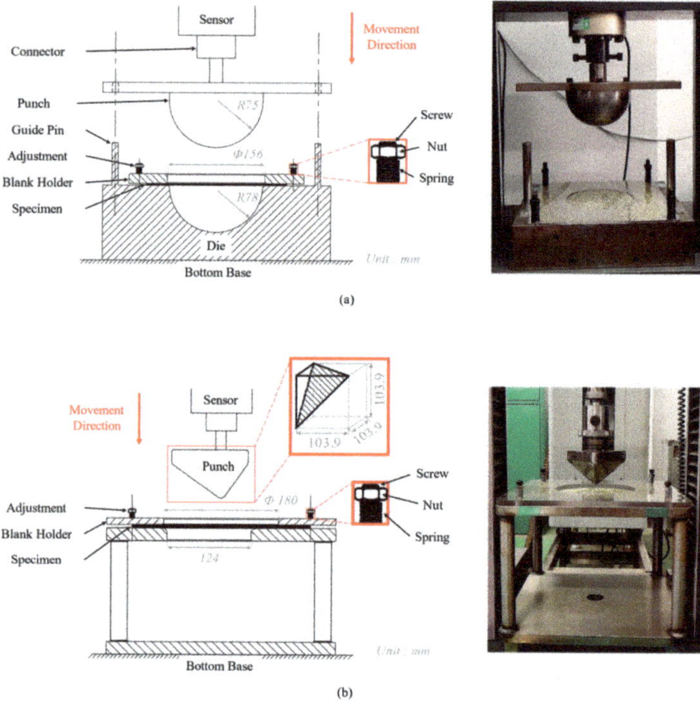

Figure 4. The hemispherical and tetrahedral preforming device: (**a**) experimental set-up for hemispherical preforming; (**b**) experimental set-up for tetrahedral preforming.

The specimens used in the experiment of hemispherical preforming and tetrahedral preforming are same (280×280 mm^2). The rectangular with a side length of 20 mm was cut from the four corners of the specimen to prevent interference with the adjustments (Figure 5). For woven structures, it is essential to study the effect of initial yarn orientation (such as $0°/90°$ and $\pm45°$) on the preforming properties [39], especially for forming complex shapes. Therefore, two initial yarn orientations ($0°/90°$ and $\pm45°$) will be considered in this study. Additionally, the quality of the final preform is also affected by the tension force imposed on the specimen during the preforming process, which can cause other defects [40,41]. Therefore, several low (0.08 Kpa for hemispherical preforming and 0.25 Kpa for tetrahedral preforming) and high (3.7 Kpa for hemispherical preforming and 4 Kpa for tetrahedral preforming) blank holder pressures were chosen for single-layer preforming experiments in this study. After preforming, the deformed reinforcement needs to be cured with the structural glue to avoid spring-back after removing the blank holder. This structural glue contains acrylic resin and a curing agent. After the two are mixed, they can be quickly cured at room temperature (usually only a few minutes). The specimen can be removed from the die after the curing of the glue for further measurement.

After the preforming process, yarn orientation and shear angle distribution are particularly essential to the permeability of the reinforcement and the mechanical behavior of the final composite product. When the preform was multi-layered, especially for that with different initial ply orientations, the occurrence possibility of defects would increase greatly [32]. In tetrahedral preforming, significant strains and additional defects occurred in the specimen due to the low punch edge radius. Therefore, tetrahedron preforming experiments have been carried out to analyze the interaction between different plies and their effect on the defects. These experiments were performed with low blank holder pressure (0.25 Kpa) and a punch loading speed of 10 mm/min. The influence of the ply orientation was also considered, and the configurations of multi-layer preforming were shown in Figure 5. The center position of the sample was marked with red marker lines for positioning. The two-layer preforms were fixed with resin after preforming. However, for four-layer preforms, it was tricky to observe the deformation of the inner layer due to the solidification. Therefore, it was necessary to carefully remove the upper layer, and the remaining ply was measured and analyzed after each removal. Then, the preforming results of different configurations were compared in order to analyze the influence of different ply orientations on the multi-layer preforming.

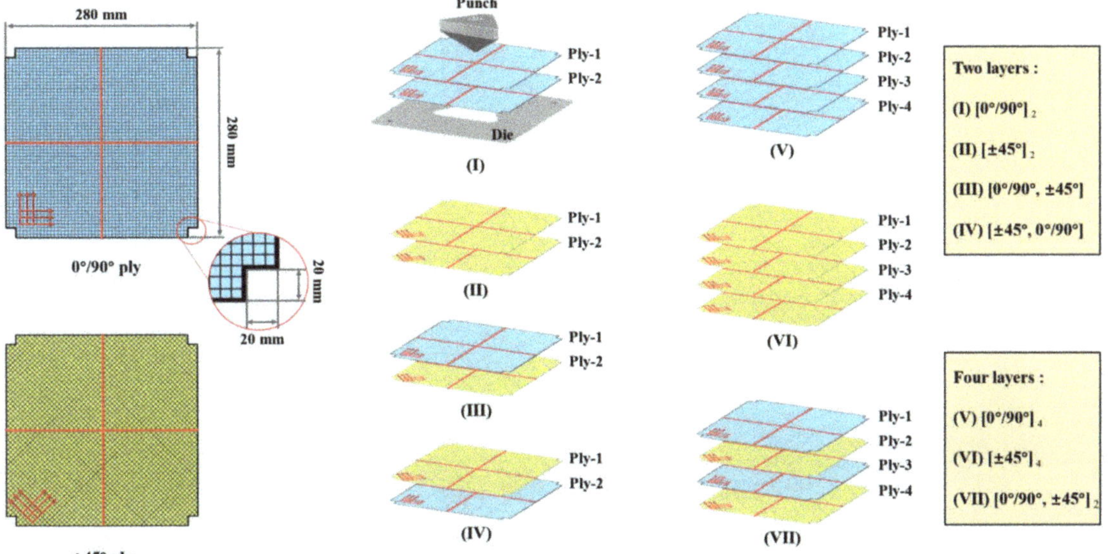

Figure 5. Schematic of specimen size (suitable for hemisphere preforming and tetrahedral preforming) and multi-layer configurations (only for tetrahedral preforming).

3. Results and Discussion

3.1. Mechanical Behavior Characterization

3.1.1. Tensile Behaviors

Figure 6 presents the curves of the tensile force versus strain at different loading speeds. The characteristics of tensile behavior are significantly different from those of reinforcement with a single fiber type. In the curve, a unique "double-peak" feature was reflected, which was related to the different properties of the two kinds of yarns. The fracture of carbon fiber yarns was directly related to the formation of the first peak, and the subsequent behavior largely depended on Kevlar yarns.

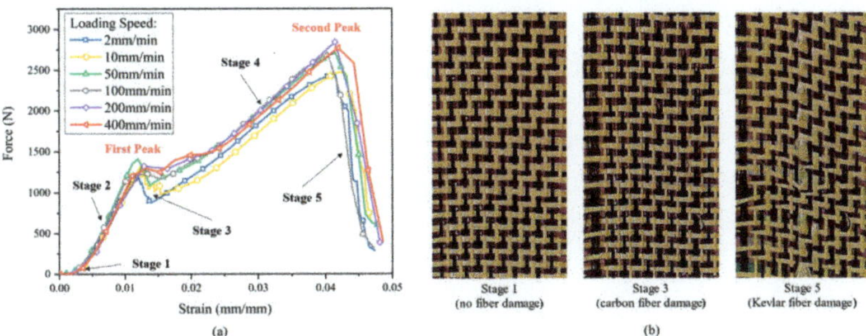

Figure 6. Tensile test results of hybrid woven reinforcement: (**a**) the force-strain relationship under different loading speeds; (**b**) the deformation and damage behaviors in some typical stages.

The whole curve can be divided into five stages (Figure 6). In stage 1, since the wave-shaped yarn was straightened in the loading direction, the stress increase was relatively low. In stage 2, the tensile curve showed a linear behavior. Both carbon yarns and Kevlar yarns were straightened with the increase in strain. In stage 3, carbon yarns firstly came to be broken due to their small ultimate strain. The curve reached the first peak. Severe fluctuations in the curve can be seen during this stage, which is due to uncertain defects existing in certain carbon yarns, and the fracture of carbon yarns may not occur at the same time. In addition, the fracture of carbon yarn will lead to the redistribution of force in the remaining yarns and a stiffness reduction in the reinforcement. Thus, there has a force decreasing as shown in stage 3. In stage 4, with the strain increased continually, the Kevlar yarns were still undamaged and continued to be stretched. Finally, the curve reached the second peak. The Kevlar yarns began to be damaged after reaching their ultimate strain, which represented the complete failure of the whole material, as shown in stage 5.

The tensile behavior of carbon-Kevlar hybrid woven reinforcement can be further elucidated by that of single carbon yarn and single Kevlar yarn. The tensile force-strain curves of both carbon yarn and Kevlar yarn are shown in Figure 7. It is worth noting that stage 1 in Figure 6 was not presented in Figure 7 since the yarn was straight at the beginning. As can be seen in Figure 7, carbon fiber provides better initial tensile stiffness, while Kevlar fiber provides better tensile strength for the hybrid woven reinforcement. Therefore, hybrid woven reinforcement materials have excellent energy absorption properties and can be applied on occasions with anti-impact requirements. In addition, the carbon yarn and the Kevlar yarn exhibited different failure modes, which can be seen in Figure 8. The fracture of carbon yarn appeared an obvious brittleness, and the energy was released instantly when the yarn was broken and then exploded outward. In contrast, the Kevlar yarn had larger toughness and formed a fluffy shape after breaking.

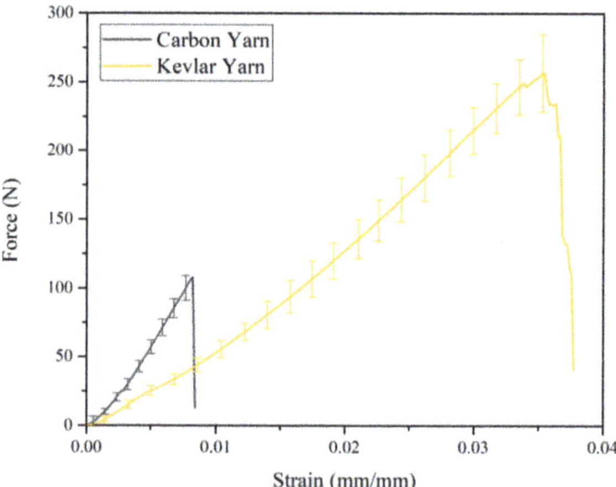

Figure 7. Tensile force versus strain curves of carbon yarn and Kevlar yarn at a loading speed of 10 mm/min.

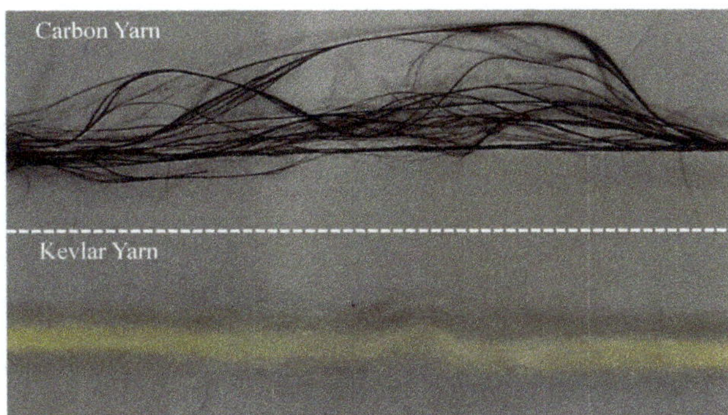

Figure 8. Failure modes of two different yarns.

3.1.2. In-Plane Shear Behaviors

Figure 9 presents the curves of the clamp force versus shear angle. It is clearly indicated that the shear behavior was also insensitive to the loading speed. Two obvious deformation phenomena can be noted. At the initial stage, the force resistance primarily comes from the friction between yarns, and yarn free rotation can be seen. As the rotation continues, the gap between adjacent yarns decreases and yarns begin to contact and compress with each other, and the load increases significantly with the increase in shear angle. Significant yarn slippage along the longitudinal direction occurred during the bias extension due to the very loose weaving structure. Furthermore, the slippage occurred at four corners of the specimen due to the yarns at this position being subjected to the least friction. Figure 10 shows the evolution of shear angles under different displacements, and three different zones corresponding to the theoretical kinematics of the test (Figure 3b) were clearly observed in the deformed specimen. Additionally, the shear stiffness of the reinforcement was low, with slight wrinkles occurring at a shear angle of about $60°$. The experimental results indicate that the reinforcement used in this study can adapt to large shear deformation with high locking angle ($55°\sim60°$). Thus, the hybrid reinforcement has suitable formability. Nevertheless, yarn slippage is a major concern, especially in the process of shaping a deep-draw shape.

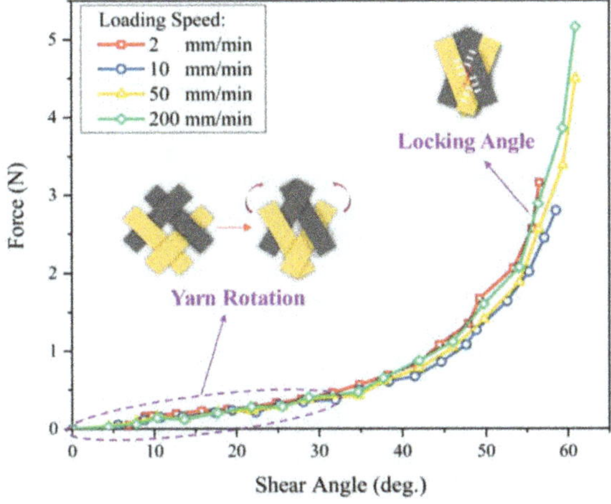

Figure 9. Force-shear angle curves at different loading speeds in bias-extension test.

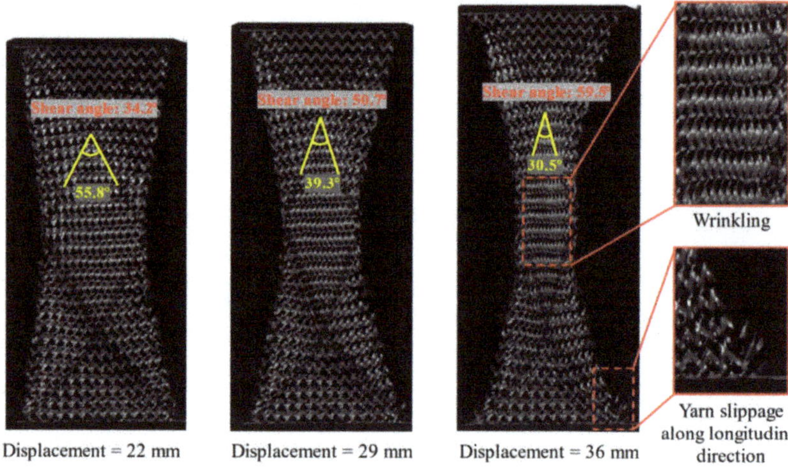

Figure 10. Shear angle evolution in the bias-extension test. The wrinkles occur at a shear angle of about 60°.

3.2. Preforming of Hybrid Woven Reinforcement

Since the reinforcement rarely breaks during the RTM processing, the double-peak tensile properties have little effect on the preforming behavior of hybrid woven reinforcements; generally, only the deformation stage before the damage of carbon yarn needs to be considered. The in-plane shear deformation is one of the main deformation modes of reinforcement in the preforming process. From the bias-extension test, it is known that yarn slippage is inclined to occur since the weaving structure of the carbon-Kevlar reinforcement is relatively loose. Furthermore, the friction behaviors in multi-layer preforming are more complex because of the relative motion between plies, which can lead to the possible enlargement of defect occurrence. In this section, we will analyze and discuss the key factors that affect the quality of preform when preforming with single-layer and multi-layers.

3.2.1. Single-Layer Preforming

Deformation of Preforms with Different Initial Yarn Directions

The feasibility of hybrid woven reinforcement forming into complex curved parts with two different initial yarn orientations was studied by using hemispherical and tetrahedral punches, respectively. The yarn orientation after preforming was analyzed firstly since it directly affects the mechanical properties of the final part and the permeability of the preforming. In the experiment, preforms with suitable shape quality were obtained at a punch loading speed of 10 mm/min and a low blank holder pressure (of 0.08 Kpa for hemispherical preforming and 0.25 Kpa for tetrahedral preforming). Figure 11 schematically depicts the final yarn orientation in the useful area (i.e., the area not covered by the blank holder) after preforming. Due to the symmetry of the hemisphere punch, there is almost no difference for the material draw-in for the preforms with different initial yarn orientations, indicating that the initial yarn orientation did not affect the quality of hemispherical preforms. Furthermore, the shear angle distribution of hemispherical preform in the useful area is shown in Figure 12. The shear angle varied continuously, and clearly distinguished shear zones cannot be observed. The area near the blank holder had a higher shear angle due to larger material draw-in, and the maximum shear angle can reach 55°. However, no obvious wrinkling was observed. In addition, yarns usually undergo in-plane orientation change at the intersectional position with the central warp and weft yarn, resulting in a noticeable curve, shown as a red line in Figure 12, and near which the shear angle was usually small (close to 0°).

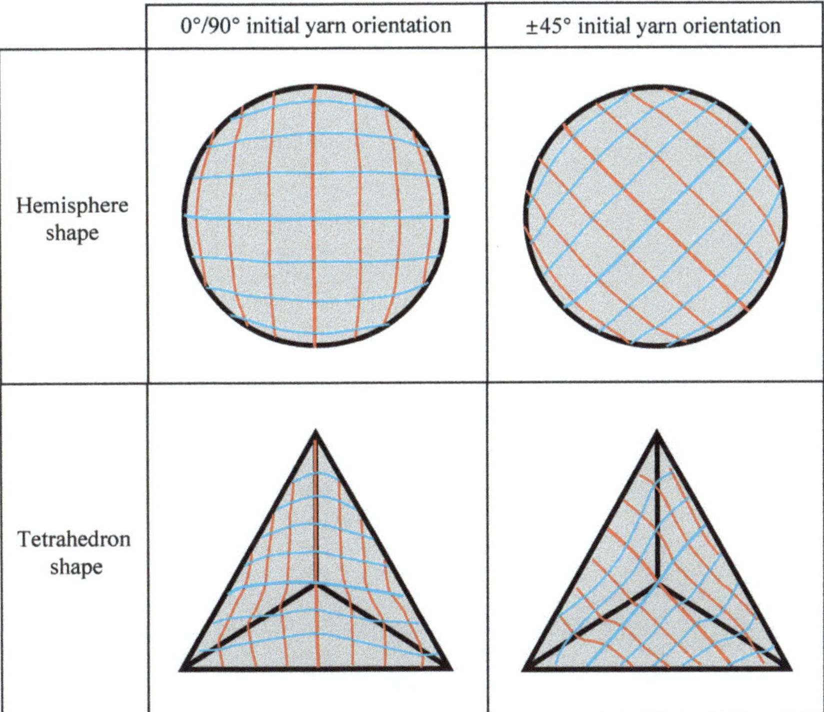

Figure 11. Yarn orientations after preforming for the hemispherical and tetrahedral preforms. Red lines for weft yarns; blue lines for warp yarns.

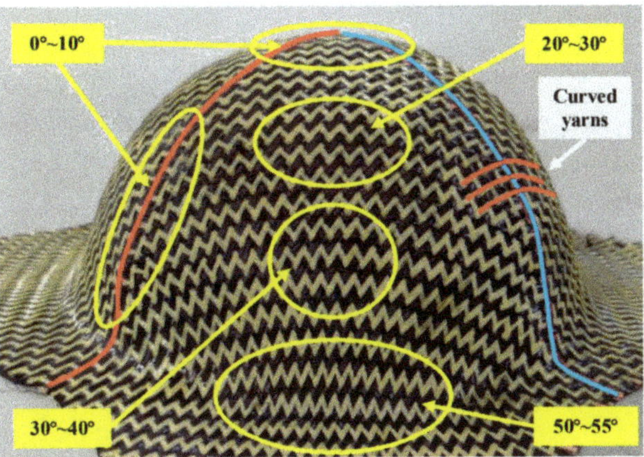

Figure 12. Shear angle distribution after hemispherical preforming. Red lines for weft yarns; blue lines for warp yarns; yellow lines for shear zones.

However, for the tetrahedron, the initial yarn orientation has a great influence on the behavior of the preform, and the shear deformation was more complex. In addition, changes in yarn orientation also occurred at the intersection edges of different planes. As can be seen in Figure 11, the preform with $\pm 45°$ initial yarn orientation underwent more yarn orientation changes. Furthermore, Figure 13 gives the shape and shear angle distribution of the tetrahedral preforms with different initial yarn orientations. Unlike hemispheric preforms, clearly distinguished shear zones could be observed in tetrahedral preforms. The dividing line between these shear zones was usually the central warp and weft yarn. It also existed the characteristic that the shear angle at the central warp and weft yarn position was close to $0°$. The yarns in the upper triangular pyramid section of the preform have a suitable alignment structure, but the shear angle becomes bigger at the bottom of the preform, especially at the edge with small curvature. The maximum shear angle can achieve $50°\sim55°$ in the preform with $0°/90°$ initial yarn orientation, which is a little bigger than that of $\pm 45°$ initial yarn orientation. Similarly, with the hemispherical preform, no wrinkles were observed in tetrahedral preforming with different initial yarn orientations. The main reason was that the loosely weaving structure prevents wrinkling even under high shear deformation. In addition, the yarn experienced a large orientation variation at the intersection of central warp and weft while the shear angle in this region was close to $0°$.

Figure 14 gives the punch force of reinforcements in $0°/90°$ and $\pm 45°$ initial yarn orientations at different forming speeds. For hybrid reinforcement with the same initial yarn orientation, the punch loading speed had little effect on the punch force. The deformation modes of the reinforcement were similar under constant blank holder pressure with different punch loading speeds. However, the punching force of the reinforcement in the $\pm 45°$ initial yarn orientation was larger. Since the punching force is to overcome sliding friction and shear forces, it means that the reinforcement in the $\pm 45°$ initial yarn orientation underwent larger sliding or shear deformation. In general, the effect of initial yarn orientation is on the punch shape. For hemisphere preforming, the results show that the initial yarn orientation has little effect on the preforming. However, when it comes to the more complex tetrahedral shape, the initial yarn orientation has a great influence on the quality of the preform. Variations of the initial yarn orientation may not cause significant defects but will result in changes in final yarn orientation [41], which directly affect the local permeability of the reinforcement and the mechanical properties of the composite parts.

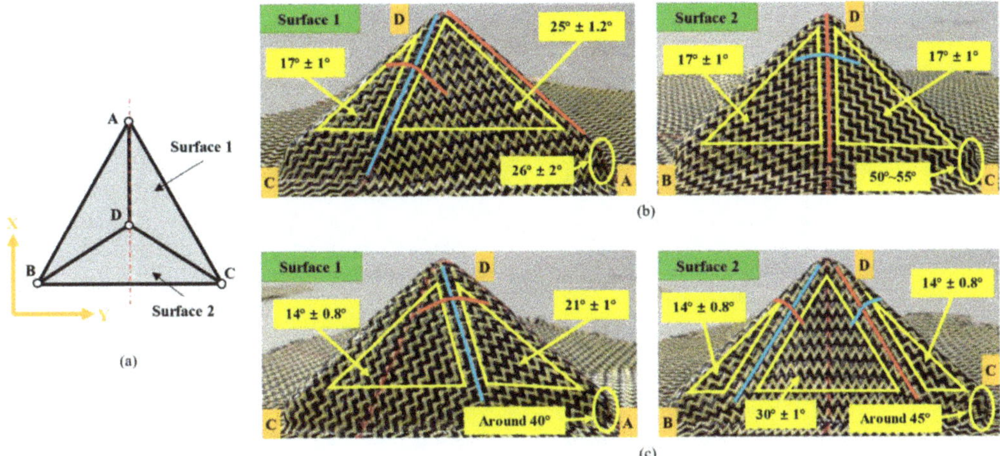

Figure 13. Shear angle distribution after tetrahedral preforming: (**a**) schematic diagram of tetrahedral preform description; (**b**) surface 1 (left) and surface 2 (right) of the tetrahedral preform with 0°/90° initial yarn orientation; (**c**) surface 1 (left) and surface 2 (right) of the tetrahedral preform with ±45° initial yarn orientation. Red lines for weft yarns; blue lines for warp yarns; yellow lines for shear zones.

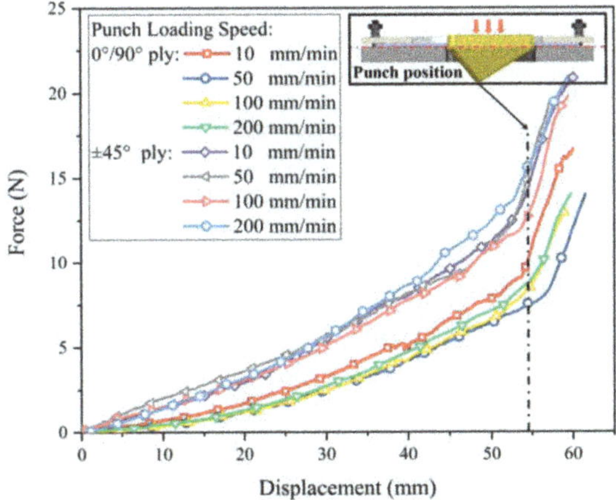

Figure 14. Punch force of reinforcements with 0°/90° and ±45° initial yarn orientations at different forming speeds.

Preforming Defect Description of Hybrid Woven Reinforcement

During the preforming process of single-layer hybrid woven reinforcement, there mainly exist two kinds of defects: yarn slippage and out-of-plane yarn buckling. For tetrahedral preforms, a typical defect is yarn slippage along the radial direction, occurring at the bottom corner of the preform, as shown in Figure 15. This random slippage made it possible to observe irregular gaps at the bottom corners of the preform and will have a strong influence on the local permeability of the preform. In single-layer preforms, such defects were located in non-useful areas, which would be cut off after forming process and not affect the quality of preform. However, this kind of defect should be taken into account when it comes to multi-layer preforming because the interaction between plies may magnify the defect. In addition, no yarn slippage was observed in the preform with an initial yarn direction of $0°/90°$ (Figure 15a, shown in the green zone), which indicated that yarn slippage can be reduced by properly adjusting the initial yarn direction.

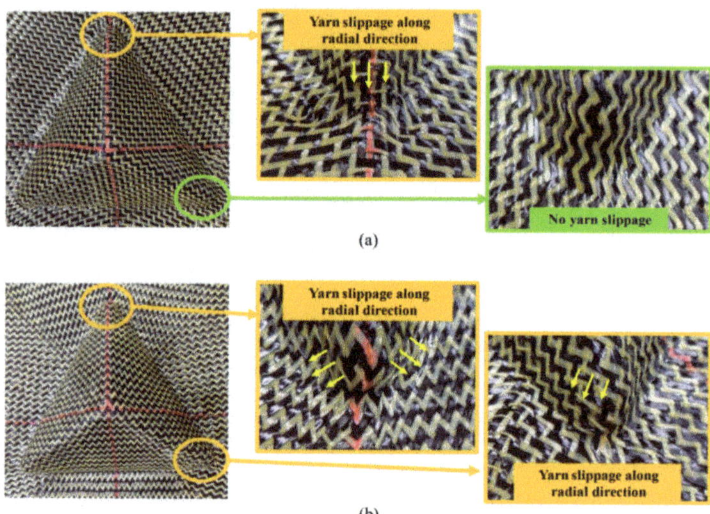

Figure 15. Yarn slippage along radial direction occurred at the bottom corner of the tetrahedral preform: (**a**) tetrahedral preform with $0°/90°$ initial yarn orientation; (**b**) tetrahedral preform with $\pm45°$ initial yarn orientation.

Another kind of defect, out-of-plane yarn buckling, occurred in both tetrahedral and hemisphere preforms. Such defects converge from the bottom to the top of the preform, and their position depends on the initial yarn direction. To be exact, they were located near the center warp and weft yarn. Buckling is a yarn scale phenomenon that usually only involves individual yarns and does not result in any membrane strain [31]. As shown in Figure 16, the yarns, which were intersected with the central warp and weft yarn, were subject to out-of-plane yarn buckling since they were impossible to accommodate in-plane bending in a small area. The magnitude of yarn buckling depended on the curvature of the in-plane bending and was proportional to the in-plane bending. These local buckling would cause a change in the thickness of the material and also affect the permeability of the reinforcement because of the conservation of yarn volume. Buckling is an unacceptable defect; thus, appropriate forming process parameters should be studied to reduce the occurrence of such defects.

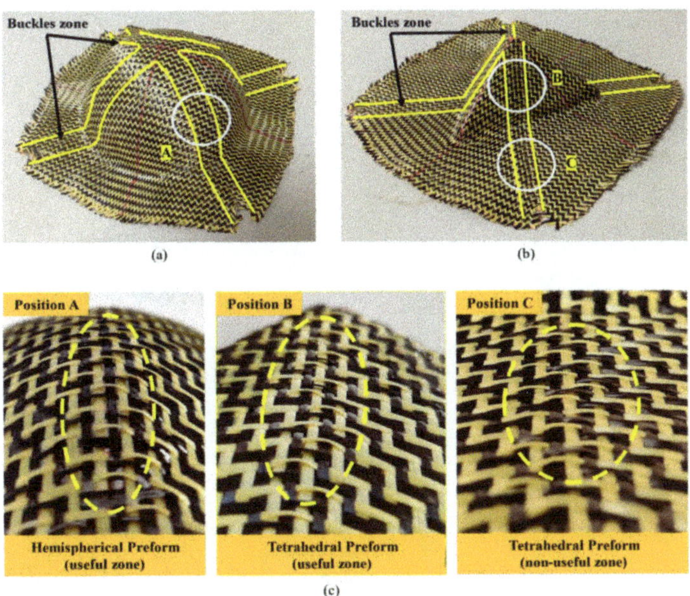

Figure 16. Out-of-plane yarn buckling in the preform and its location. Located near the central warp and weft yarn: (**a**) buckles zone on tetrahedral preform; (**b**) buckles zone on hemispherical preform; (**c**) zoom of the buckles.

Influence of Blank Holder Pressure on Defects

The blank holder pressure plays a very important role in the composite-forming process [42]. The blank holder pressure imposes tension on the reinforcement, which is also called preloading. The increase in blank holder pressure tends to delay the appearance of wrinkles. This means that if tension is applied to the reinforcement and even the shear angle is higher than the "locking angle", it is possible to obtain high shear deformation without wrinkles. For the carbon-Kevlar hybrid woven reinforcement used in this study, slight wrinkling could be observed when the shear angle reached 60° in the bias-extension test. However, wrinkles did not occur in the preforming. The influence of different blank holder pressure on the defects was quantified in this study. The preforms obtained by a tetrahedron preforming under high blank holder pressure (4 Kpa) are shown in Figure 17. The yarn slippage existed at the bottom corners of the preform, and the "weave pattern heterogeneity" phenomenon caused by yarn slippage along radial direction can be observed in the useful area. When the friction between the yarn/yarn and the yarn/punch is not enough to resist this tension, yarn slippage occurs. This type of defect is obviously unacceptable. In addition, out-of-plane yarn buckling would still exist when the pressure on the blank holder was large enough. Studies have shown that out-of-plane yarn buckling can be avoided by increasing the pressure on the blank holder [43]. However, for the carbon-Kevlar hybrid woven reinforcement in this study, this method did not seem to work since out-of-plane yarn buckling was noted even at high blank holder pressure (4 Kpa). In future work, related research includes optimizing the structure of the reinforcement should be carried out to reduce such defects.

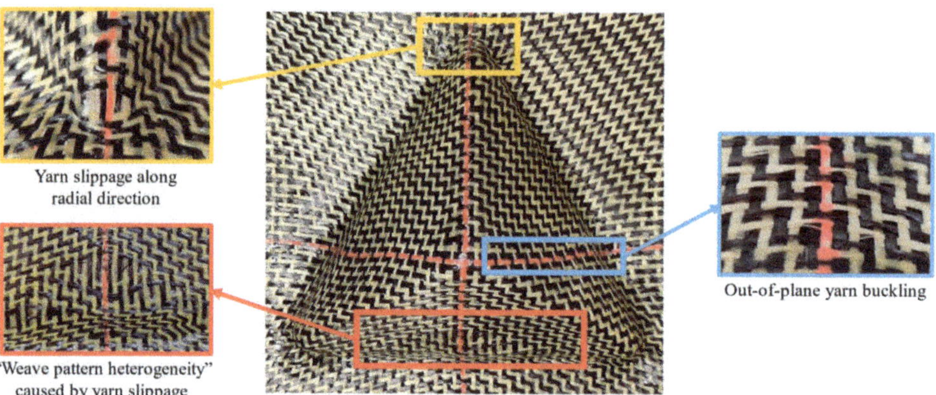

Figure 17. The preforms shape obtained from a tetrahedron preforming under high blank holder pressure (4 Kpa). Excessive yarn tension leads to more serious yarn slippage along the radial direction, but out-of-plane yarn buckling does not disappear.

Another interesting phenomenon that occurred in the non-useful area caused by blank holder pressure was also analyzed. As shown in Figure 18, in the tetrahedral preform with $0°/90°$ initial yarn orientation, less material draw-in under high blank holder pressure can be observed because the movement of the yarn was restricted by the pressure of the blank holder (Figure 18a). The preform obtained under low blank holder pressure had a suitable symmetrical profile while existing little deviations. One of the reasons was that it was difficult to ensure the specimen was placed on the mold with complete symmetry. On the contrary, the profile of the preform obtained under high blank holder pressure was more distorted, which indicated that the yarn movement was discontinuous. In addition, the randomness of shear angle distribution could also be observed in non-useful areas. Since the increase in shear angle can cause the local thickness changes, it makes the measurement of shear angle difficult. When the material thickness changes, the pressure exerted by the blank holder on the material is uniform, which may affect the friction coefficient between local materials and tools. This non-uniformity of friction coefficient can lead to different yarn slippages. Since the yarn cannot be extended, it can affect the shear angle in the useful area and even produce buckling defects. It can be seen that the influence of blank holder pressure on preform quality is very complex. In order to further understand this phenomenon, it is necessary to develop a specific test independent of the process. In actual production, this process is difficult, and corresponding optimization research can be carried out through numerical simulation on the mesoscopic scale [44].

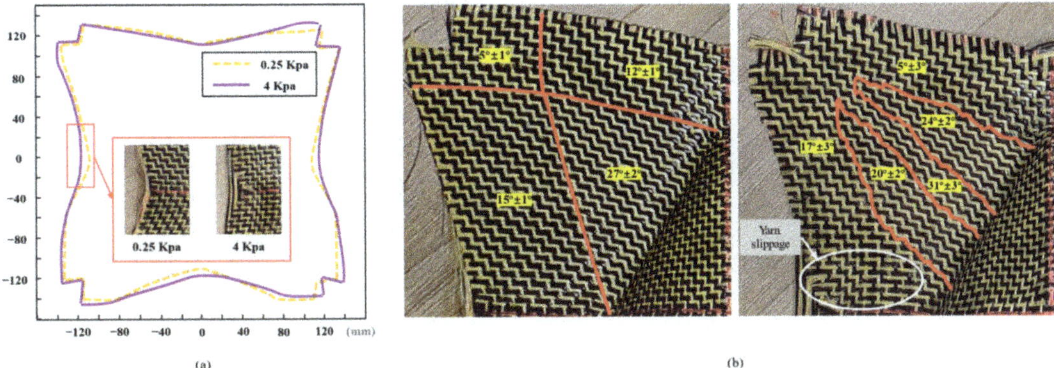

Figure 18. 0°/90° tetrahedral preforms under different blank holder pressures. A decrease in material draw-in and an uncertain change in shear angle distribution occur at high pressure: (**a**) material draw-in comparison; (**b**) shear angle distribution of non-useful areas under 0.25 Kpa (left) and 0.4 Kpa (right) blank holder pressure.

3.2.2. Multi-Layer Preforming

Defects in Multi-Layer Preforms with Different Stacking Configurations

In order to study the deformation behavior and possible defects of the multi-layer preforming hybrid woven reinforcements, tetrahedral preforming experiments with different ply configurations were carried out. The test was performed at low blank holder pressure (0.25 Kpa) and punch loading speed of 10 mm/min. The multi-layer configurations used in this study are shown in Figure 5, and the corresponding sequences are used to indicate the layering situation in the following description (for example, "Configuration I" represents the two plies are both 0°/90°, see Figure 5 for more details). In the multi-layer preforming, all the defect types observed in the previous single-layer preforms can also be seen and even more complex. Similarly, wrinkling can be completely suppressed under low blank holder pressure, as analyzed in single-layer preforming. However, there were several key locations worth noting where defects occurred in single-layer preforms (as mentioned in Section "Preforming Defect Description of Hybrid Woven Reinforcement"). When it comes to a multi-layer preforming, defects will be magnified due to the inter-ply friction. Although the yarn slippage occurred in non-useful areas in single-layer preforms, it should also be noted that the influence of these defects will not spread to useful areas in multi-layer preforms.

Figure 19 shows the defects observed in the multi-layer preform (configuration II) with consistent relative orientation between plies. It can be observed that the preform had no yarn slippage at the bottom corner, which is due to the difficulty of yarn slip caused by inter-ply friction at this position. In addition, the local shear angle of the inner layer (ply-1) was larger than that of the outer layer (ply-2). The difference in radius was one of the possibilities that led to this phenomenon. Due to the volume conservation in the in-plane shear process, the thickness of the reinforcement increases [45]. In the multi-layer preform containing 0°/90° ply orientation, although the yarn slippage at the bottom corner did not disappear, they did not affect the useful area. Therefore, this defect can be ignored. However, the fiber distortion can be observed near the buckles zone. This phenomenon was induced by the influence of the interaction of the plies since the buckles zone between the plies were at the same location. In this position, the local volume fraction of the preform changed greatly, which would affect the mechanical properties. In addition, the range of fiber distortion in the inner layer (ply-1) was narrower than that in the outer layer (ply-2). This may be due to the compaction. The compaction provided by the outer layer can reduce some defects in the inner layer to a certain extent. However, this compaction did not cause the loss of buckling. In general, for multi-layer preforms with consistent relative orientation between plies, the increase in inter-ply friction caused by out-of-plane yarn buckling was the main influence on the occurrence of defects in multi-layer structures. In some cases, inter-ply friction is more important than compaction, especially when out-of-plane deformation occurs.

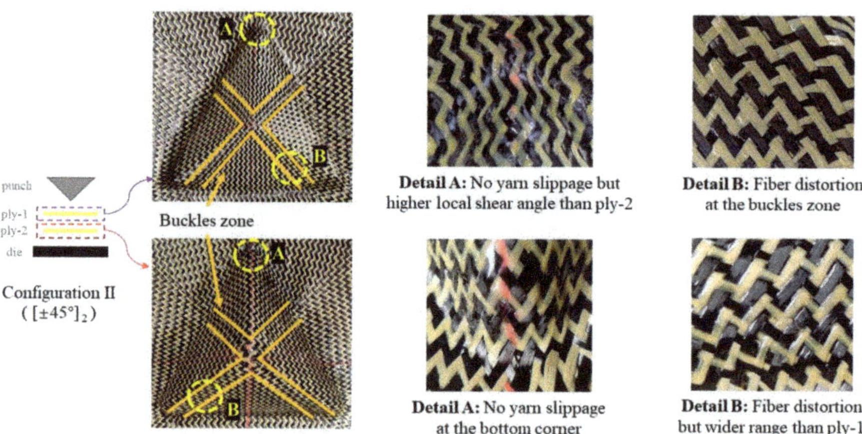

Figure 19. Defects in preform with the consistent orientation between plies.

For multi-layer preforms with inconsistent relative orientation between plies, different deformation modes can lead to more serious defects. Figure 20 lists the deformation of each ply in configuration VII. Detail A and detail B correspond to the two bottom corners of the preform, which is the location where yarn slippage along radial direction was easy to occur. It can be seen that under the superposition of multi-layer defects, the deformation of the preform in this place was unsatisfactory, even affecting the useful area. Detail C corresponds to the location where buckling occurs in each ply. Although the buckling positions of adjacent plies were different, they would still affect each other. Under the influence of different deformation modes, the relative slippage between plies was more severe. It can be observed that the yarns were pulled out of the weaving structure at the buckles zone. From the whole preforming results, the location of defects produced by multi-layer preforms was the same as that of single-layer preforms while the range was larger. The factors causing defects were mainly attributed to the inter-ply friction. For the surface with only in-plane shear deformation, the friction is relatively small. The inter-ply friction will not affect the preforming quality, and even better deformation behavior is represented under the action of compaction. However, when out-of-plane deformation occurs, the friction would increase rapidly, which would directly affect the quality of preforms. Fully understanding these friction phenomena is a prerequisite condition for obtaining ideal preforms. Thus, the research on inter-ply friction needs to be further studied.

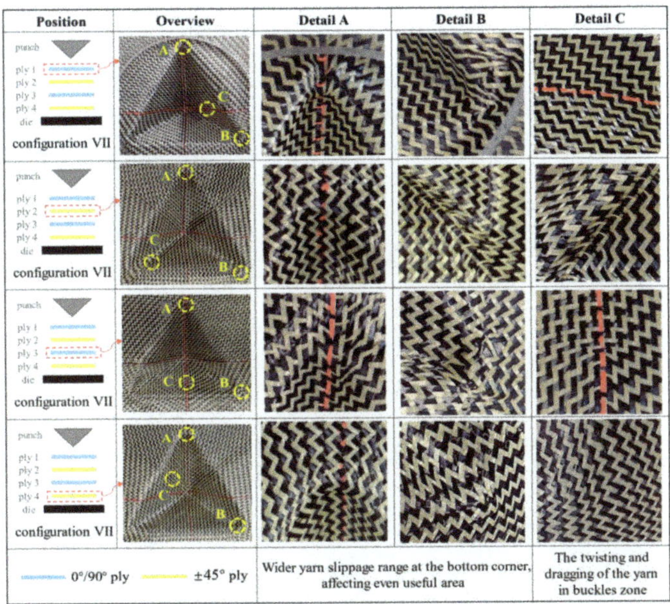

Figure 20. The deformation of each ply in configuration VII. Differences in relative deformation patterns between plies lead to a wider range of yarn slippage at the bottom corner, and yarn pull-out was observed in the buckles zone.

Inter-Ply Friction Mechanism in Multi-Layer Preforming

In multi-layer preforming, the friction behavior between yarns is the superposition of sliding friction and shock between yarns, as shown in Figure 21. In the current work, the sliding friction between yarns does not cause defects. Another phenomenon is the shock between yarns, especially when out-of-plane deformation occurs. Different plies of yarns would be embedded. When relative slip occurs between plies, these embedded yarns squeeze with each other, resulting in a sharp rise in friction. Compared to the multi-layer configuration with consistent relative orientation between plies, the configuration with inconsistent relative orientation between plies has a more significant relative slip due to the difference in the deformation mode of each ply. Nevertheless, some out-of-plane deformation is inevitable, so the possibility of defects is greater. Figure 22 shows the punch force under different configurations in multi-layer preforming. In the two-layer preforming, the punch force of configuration I is the lowest on the whole, and that of configuration IV is the highest. The punch force of configuration III at some positions even exceeds configuration IV. This proves the rationality of the above inference. Finally, in the four-layer preforming configuration VII, the punch force reaches the highest value. This is obviously not only due to the sliding friction between yarns, but the shock between yarns contributed much friction. Hence, the relative movement of each ply should be considered and reduced as much as possible. This reduction in friction is critical to the quality of the preform. Controlling the shock between yarns in the preforming process is one of the methods to reduce defects. Furthermore, introducing a matted fabric between plies can also reduce friction [46].

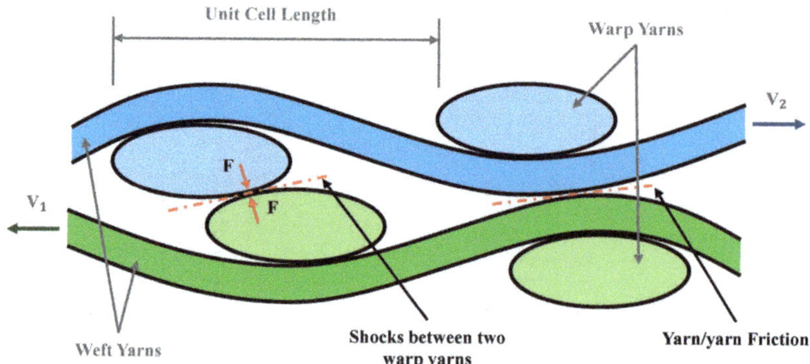

Figure 21. Two different friction behaviors between yarns.

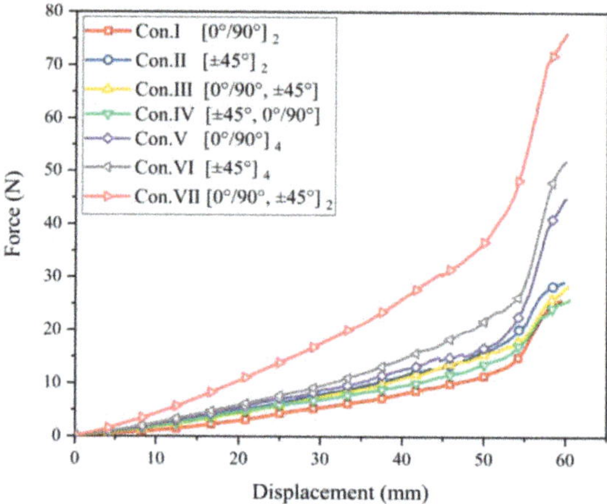

Figure 22. Punch force under different configurations in multi-layer preforming.

4. Conclusions

This paper investigated the mechanical behaviors of the carbon-Kevlar hybrid woven reinforcement and its deformation behaviors during the preforming process. The tensile properties of hybrid woven reinforcements exhibited a unique "double peak" phenomenon caused by the different properties of two kinds of fibers. The shear stiffness of the reinforcement was very low, and small wrinkles occurred at the shear angle of about 60°. Neither the tensile nor shear behaviors presented sensitivity to loading rate. In the preforming experiments, wrinkling and yarn breakage were not observed in the useful area of the preform, while significant yarn slippage and out-of-plane yarn buckling were observed in all the preforms. These defects were closely related to the punch shape and the initial yarn orientation and cannot be eliminated by just increasing blank holder pressure. In addition, in multi-layer preforming, the compaction between plies and the friction between yarns simultaneously affect the preform quality. Both factors resulted in a wider range of defects observed in multi-layer preforms than in single-layer preforms, while the locations and defect types were consistent in them. The results obtained in this paper can provide useful guidance for determining the optimized preforming parameters for the hybrid woven reinforcement. In the future, based on the knowledge obtained in this paper, numerical modeling of the preforming process of carbon-Kevlar hybrid woven reinforcement will be conducted to better understand the defect formation mechanism in the preforming.

Author Contributions: Experiment design, S.G., Z.Q. and Y.Z.; methodology, S.G., Z.Q., J.J. and Y.Z.; testing of mechanical properties and preforming, Z.Q. and Y.Z.; data analysis, S.G. and Z.Q.; writing and editing the manuscript, S.G., Z.Q., Y.Z. and J.J.; supervision, S.G. and J.J.; project administration, S.G.; language correction, S.G., Z.Q. and Y.Z. All authors have read and agreed to the published version of the manuscript.

Funding: This research was funded by the Young fund of Natural Science Foundation of Shaanxi province, grant number 2020JQ-701, and the Open Fund of State Key Laboratory of Applied Optics, China, grant number SKLAO2020001A09.

Data Availability Statement: The data presented in this study are available on request from the corresponding author.

Conflicts of Interest: The authors declare no conflict of interest.

References

1. Meola, C.; Boccardi, S.; Carlomagno, G.M. (Eds.) Chapter 1-Composite Materials in the Aeronautical Industry. In *Infrared Thermography in the Evaluation of Aerospace Composite Materials*; Woodhead Publishing: Philadelphia, PA, USA, 2017; pp. 1–24.
2. Xiao, S.; Lanceron, C.; Wang, P.; Soulat, D.; Gao, H. Mechanical and thermal behaviors of ultra-high molecular weight polyethylene triaxial braids: The influence of structural parameters. *Text. Res. J.* **2018**, *89*, 3362–3373. [CrossRef]
3. Omrani, F.; Wang, P.; Soulat, D.; Ferreira, M.; Ouagne, P. Analysis of the deformability of flax-fibre nonwoven fabrics during manufacturing. *Compos. Part B Eng.* **2017**, *116*, 471–485. [CrossRef]
4. Dufour, C.; Wang, P.; Boussu, F.; Soulat, D. Experimental Investigation About Stamping Behaviour of 3D Warp Interlock Composite Preforms. *Appl. Compos. Mater.* **2014**, *21*, 725–738. [CrossRef]
5. Shen, H.; Wang, P.; Legrand, X. In-plane shear characteristics during the forming of tufted carbon woven fabrics. *Compos. Part A Appl. Sci. Manuf.* **2021**, *141*, 106196. [CrossRef]
6. Kim, D.-J.; Yu, M.-H.; Lim, J.; Nam, B.; Kim, H.-S. Prediction of the mechanical behavior of fiber-reinforced composite structure considering its shear angle distribution generated during thermo-compression molding process. *Compos. Struct.* **2019**, *220*, 441–450. [CrossRef]
7. Sevkat, E.; Liaw, B.; Delale, F. Drop-weight impact response of hybrid composites impacted by impactor of various geometries. *Mater. Des.* **2013**, *52*, 67–77. [CrossRef]
8. Bunea, M.; Cîrciumaru, A.; Buciumeanu, M.; Bîrsan, I.G.; Silva, F.S. Low velocity impact response of fabric reinforced hybrid composites with stratified filled epoxy matrix. *Compos. Sci. Technol.* **2019**, *169*, 242–248. [CrossRef]
9. Hashim, N.; Majid, D.L.A.; Mahdi, E.-S.; Zahari, R.; Yidris, N. Effect of fiber loading directions on the low cycle fatigue of intraply carbon-Kevlar reinforced epoxy hybrid composites. *Compos. Struct.* **2019**, *212*, 476–483. [CrossRef]
10. Woo, S.-C.; Kim, T.-W. High strain-rate failure in carbon/Kevlar hybrid woven composites via a novel SHPB-AE coupled test. *Compos. Part B Eng.* **2016**, *97*, 317–332. [CrossRef]
11. Liang, B.; Hamila, N.; Peillon, M.; Boisse, P. Analysis of thermoplastic prepreg bending stiffness during manufacturing and of its influence on wrinkling simulations. *Compos. Part A Appl. Sci. Manuf.* **2014**, *67*, 111–122. [CrossRef]
12. Gao, J.; Liang, B.; Zhang, W.; Liu, Z.; Cheng, P.; Bostanabad, R.; Cao, J.; Chen, W.; Liu, W.K.; Su, X.; et al. *Multiscale Modeling of Carbon Fiber Reinforced Polymer (CFRP) for Integrated Computational Materials Engineering Process*; Ford Motor Company: Detroit, MI, USA, 2017.
13. Liang, B.; Colmars, J.; Boisse, P. A shell formulation for fibrous reinforcement forming simulations. *Compos. Part A Appl. Sci. Manuf.* **2017**, *100*, 81–96. [CrossRef]
14. Zhang, W.; Ren, H.; Liang, B.; Zeng, D.; Su, X.; Dahl, J.; Mirdamadi, M.; Zhao, Q.; Cao, J. A non-orthogonal material model of woven composites in the preforming process. *CIRP Ann.* **2017**, *66*, 257–260. [CrossRef]
15. Boisse, P.; Hamila, N.; Vidal-Sallé, E.; Dumont, F. Simulation of wrinkling during textile composite reinforcement forming. Influence of tensile, in-plane shear and bending stiffnesses. *Compos. Sci. Technol.* **2011**, *71*, 683–692. [CrossRef]
16. Boisse, P.; Hamila, N.; Guzmán Maldonado, E.; Madeo, A.; Hivet, G.; dell'Isola, F. The bias-extension test for the analysis of in-plane shear properties of textile composite reinforcements and prepregs: A review. *Int. J. Mater. Form.* **2017**, *10*, 473–492. [CrossRef]
17. Nguyen, Q.T.; Vidal-Sallé, E.; Boisse, P.; Park, C.H.; Saouab, A.; Bréard, J.; Hivet, G. Mesoscopic scale analyses of textile composite reinforcement compaction. *Compos. Part B Eng.* **2013**, *44*, 231–241. [CrossRef]
18. Yao, Y.; Peng, X.; Youkun, G. Influence of tension–shear coupling on draping of plain weave fabrics. *J. Mater. Sci.* **2019**, *54*, 6310–6322. [CrossRef]
19. Zhang, W.; Bostanabad, R.; Liang, B.; Su, X.; Zeng, D.; Bessa, M.A.; Wang, Y.; Chen, W.; Cao, J. A numerical Bayesian-calibrated characterization method for multiscale prepreg preforming simulations with tension-shear coupling. *Compos. Sci. Technol.* **2019**, *170*, 15–24. [CrossRef]
20. Hallander, P.; Akermo, M.; Mattei, C.; Petersson, M.; Nyman, T. An experimental study of mechanisms behind wrinkle development during forming of composite laminates. *Compos. Part A Appl. Sci. Manuf.* **2013**, *50*, 54–64. [CrossRef]

21. Zhu, B.; Yu, T.X.; Zhang, H.; Tao, X.M. Experimental investigation of formability of commingled woven composite preform in stamping operation. *Compos. Part B Eng.* **2011**, *42*, 289–295. [CrossRef]

22. Labanieh, A.R.; Garnier, C.; Ouagne, P.; Dalverny, O.; Soulat, D. Intra-ply yarn sliding defect in hemisphere preforming of a woven preform. *Compos. Part A Appl. Sci. Manuf.* **2018**, *107*, 432–446. [CrossRef]

23. Azzouz, R.; Allaoui, S.; Moulart, R. Composite preforming defects: A review and a classification. *Int. J. Mater. Form.* **2021**, 1–20. [CrossRef]

24. Peng, X.; Rehman, Z.U. Textile composite double dome stamping simulation using a non-orthogonal constitutive model. *Compos. Sci. Technol.* **2011**, *71*, 1075–1081. [CrossRef]

25. Peng, X.; Guo, Z.; Du, T.; Yu, W.-R. A simple anisotropic hyperelastic constitutive model for textile fabrics with application to forming simulation. *Compos. Part B Eng.* **2013**, *52*, 275–281. [CrossRef]

26. Aimene, Y. Approche Hyperélastique pour la Simulation des Renforts Fibreux en Grandes Transformations. Ph.D. Thesis, INSA Lyon, Lyon, France, 2007.

27. Youkun, G.; Xu, P.; Peng, X.; Wei, R.; Yao, Y.; Zhao, K. A Lamination Model for Forming Simulation of Woven Fabric Reinforced Thermoplastic Prepregs. *Compos. Struct.* **2018**, *196*, 89–95.

28. Gatouillat, S.; Bareggi, A.; Vidal-Sallé, E.; Boisse, P. Meso modelling for composite preform shaping–Simulation of the loss of cohesion of the woven fibre network. *Compos. Part A Appl. Sci. Manuf.* **2013**, *54*, 135–144. [CrossRef]

29. Capelle, E.; Ouagne, P.; Soulat, D.; Duriatti, D. Complex shape forming of flax woven fabrics: Design of specific blank-holder shapes to prevent defects. *Compos. Part B Eng.* **2014**, *62*, 29–36. [CrossRef]

30. Shanwan, A.; Allaoui, S.; Gillibert, J.; Hivet, G. Development of an experimental approach to study preforming mesoscopic defects of woven fabrics. In Proceedings of the ESAFORM 2021, 24th International Conference on Material Forming, Liège, Belgium, 14–16 April 2021.

31. Allaoui, S.; Cellard, C.; Hivet, G. Effect of inter-ply sliding on the quality of multilayer interlock dry fabric preforms. *Compos. Part A Appl. Sci. Manuf.* **2015**, *68*, 336–345. [CrossRef]

32. Guzman-Maldonado, E.; Wang, P.; Hamila, N.; Boisse, P. Experimental and numerical analysis of wrinkling during forming of multi-layered textile composites. *Compos. Struct.* **2019**, *208*, 213–223. [CrossRef]

33. Nosrat Nezami, F.; Gereke, T.; Cherif, C. Analyses of interaction mechanisms during forming of multilayer carbon woven fabrics for composite applications. *Compos. Part A Appl. Sci. Manuf.* **2016**, *84*, 406–416. [CrossRef]

34. Vanclooster, K.; Lomov, S.; Verpoest, I. Simulation of Multi-layered Composites Forming. *Int. J. Mater. Form.* **2010**, *3*, 695–698. [CrossRef]

35. Thompson, A.; Belnoue, J.; Hallett, S. Modelling defect formation in textiles during the double diaphragm forming process. *Compos. Part B Eng.* **2020**, *202*, 108357. [CrossRef]

36. Huang, J.; Boisse, P.; Hamila, N. Simulation of the forming of tufted multilayer composite preforms. *Compos. Part B Eng.* **2021**, *220*, 108981. [CrossRef]

37. Wagih, A.; Sebaey, T.A.; Yudhanto, A.; Lubineau, G. Post-impact flexural behavior of carbon-aramid/epoxy hybrid composites. *Compos. Struct.* **2020**, *239*, 112022. [CrossRef]

38. Guled, F.; Chittappa, H. Influence of interply arrangement on inter-laminar shear strength of carbon-Kevlar/epoxy hybrid composites. In Proceedings of the AIP Conference Proceedings, Surathkal, India, 15–16 December 2018.

39. Jacquot, P.-B.; Wang, P.; Soulat, D.; Legrand, X. Analysis of the preforming behaviour of the braided and woven flax/polyamide fabrics. *J. Ind. Text.* **2015**, *46*, 698–718. [CrossRef]

40. Hamila, N.; Boisse, P.; Chatel, S. Semi-Discrete Shell Finite Elements for Textile Composite Forming Simulation. *Int. J. Mater. Form.* **2009**, *2*, 169–172. [CrossRef]

41. Allaoui, S.; Hivet, G.; Soulat, D.; Wendling, A.; Ouagne, P.; Chatel, S. Experimental preforming of highly double curved shapes with a case corner using an interlock reinforcement. *Int. J. Mater. Form.* **2014**, *7*, 155–165. [CrossRef]

42. Lee, J.S.; Hong, S.J.; Yu, W.-R.; Kang, T.J. The effect of blank holder force on the stamp forming behavior of non-crimp fabric with a chain stitch. *Compos. Sci. Technol.* **2007**, *67*, 357–366. [CrossRef]

43. Ouagne, P.; Soulat, D.; Moothoo, J.; Capelle, E.; Gueret, S. Complex shape forming of a flax woven fabric; analysis of the tow buckling and misalignment defect. *Compos. Part A Appl. Sci. Manuf.* **2013**, *51*, 1–10. [CrossRef]

44. Iwata, A.; Inoue, T.; Naouar, N.; Boisse, P.; Lomov, S.V. Coupled meso-macro simulation of woven fabric local deformation during draping. *Compos. Part A Appl. Sci. Manuf.* **2019**, *118*, 267–280. [CrossRef]

45. White, K.D.; Campshure, B.; Sherwood, J.A. Effects of Thickness Changes and Friction during the Thermoforming of Composite Sheets. In Proceedings of the ESAFORM 2021, 24th International Conference on Material Forming, Liège, Belgium, 14–16 April 2021.

46. Shanwan, A.; Allaoui, S. Different experimental ways to minimize the preforming defects of multi-layered interlock dry fabric. *Int. J. Mater. Form.* **2019**, *12*, 69–78. [CrossRef]

polymers

MDPI

Article

Layered-Fabric Materiality Fibre Reinforced Polymers (L-FMFRP): Hysteretic Behavior in Architectured FRP Material

Arielle Blonder * and Maurizio Brocato

Laboratoire GSA, Ecole National Supérieure d'Architecture Paris-Malaquais, Université PSL, 75006 Paris, France; maurizio.brocato@paris-malaquais.archi.fr
* Correspondence: arielle.blonder@gmail.com

Abstract: L_FMFRP is an architectural fiber composite surface element with an airy internal structure and variable section. This architectured material is the product of an alternative design and fabrication process that integrates *fabric materiality*, suggesting moldless shaping of the material through pleating and layering. Initial study of the mechanical properties of the element showed a structural behavior that would satisfy the requirement for schematic architectural cladding configurations, indicating a unique hysteretic behavior of the material. This paper further investigates the hysteretic capacities of L-FMFRP, examining the behavior under repeated loading and the effect of its internal material architecture. Parallels to entangled materials are suggested for a deeper understanding of the phenomenon, and the potential future application as an energy-absorbent material for façade cladding is outlined.

Keywords: fiber composites; FRP; architecture; architectured materials; hysteresis

Citation: Blonder, A.; Brocato, M. Layered-Fabric Materiality Fibre Reinforced Polymers (L-FMFRP): Hysteretic Behavior in Architectured FRP Material. *Polymers* **2022**, *14*, 1141. https://doi.org/10.3390/polym14061141

Academic Editor: Rajesh Mishra

Received: 26 January 2022
Accepted: 2 March 2022
Published: 12 March 2022

Publisher's Note: MDPI stays neutral with regard to jurisdictional claims in published maps and institutional affiliations.

1. Introduction

Fiber-reinforced polymers (FRPs) are the composite outcome of advanced fibers and polymer resins, making a family of high-performance materials that offers high strength-to-weight ratio, durability and versatility. In the past decades, numerous types of FRPs have been developed and introduced into extensive use across industries, from aeronautics and space, to infrastructure, automotive and consumer products. In the construction industry, FRPs are widely used for civil infrastructure [1], rehabilitation and reinforcement, light and moveable constructions, footbridges, profiles and decks, under various forms and fabrication processes [2]. Their architectural use is re-emerging [3], expanding from experimental pavilions and academic research [4] to wider commercial applications by world-leading architects [5].

Elements of FRPs are traditionally manufactured based on rigid molds. A compact laminate is made by pressing a number of fiber layers, mainly under the form of textile fabrics, over the mold. The composite piece is shaped by the mold's morphology, assuring conformity to the requested designed shape as well as material homogeneity [6]. While being substantial to applications in fields such as the automotive, security or aeronautics sectors, the dependence on rigid molds can be restrictive in the case of architectural applications [7]. The typical size-of-element and its one-off nature, together with contemporary architectural practices that promote complex morphologies and high variability, stand in contradiction to the practice and economy of mold-based manufacturing. Standard industrial composite-forming processes therefore represent a significant factor in the barrier to the wider application of the material in the architectural field [8]. This barrier is reflected in numerous contemporary research projects, seeking for alternative, adaptive, reconfigurable moldless forming processes and material systems in architecture [9–11]. In fiber composites, a major approach to alternative shaping is tackled at the fiber-bundle level, through robotic filament winding fabrication [12,13].

This research tackles the alternative shaping of FRPs on the fabric level, enhancing the textile qualities of the fiber constituent of the composite. The term *"fabric materiality FRP"* (FM-FRP) was coined to represent the integration of textile-related practices, techniques, design paradigms and material qualities into the world of fiber composites, as novel FRP material systems [14]. FM-FRP typically would inherit key assets from the world of textiles, such as parametric variability, self-organization and resilience, resulting in novel architectural material systems. Fabric manipulations and self-organization capacities substitute the extensive use of molds, thus suggesting alternative architectural outcomes. The FM-FRP material system that was developed based on the layering of pleated surfaces resulted in an airy element with a variable and intricate section; blurring the boundaries between structure and matter, it resembles a thick panel [14] (Figure 1). The combination of matter and voids as an internal material structure and the creation of porous systems with enhanced capacities for multifunctional applications [15] is currently of high interest, developing *architectured materials*, or *metamaterials* [16]. The voids can be distributed periodically (such as in lattices [17]) or in relative disorder (such as in entangled materials [18]).

Figure 1. Layered FM-FRP material system (L-FMFRP) resulting in architectured material shaped as a thick panel.

In the field of composites, the integration of voids in the material architecture is mainly applied as tubular hollow composites, based on techniques such as braiding, knitting, or spacer [19]. Such hollow-structured elements can also be incorporated as nested inserts in more complex sandwich structures when seeking to further optimize performance and light weight [20]. Out of the various applications of tubular composites, from sports equipment and printing rollers to rocket structures and helicopter landing gears, various applications make use of the crashworthiness and energy absorption properties of hollow composites, as energy-absorbent composite structures (EACS). Composites can absorb a substantial amount of energy per unit mass in comparison to metals. Other advantages of EACS include higher strength, lower weight, higher specific-stiffness, better potential in terms of vibration control and noise reduction [21]. Due to the brittle nature of composites, the energy is absorbed in FRP structures mostly through the conversion of kinetic energy to a form of deformation absorbed energy—a complex fracture mechanism of cracking, delamination and fiber breakage [21]. The level of energy absorption is affected by a variety of parameters of geometrical and material nature, depending on the fiber architecture as well as the resin and matrix material characteristics [22]. The unique energy-absorbing qualities of composite structures gradually make them a preferred choice for crashworthiness applications, along with increasing research interest and technological progress in the field [21].

The preliminary testing of L-FMFRP material demonstrated its general suitability for service as an architectural façade element [23]. In particular, compression tests revealed a capacity of the panel for large quasi-elastic deformation and a good recovery of the material after extensive compression, indicating the potential energy-absorption qualities of the material.

In the past two decades, there has been a growing concern for safety and security in the built environment. From the scale of urban planning to material design, efforts are being invested in the creation of resilient built environments that would actively contribute toward facing the challenges of terrorism and of natural disasters [24]. The growing threat of terrorist attacks in city centers urges the development of blast-resistant façade systems [25] and energy-absorbent material systems [26]. In parallel, climate change accentuates the occurrence of natural disasters such as cyclones, hurricanes and typhoons, causing tremendous casualties and physical damage. Studies have shown that the wind-borne debris of storms, generated either by unsecured items or by the progressive failure of the built environment, plays a major part in causing damage. Failure of the façade elements prove to be hazardous for the surroundings, the structure itself and its occupants, and can ultimately lead to the general failure of the structure by changing the internal–external balance of air pressure [27]. Aiming to improve performance with regard to both natural disasters and terrorist attacks, façade materials are re-evaluated with regard to updated codes and regulations [28], and structural concepts such as sacrificial façade systems are developed [29]. Fiber composites can potentially answer the need for cladding systems that are lightweight and perform as efficient energy absorbers [26]. The typical plastic deformation of extensive micro-cracking, rather than general buckling failure, together with the high ratio of density to flexural stiffness make FRP materials a potentially suitable domain for the development of such solutions [30].

Following the initial indications for the potential energy-absorption capacities and interesting properties of L-FMFRP under compression, three issues were defined for further investigation through compressive tests:

(a) Test the unloading behavior: verify whether or not the unloading curve is inversely similar to the loading curve and identify the elasto-plastic characteristics of the material behavior.

(b) Test repeated cycles: observe the element during multiple loading–unloading cycles to identify a possible drift in the cyclic curves.

(c) Test various folding patterns: investigate the effect of different folding patterns on the overall elasto-plastic behavior of the element.

Two experimental campaigns were carried out for the investigation of the above three issues. The main results of these tests are the observation of a phenomenon of hysteresis and some correlations between the shape of the loading–unloading paths and the meso-architecture of the specimens.

2. Materials and Methods

2.1. Materials and Sample Preparation

Testing was realized on L-FMFRP specimens, all made of fiberglass–epoxy prepreg 0/90 satin weave of 300 g/m^2: Prepreg E-glass 7781: fabric thickness 0.23 mm, weave pattern 8HS; resin content 30% +/−3%, Tg 124°.

The samples were fabricated according to a protocol that was developed in the framework of previous research [23], which consists of the pleating of single sheets and their stacking with light peripheral constraint, in order to form the equivalent of a laminate, a volumetric element (Figure 2):

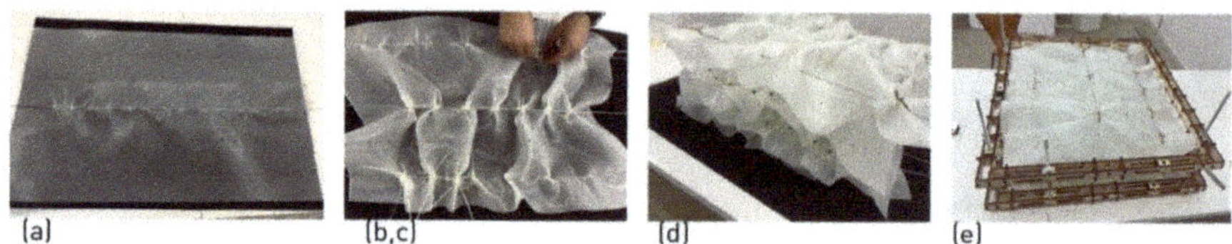

Figure 2. L-FMFRP fabrication process: (**a**) introducing metal rods in pre-cut holes in the fabric; (**b**) contracting the fabric at points along the rod; (**c**) assuring the local contractions with a temporary string; (**d**) super-imposing the manipulated fabric; (**e**) contracting the layered assembly with a jig for oven curing.

Pleating of the single sheet:

(a)   Perforation of the prepreg sheet according to pattern (number and placing of lines over the sheet, spacing of holes along the lines).

(b)   Sliding metal rods into the perforated lines of holes to serve as pleating guides.

(c)   Gathering the pleats by temporary knots along the pleating guides (the metal rod).

Forming the 'laminate':

(a)   Stacking the pleated sheet by superposition.

(b)   Constraining the assembly by a jig.

(c)   Curing.

All samples were composed of two prepreg layers, 600/200 mm each. The stacked assembly was oven-cured at a temperature of 125 °C for 2 h.

2.1.1. First Experimental Campaign

The testing was carried over three samples of identical pattern and similar internal structure (Figure 3a). The pleating pattern is composed of three rows, with metal rods aligning five pleats (gathering points) along its central row. As the product is not molded, and the manipulated layers are just lightly compressed in the curing jig, the different samples exhibit variations; the final dimensions of the panel were averagely 300–320 L/190 W/40–55 H (mm).

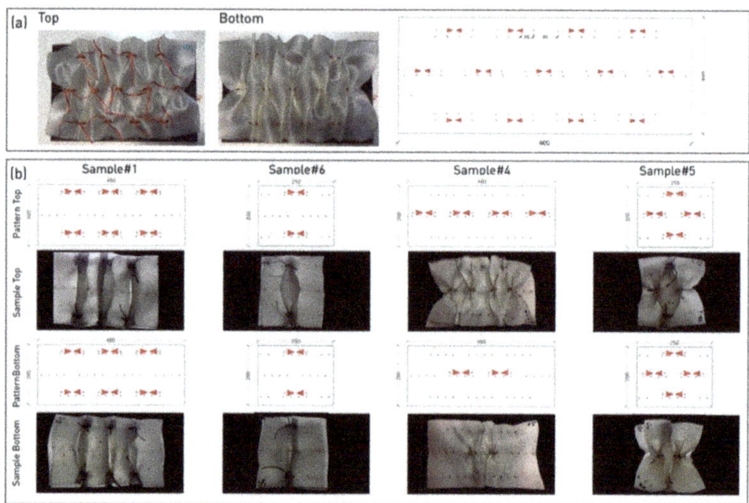

Figure 3. Samples of two experimental campaigns. (**a**) First campaign: typical sample top (left), and bottom view (middle), pleating pattern for both layers (right). (**b**) Second campaign: four tested samples, views and pleating patterns of top and bottom layers.

2.1.2. Second Experimental Campaign

Testing was carried over four samples, each with a different pleating pattern. Two pleat types were used: a 'simple pleat' pinched along two rods (top and bottom) and a 'full pleat' pinched along three rods (top, center, bottom) (Figure 3b). Each sample comprised two layers of similar pleating type and a variable number of pleats. The final dimensions of the samples varied according to the pleat types (Table 1).

Table 1. Pleating types–second experimental campaign: samples' data.

Sample Num.	Pleat Type	Num. Pleats Layer 1	Num. Pleats Layer 2	Final Dimensions (mm)
1	simple	3	3	180 L/195 W/35 H
6	Simple	1	1	205 L/200 W/30 H
5	Full	1	1	330 L/195 W/40 H
4	Full	2	1	300 L/190 W/45 H

2.2. Mechanical Testing

2.2.1. First Experimental Campaign

The testing was realized in an electromechanical tension–compression MTS testing machine of 100 kN capacity, with the samples placed between two steel plates (20 mm thickness) for levelling and load distribution over the surface. An average of 20 loading cycles were realized per sample, at constant displacement rate (between 8 and 40 mm/min, varying between test sets). The maximal load was averagely 3000 N (excluding the first cycles of the setting and the last cycles of the rupture).

2.2.2. Second Experimental Campaign

The testing was realized in an electromechanical tension–compression Instron testing machine of 10 kN capacity, with the samples placed between two steel plates (20 mm thickness) for levelling and load distribution over the surface (Figure 4).

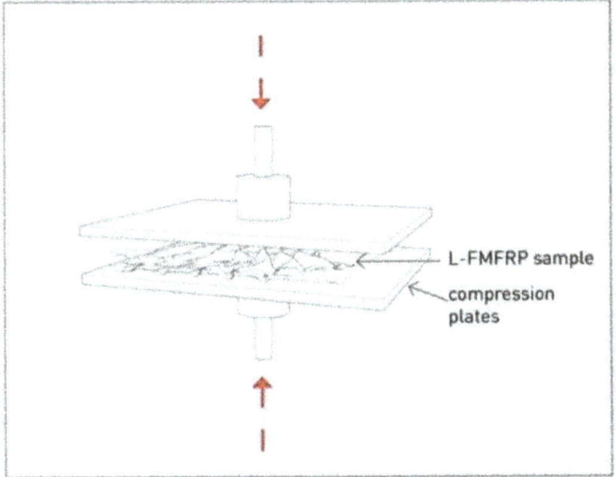

Figure 4. Experiment setup: sample positioned between two steel levelling plates.

3. Results

3.1. Preliminary Testing

A preliminary study of L-FMFRP's structural characteristics was carried out in the framework of previous research, testing the surface element under tension, compression and bending. The results showed a structural behavior that would satisfy the requirements for different theoretical schematic architectural cladding applications [23]. The compression test set (distributed load perpendicular to the panel surface) indicated a capacity of the panel for large quasi-elastic deformations. The generation of large displacement under loading (compressing the system to less than half of its initial thickness) did not lead to failure, but for the appearance of a few minor localized cracks. Furthermore, the initial thickness of the panel was almost entirely recovered after unloading (Figure 5). The observed behavior indicated the potential properties of hysteresis, and the capacity of energy absorption of the system, which led to the following two experimental campaigns.

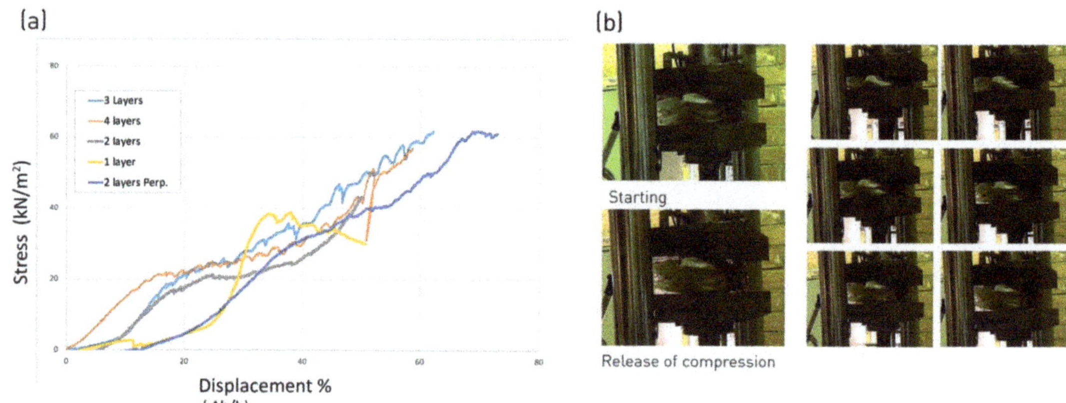

Figure 5. Compressive tests, distributed load. (**a**) Stress-displacement curves over five samples of different layer composition. (**b**) Sample during test: application of force and release of compression stress.

3.2. First Experimental Campaign: Cyclic Loading

The results of the first experimental campaign showed a significant difference between the loading and the unloading curves. When plotting the force against deformation, the two curves form a loop with an area enclosed between the curves (Figure 6). The loading curve shows positive hardening (a greater force required as the deformation increases) and the difference between the curves indicates a different path taken by the material upon loading/unloading. This typical curve, resembling exponential progression, is different from the curves that were plotted in the initial compression tests (operated on samples of various architectures (Figure 5)), which showed a more linear progression of force as the deformation advances. Looking at the first cycles of compression in the repeated loading of the current experiments, we see an evolution of the curve, from a quasi-linear slope in the first compression cycle, which resembles the previous initial compression tests, to the exponential type of curve by the third compression cycle. The evolution between the test sets of a single sample shows a settling of the material during the first loading cycles, and as the cycles are repeated, the response gradually stabilizes (Figure 6).

Increasing the applied force beyond the previously achieved threshold generates an irreversible plastic deformation, visible in the dented curve section. The repeated loading thereafter within the newly achieved threshold keeps a reversible elastic behavior with smooth curves. A small shift in the zero-state at each loading cycle is noticeable (within a similar threshold), showing an accumulated minimal irreversible deformation (about 6% overall). However, as the number of cycles increases under the same threshold, the accumulated irreversible deformation decreases (Figure 7).

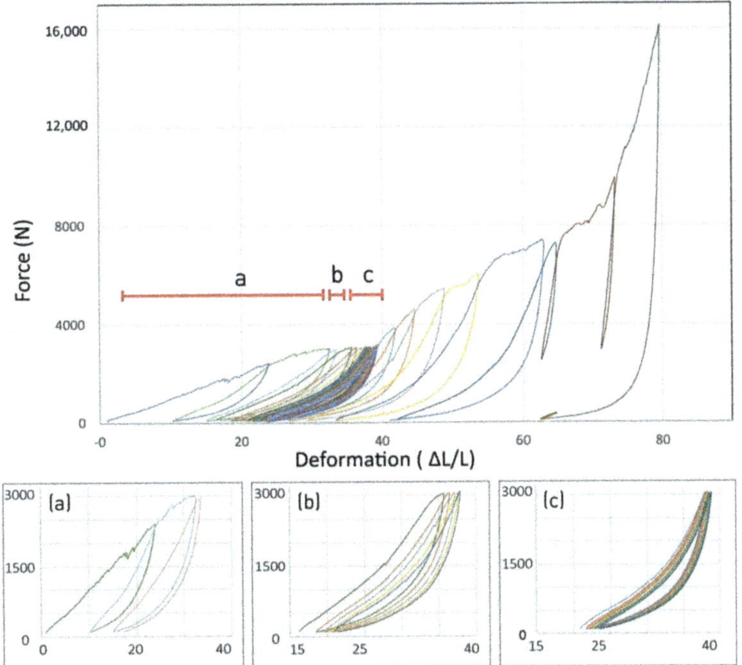

Figure 6. Typical loading/unloading curve: evolution along test sets. Sections in detail below: (**a**) settling, (**b**) first hysteretic cycles, (**c**) steady repeated cycles (Experimental campaign1_sample01).

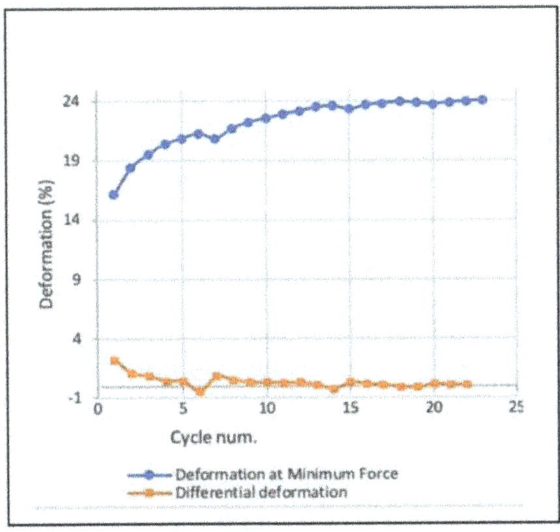

Figure 7. Accumulated irreversible deformation across loading cycles: plotted deformation at minimal applied force (blue) and differential deformation (orange).

A similar loop curve appears among all three samples, despite their relative differences due to manual fabrication and material self-organization. A superposition of loops of different samples under identical loading thresholds strongly resembles the curves and the enclosed area between the loading–unloading (Figure 8).

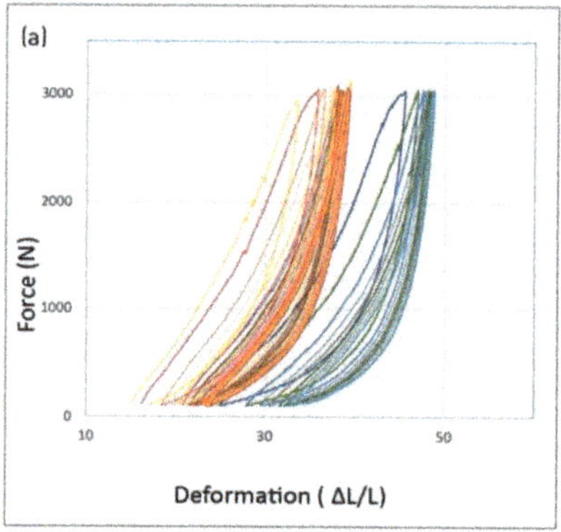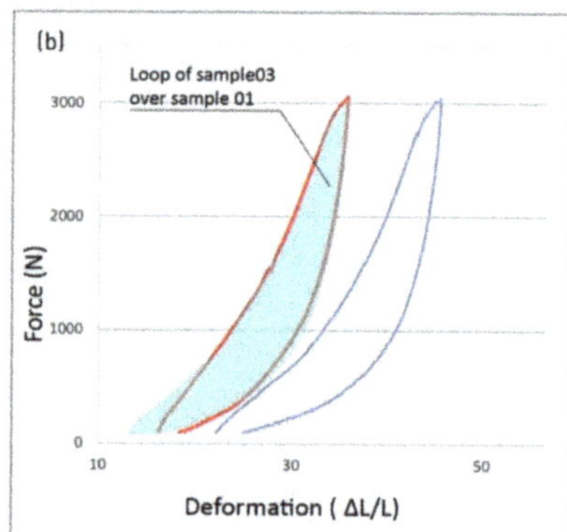

Figure 8. Similarity of loading–unloading loops between two samples of identical pleating pattern. (**a**) Repeated loading of sample 01 (red) and sample 03 (blue). (**b**) Comparison of loops between samples: enclosed area of sample 03 (blue stain) superimposed over loop curve of sample 01 (red).

3.3. Second Experimental Campaign: Pleating Pattern

The results of the second experimental campaign show the variation of the hysteretic loop between the samples of different pleating patterns. General similarity remains between the samples, as all demonstrate the hysteretic loop. However, the enclosed area of the loops as well as the overall steepness of the curves vary between samples. The plotting force against the normalized displacement of all samples combined shows that the samples of the full pleats (red, purple) go through larger relative deformation compared to the simple pleats (blue, green). Accordingly, the enclosed area of their loops is larger, showing a stronger difference between the loading and unloading paths (Figure 9). A comparative look at two samples of a single pleat in the two configurations (simple/full) shows the steeper curves of the single pleat compared to the full one, as well as a larger area enclosed in the loading–unloading loop.

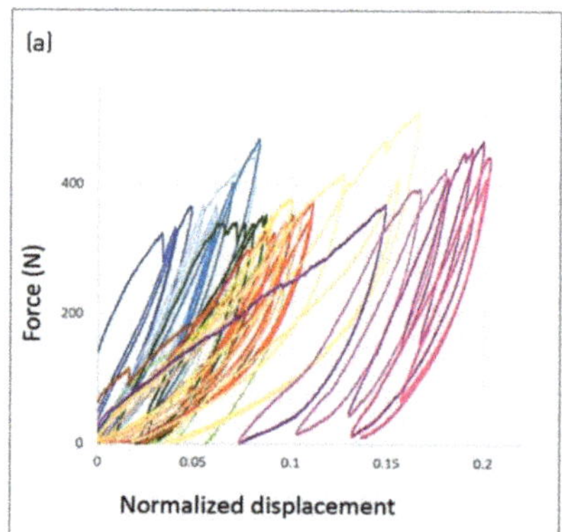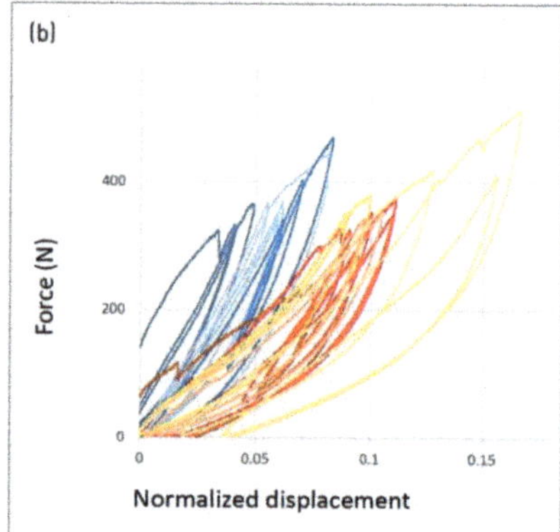

Figure 9. Effect of pleat types force plotted against normalized deformation and grouped by color shades. (**a**) All samples combined: one simple pleat (blue); one full pleat (red/yellow); three simple pleats (green); three full pleats (purple). (**b**) Comparison between simple and full pleat types: simple pleat (blue), showing steeper curves and smaller enclosed area in path; full pleat (red), showing larger enclosed area and shallower curves.

The comparison between single and triple pleats of the same kind shows a similar behavior for both simple and full pleats (Figure 10). The higher displacement values for a larger number of pleats indicate lower stiffness. The enclosed area of the hysteresis loops rises with the number of pleats, indicating higher energy absorption.

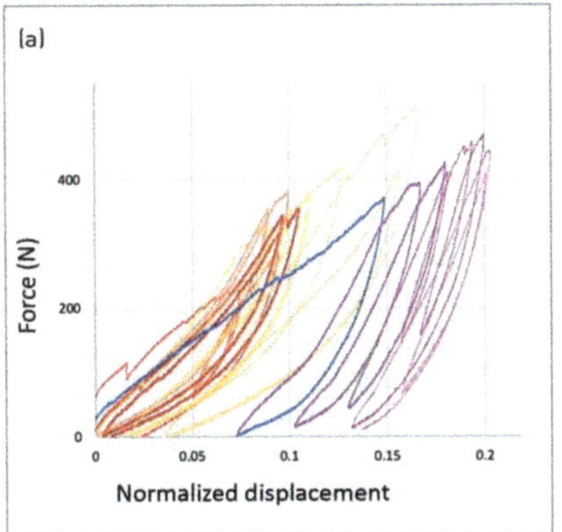
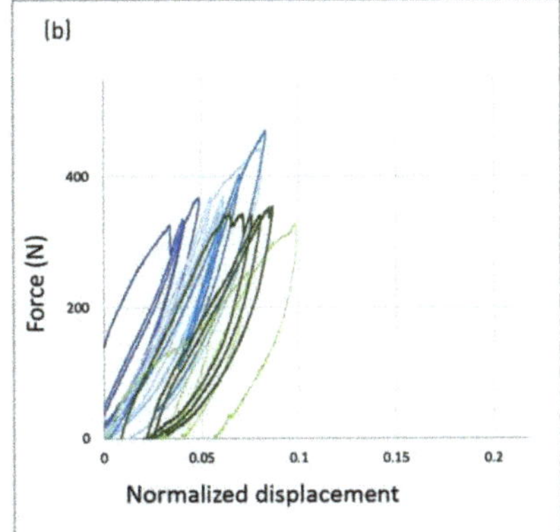

Figure 10. Effect of pleat number—increasing the number of pleats reduces stiffness and increases energy absorption. (**a**) Full pleats: single (purple) and triple (red). (**b**) Simple pleats: single (blue) and triple (green).

4. Discussion

The experiments show a loop of a defined deformation cycle (loading and unloading curves) enclosing an area, which indicates a loss of elastic energy in the process, allegedly attributed to friction. Such behavior is known as hysteresis, occurring in a variety of materials, structures and systems, where the dependence of a physical system upon its history is expressed as a non-linear behavior with a retardation of the effect when external forces acting upon the system are changed. This phenomenon is typically associated with materials of visco-elastic behavior such as elastomers [31], and is not particularly typical of FRP, where no intrinsic dissipation mechanisms exist in statics before failure [32]. On the contrary, when in dynamics, FRP may have damping capacities by dissipating some elastic energy through the visco-elastic behavior of their resin, or by various fracture mechanisms [20].

The similarity of load–unload loops between the different samples clearly demonstrates that the hysteretic behavior is a typical property of the L-FMFRP architectured material. While the samples show significant variations in their specific material configuration, due to manual manufacturing and material self-organization, the strong similarity of hysteretic loops between the samples indicates that this property is not sensitive to specific internal configuration (Figure 8). Rather, it seems to be dictated by the overall material architecture, i.e., the pleating pattern and the connections between the layers. A parallel can be made to other architectured materials with randomly disordered structure at the meso scale, such as entangled materials [33] or the jamming of aggregate systems [18]. Such systems demonstrate hysteretic behaviors that depend on their internal architectural parameters. Parameters such as the number of connections between the discrete components of the system (such as fibers or granules), their orientation or aspect ratio determine the level of energy dissipation of the system [34].

The effect of different parameters of the material architectures of L-FMFRP and its behavior can be initially observed through the comparison between different pleating patterns (campaign 02). Although the isolation and modelling of the effect of each parameter requires further study, the difference between the full and simple pleats is noticeable in the typical loading–unloading paths (Figure 9). The two different patterns generate surfaces of different geometrical complexity; the manipulation of the simple pleat develops into a barrel-like curved surface, and the full pleat develops into conical-type formations with a higher degree of geometrical complexity. The elements generated by patterns that are more geometrically complex show reduced stiffness (indicated by the reduced overall steepness of the loading curve). The connection between the folding complexity of the surface and its reduced stiffness could be compared to the behavior of knitted hollow composites, where the elastic stage of the compression curve corresponds to the flattening and ovalization of the hollow channels, and is correlated with the channel's radius of curvature [24]. Whether the model describing the compression curve of hollow composites could be suitable for our case should be further investigated as well (transitioning from elastic flattening of the structure to deformation at the joining points, and finally the densification of fibers and load transfer to matrix). In parallel to reduced stiffness, a larger area is enclosed in the hysteretic loop of the load–unload curves, with increased surface complexity. This is visible both in the comparison between simple and full pleats, as well as between surfaces of single or multiple pleats. This could be explained by the role friction plays in the hysteretic process, determined by parameters such as the number of contact points, the tangency of surfaces and areas of contact, which are generated by the pleating pattern and will be further investigated. A parallel can be drawn with knitted tubular lattice composites, where energy absorption improves with a higher number of cells to be deformed [35].

The role of the connection between the layers is to be further studied. The contact points between the pleated layers of the laminate are essential for the material to perform as a constrained unit under load and prevent immediate sliding of the layers. However, the displacement of the layers during compression shows a relative sliding of the layers. This sliding is within the elastic range of the material and its contact points, since the meso-scale architecture of the material is not destroyed through the extensive loading, and the layers within each sample remain connected.

Looking at the analysis of entangled materials, the connection between the components' contact length and permanence of contact seems to be a critical architectural parameter in the material performance [34]. A fundamental difference was found between entangled materials with initial permanent connections between components (epoxy cross-links between fibers) and without rigid connections (loose fibers). The type of stress–strain curves of the two materials is different and indicates an evolution of connection points along loading; the unlinked material starts with low initial stiffness, which increases with deformation as new contact points are created by compression and friction augments. This typically exponential stress–strain curve resembles the curve of L-FMFRP, where stiffness increases with deformation. The material of permanent links goes through a three-step process, starting with elastic behavior, going through a plateau where permanent connections are destroyed and new contact points are constructed, and ending with pure densification of the contact points [34]. This type of curve resembles the curve of the first loading cycle (Figure 5). The evolution of the stress–strain curve in cyclic loading, from linear-like curve to exponentially growing curve, indicates the evolution of the contact points and friction level of the material during increased compression. Contact points between the pleated layers of L-FMFRP are permanent and persist during compression, indicating that these contacts perform within the elastic range of the resin incorporating them. However, we have no indication of the number of contacts and the extension of the contact areas before and after the loading, and it is probable that a certain number of connections are at least partly destroyed during the first cycle of compression. This might explain the resemblance between the curve of the first compression cycles and the typical curves of the permanently linked entangled systems, with a three-step evolution of contact points.

The role of friction in the hysteresis of entangled material was studied previously [36,37], indicating its direct relation to the hysteresis in stretching and bending energy; it was found to be determined by the friction coefficient and average number of contact points between the elements. As L-FMFRP only partially resembles entangled materials, the relevance of its friction model to describe the behavior of our system requires further study.

5. Conclusions

The L-FMFRP material system product resembles a thick panel with highly articulated surfaces of extreme low density. It stands in contrast to typical FRP panels: its section is not uniform and dense, but variable and airy; it has an intricate internal configuration set by the pleats and the partial adherence between the layers. This, in turn, affects and determines the characteristics of the overall resulting element, its stiffness and strength. L-FMFRP blurs the boundaries between material and structure, and thus can be considered an 'architectured material', for which the overall macro-scale behavior is related to two underlying scales: the micro-scale, where the density and the direction of the fibers play a principal role in defining the textile's properties; and the meso-scale, where a complex geometry and a set of internal bonds give rise to a particular, airy, structure.

The finite deformation properties of L-FMFRP and the hysteresis phenomena demonstrated in the experiments indicate a potential capacity of the material system for energy absorbance. The experiments show that this property is inherent to the material's architecture, and is affected by parameters such as pleat type and number. Such a capacity for energy absorbance, if further investigated and developed, could prove to be beneficial for architectural cladding applications, as retrofit or for new structures. Other energy absorbent composite structures designed for crashworthiness applications could serve as a reference [26]. While these are mainly implemented in moving structures (transportation), here, another application in architectural structures is suggested. Possible implementation could be a development of L-FMFRP, either as part of a composite sandwich construction or as a single "thick skin" with cellular characteristics.

An initial estimation is to be carried out to assess the capacities of the L-FMFRP system to withstand the range of loads typically considered for windborne debris, blast mitigation by sacrificial façade systems and general energy dissipation for fluctuating air pressure. Referring to its crashworthiness properties, its characterization should refer to indicators of performance such as the specific energy absorption (SEA), energy absorption (EA) capacity, crush force efficiency (CFE), mean crushing force (MCF) and sound transmission loss (STL) [26]. With that, a better understanding is required as to the capacity of the material system to be enhanced by variation of the fabrication parameters or the association of additional elements to the system.

The further investigation of L-FMFRP calls for a deeper understanding of the internal physical phenomena of energy dissipation in the system and its determining parameters. Wishing to augment its capacities for energy absorbance, the relation between friction, structure and geometrical pattern is to be evidenced. Monitoring the heat loss and the sound emitted throughout the full loading cycle could indicate the fractional energy loss that takes place. The role of contact points, their evolution along compression cycles, their effect on friction, their relation to the folding pattern and the effect of those on the hysteresis level should be further investigated in order to optimize the energy absorbance of the system. Through further physical experimentations and mathematical modelling, control of the system's performance could be obtained through the variation of the geometrical folding pattern and raw material properties.

Author Contributions: Conceptualization, A.B.; Data curation, A.B.; Formal analysis, M.B.; Investigation, A.B.; Supervision, M.B.; Visualization, A.B.; Writing—original draft, A.B.; Writing—review & editing, M.B. All authors have read and agreed to the published version of the manuscript.

Funding: This research received no external funding.

Data Availability Statement: The data that support the findings of this study are openly available in the Figshare repository: https://doi.org/10.6084/m9.figshare.18515129 (accessed on 25 January 2022).

Acknowledgments: The research was realized in the framework of postdoctoral scholarship by the Israeli Council for Higher Education, PBC (Vatat). The mechanical tests were realized at the Laboratoire Navier at Université Paris-Est, Ecole des Ponts ParisTech. Special thanks to Gilles Foret for his assistance and guidance.

Conflicts of Interest: The authors declare no conflict of interest.

References

1. Hollaway, L.C. A Review of the Present and Future Utilisation of FRP Composites in the Civil Infrastructure with Reference to Their Important In-Service Properties. *Constr. Build. Mater.* **2010**, *24*, 2419–2445. [CrossRef]
2. Bakis, C.; Bank, L.; Brown, V.; Cosenza, E.; Davalos, J.; Lesko, J.; Machida, A.; Rizkalla, S.; Triantafillou, T. Fiber-Reinforced Polymer Composites for Construction—State-of-the-Art Review. *J. Compos. Constr.* **2002**, *6*, 73–87. [CrossRef]
3. Bell, M.; Buckley, C. (Eds.) *Permanent Change: Plastics in Architecture and Engineering*; Princeton Architectural Press: New York, NY, USA, 2014; ISBN 978-1-61689-166-4.
4. Lynn, G.; Gage, M.F. *Composites, Surfaces, and Software: High Performance Architecture*; Yale School of Architecture: New Haven, CT, USA, 2010; ISBN 978-0-393-73333-4.
5. Faircloth, B. *Plastics Now: On Architecture's Relationship to a Continuously Emerging Material*, 1st ed.; Routledge: London, UK; New York, NY, USA, 2015; ISBN 978-1-138-80451-7.
6. Mallick, P.K. *Fiber-Reinforced Composites: Materials, Manufacturing, and Design*, 3rd ed.; CRC Press: Boca Raton, FL, USA, 2008; ISBN 978-0-8493-4205-9.
7. Moskaleva, A.; Safonov, A.; Hernández-Montes, E. Fiber-Reinforced Polymers in Freeform Structures: A Review. *Buildings* **2021**, *11*, 481. [CrossRef]
8. Blonder, A.; Grobman, Y.J. Design and Fabrication with Fibre-Reinforced Polymers in Architecture: A Case for Complex Geometry. *Archit. Sci. Rev.* **2016**, *59*, 257–268. [CrossRef]
9. Veenendaal, D.; Block, P. Design Process for Prototype Concrete Shells Using a Hybrid Cable-Net and Fabric Formwork. *Eng. Struct.* **2014**, *75*, 39–50. [CrossRef]
10. Raun, C.; Kristensen, M.K.; Kirkegaard, P.H. Dynamic double curvature mould system. In *Computational Design Modelling*; Gengnagel, C., Kilian, A., Palz, N., Scheurer, F., Eds.; Springer: Berlin/Heidelberg, Germany, 2012; pp. 291–300. ISBN 978-3-642-23434-7.
11. Hawkins, W.J.; Herrmann, M.; Ibell, T.J.; Kromoser, B.; Michaelski, A.; Orr, J.J.; Pedreschi, R.; Pronk, A.; Schipper, H.R.; Shepherd, P.; et al. Flexible Formwork Technologies—A State of the Art Review. *Struct. Concr.* **2016**, *17*, 911–935. [CrossRef]
12. Reichert, S.; Schwinn, T.; La Magna, R.; Waimer, F.; Knippers, J.; Menges, A. Fibrous Structures: An Integrative Approach to Design Computation, Simulation and Fabrication for Lightweight, Glass and Carbon Fibre Composite Structures in Architecture Based on Biomimetic Design Principles. *Comput. Aided Des.* **2014**, *52*, 27–39. [CrossRef]
13. Prado, M.; Dorstellman, M.; Solly, J.; Knippers, J. Elytra filament pavilion: Robotic filament winding for structural composite building systems. In *Fabricate 2017*; Menges, A., Sheil, B., Glynn, R., Skavara, M., Eds.; UCL Press: London, UK, 2017; pp. 224–231. ISBN 978-1-78735-000-7.
14. Blonder, A. FM-FRP: New Materiality in FRP as Architectured Matter with Textile Attributes. *Archit. Sci. Rev.* **2021**, *64*, 478–489. [CrossRef]
15. Brechet, Y.; Embury, J.D. Architectured Materials: Expanding Materials Space. *Scr. Mater.* **2013**, *68*, 1–3. [CrossRef]
16. Christensen, J.; Kadic, M.; Kraft, O.; Wegener, M. Vibrant Times for Mechanical Metamaterials. *MRS Commun.* **2015**, *5*, 453–462. [CrossRef]
17. Kadic, M.; Bückmann, T.; Schittny, R.; Wegener, M. Metamaterials beyond Electromagnetism. *Rep. Prog. Phys.* **2013**, *76*, 126501. [CrossRef] [PubMed]
18. Weiner, N.; Bhosale, Y.; Gazzola, M.; King, H. Mechanics of Randomly Packed Filaments—The "Bird Nest" as Meta-Material. *J. Appl. Phys.* **2020**, *127*, 050902. [CrossRef]
19. Jamshaid, H.; Mishra, R.; Zeeshan, M.; Zahid, B.; Basra, S.A.; Tichy, M.; Muller, M. Mechanical Performance of Knitted Hollow Composites from Recycled Cotton and Glass Fibers for Packaging Applications. *Polymers* **2021**, *13*, 2381. [CrossRef] [PubMed]
20. Tarlochan, F.; Ramesh, S. Composite Sandwich Structures with Nested Inserts for Energy Absorption Application. *Compos. Struct.* **2012**, *94*, 904–916. [CrossRef]
21. Isaac, C.W.; Ezekwem, C. A Review of the Crashworthiness Performance of Energy Absorbing Composite Structure within the Context of Materials, Manufacturing and Maintenance for Sustainability. *Compos. Struct.* **2021**, *257*, 113081. [CrossRef]
22. Beard, S.J.; Chang, F.-K. Energy Absorption of Braided Composite Tubes. *Int. J. Crashworthiness* **2002**, *7*, 191–206. [CrossRef]
23. Blonder, A.; Grobman, Y.; Latteur, P. Pleated Facades: Layered Fabric Materiality in FRP Surface Elements. In Proceedings of the IASS Annual Symposia International Association for Shell and Spatial Structures (IASS), Boston, MA, USA, 16–20 July 2018; International Association for Shell and Spatial Structures (IASS): Boston, MA, USA, 2018; Volume 2018, pp. 1–8.
24. Coaffee, J.; Moore, C.; Fletcher, D.; Bosher, L. Resilient Design for Community Safety and Terror-Resistant Cities. *Proc. Inst. Civ. Eng. Munic. Eng.* **2008**, *161*, 103–110. [CrossRef]

25. Brewer, T.R.; Morrill, K.B.; Crawford, J.E. A New Kind of Blast-Resistant Curtain Wall Façade. *Int. J. Prot. Struct.* **2015**, *6*, 671–689. [CrossRef]
26. Tarlochan, F.; Ramesh, S.; Harpreet, S. Advanced Composite Sandwich Structure Design for Energy Absorption Applications: Blast Protection and Crashworthiness. *Compos. Part B Eng.* **2012**, *43*, 2198–2208. [CrossRef]
27. Minor, J.E. Windborne Debris and the Building Envelope. *J. Wind. Eng. Ind. Aerodyn.* **1994**, *53*, 207–227. [CrossRef]
28. Chen, W.; Hao, H.; Du, H. Failure Analysis of Corrugated Panel Subjected to Windborne Debris Impacts. *Eng. Fail. Anal.* **2014**, *44*, 229–249. [CrossRef]
29. Codina, R.; Ambrosini, D.; de Borbón, F. New Sacrificial Cladding System for the Reduction of Blast Damage in Reinforced Concrete Structures. *Int. J. Prot. Struct.* **2017**, *8*, 221–236. [CrossRef]
30. Zhou, H.; Wang, X.; Ma, G.; Liu, Z. On the Effectiveness of Blast Mitigation with Lightweight Claddings. *Procedia Eng.* **2017**, *210*, 148–153. [CrossRef]
31. Gorce, J.-N.; Hellgeth, J.W.; Ward, T.C. Mechanical Hysteresis of a Polyether Polyurethane Thermoplastic Elastomer. *Polym. Eng. Sci.* **1993**, *33*, 1170–1176. [CrossRef]
32. Treviso, A.; Van Genechten, B.; Mundo, D.; Tournour, M. Damping in Composite Materials: Properties and Models. *Compos. Part B Eng.* **2015**, *78*, 144–152. [CrossRef]
33. Bouaziz, O.; Masse, J.P.; Bréchet, Y. An Analytical Description of the Mechanical Hysteresis of Entangled Materials during Loading–Unloading in Uniaxial Compression. *Scr. Mater.* **2011**, *64*, 107–109. [CrossRef]
34. Masse, J.-P.; Poquillon, D. Mechanical Behavior of Entangled Materials with or without Cross-Linked Fibers. *Scr. Mater.* **2013**, *68*, 39–43. [CrossRef]
35. Hajjari, M.; Jafari Nedoushan, R.; Dastan, T.; Sheikhzadeh, M.; Yu, W.-R. Lightweight Weft-Knitted Tubular Lattice Composite for Energy Absorption Applications: An Experimental and Numerical Study. *Int. J. Solids Struct.* **2021**, *213*, 77–92. [CrossRef]
36. Poquillon, D.; Viguier, B.; Andrieu, E. Experimental Data about Mechanical Behavior during Compression Tests for Various Matted Fibres. *J. Mater. Sci.* **2005**, *40*, 5963–5970. [CrossRef]
37. Barbier, C.; Dendievel, R.; Rodney, D. Role of Friction in the Mechanics of Nonbonded Fibrous Materials. *Phys. Rev. E* **2009**, *80*, 016115. [CrossRef]

Article

Low-Velocity Impact Response of Auxetic Seamless Knits Combined with Non-Newtonian Fluids

Vânia Pais [1,2,*], Pedro Silva [1,2], João Bessa [1,2], Hernâni Dias [3], Maria Helena Duarte [3], Fernando Cunha [1,2] and Raul Fangueiro [1,2]

[1] Fibrenamics, Institute of Innovation on Fiber-Based Materials and Composites, University of Minho, 4800-058 Guimarães, Portugal; pedrosilva@fibrenamics.com (P.S.); joaobessa@fibrenamics.com (J.B.); fernandocunha@det.uminho.pt (F.C.); rfangueiro@dem.uminho.pt (R.F.)

[2] Centre for Textile Science and Technology (2C2T), University of Minho, 4800-058 Guimarães, Portugal

[3] Playvest-Nextil Group Sports Division, Rua das Austrálias, 4705-3226 Braga, Portugal; hernadias@gmail.com (H.D.); maria.duarte@playvest.pt (M.H.D.)

* Correspondence: vaniapais@fibrenamics.com

Abstract: Low-velocity impact can cause serious damage to the person or structure that is hit. The development of barriers that can absorb the energy of the impact and, therefore, protect the other side of the impact is the ideal solution for the pointed situation. Auxetic materials and shear thickening fluids are two types of technologies that have great capabilities to absorb high levels of energy when an impact happens. Accordingly, within this study, the combination of auxetic knits with shear thickening fluids by the pad-dry-cure process was investigated. It was observed that, by applying knits with auxetic patterns produced with denser materials and combined with the shear thickening fluids, high performance in terms of absorbed energy from puncture impact is obtained. The increment rates obtained are higher than 100% when comparing the structures with and without shear thickening fluids.

Keywords: auxetic; non-Newtonian fluids; low-velocity impact; personal protection

Citation: Pais, V.; Silva, P.; Bessa, J.; Dias, H.; Duarte, M.H.; Cunha, F.; Fangueiro, R. Low-Velocity Impact Response of Auxetic Seamless Knits Combined with Non-Newtonian Fluids. *Polymers* **2022**, *14*, 2065. https://doi.org/10.3390/polym14102065

Academic Editor: Rajesh Mishra

Received: 20 April 2022
Accepted: 16 May 2022
Published: 19 May 2022

Publisher's Note: MDPI stays neutral with regard to jurisdictional claims in published maps and institutional affiliations.

1. Introduction

Any structure that suffers an impact can be damaged, reversible, or irreversible, depending on factors such as the geometry, the mass, and the velocity of the striker. The frequency of the impact will also affect the result. Low-velocity impacts usually occur in the range of 1–10 m/s depending on the properties of the projectile and the defensive structure. In addition, the properties of the structure will influence the impact consequences [1,2]. These structures can act as a barrier to protect an object or a person located on the opposite side of the impact and can be applied in areas such as sports, automotive, or personal protection [2–4]. In an ideal situation, this structure should absorb as much energy as possible, so that the user or the object that is being protected feels the smallest impact possible [5].

Auxetic materials present a special behaviour when subjected to mechanical forces. These materials have a negative Poissons' ratio since, when they expand laterally, an expansion occurs also in the opposite direction. On the other side, when they are compressed laterally, compression on the opposite side also occurs [3]. Auxetic materials present exceptional properties such as increased mechanical properties, shear resistance, indentation resistance, surface fitting ability, and increased breaking tenacity and energy absorption [3,5–7]. Due to the mentioned topics, auxetic materials are widely applied in the scope of protection.

Shear thickening fluid (STF) is a non-Newtonian fluid that has low viscosity values when under normal conditions. However, when the shear rate is increased to a critical value, the viscosity also increases at a high rate. At this point, the fluid changes to a solid-state [6,8,9]. Due to this ability, the non-Newtonian fluid can have an important role in impact dynamics [10], since it positively influences the mechanical properties and increases parameters performance such as energy absorption [9].

Due to these special behaviours of auxetic materials and non-Newtonian fluids, the combined application of these two technologies in the protection has much potential. The research presented in this paper aims to investigate the mechanical behaviour of auxetic structures produced in a seamless loom as well as the protective effect of this auxetic structure when combined with non-Newtonian fluids. To develop the auxetic seamless knits, two distinct patterns produced with conditions previously optimized by the group were studied. The amount of non-Newtonian fluid, as well as its application by the pad-dry-cure process, was also studied to obtain a hybrid system with protective properties. Beyond the potential of the application of these two combined technologies, the auxetic structures produced by seamless technology are also an innovative concept and have the high advantage of their reduced weight and high flexibility—when compared to composite structures. Therefore, auxetic knits as well as their combination with STF were optimized within this study in order to obtain materials with increased mechanical performance.

2. Materials and Methods

2.1. Materials

The auxetic knits were produced on a seamless Santoni loom, model Top 2 by Playvest (Braga, Portugal). The knits are composed of polyamide (PA) and elastane. The PA is produced by air-jet spinning. Two different types of PA were studied—A 78/68 × 2 and PA 78/68 × 3 and two different knit stitch tightening were studied for each type of polyamide, with P0 being the least tight and P-15 being the tightest. The two auxetic patterns studied are represented in Figure 1. A total of six different combinations was obtained, as represented in Table 1. The non-Newtonian fluids were acquired at Polyanswer.

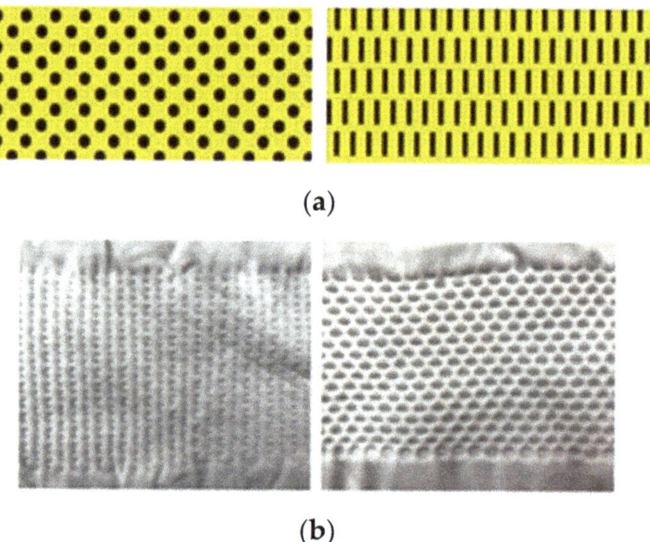

Figure 1. Technical drawing of pattern L and P (**a**) and pattern L and P on seamless knit (**b**).

Table 1. Knits studied in the present study.

Pattern	Polyamide	Knit Stitch	Reference
L	PA 78/68 × 2 × 2	P0	L_PA 78/68 × 2_P0
	PA 78/68 × 3	P-15	L_PA 78/68 × 3_P-15
		P0	L_PA 78/68 × 3_P0
P	PA 78/68 × 2 × 2	P0	P_PA 78/68 × 2_P0
	PA 78/68 × 3	P-15	P_PA 78/68 × 3_P-15
		P0	P_PA 78/68 × 2 × 2_P0

2.2. Methods

The auxetic behaviour was studied by imposing the extension of 4 cm of the knits in wales and course directions, separately, with the measurement of the extension or contraction between two specific points in the opposite direction.

The weight of the samples was calculated by dividing the mass of the sample mat by its effective area. The friction coefficients were measured using a binary sensor. This equipment was developed in the University of Minho [11] present on the Frictorq equipment. During this assay, a constant rotational motion was applied between the sample under study and a counter-fabric under a uniformly distributed contact force. The air permeability was measured using an air permeability tester from TEXTEST instruments (Zurich, Switzerland, model FX 3300 and a pressure of 200 Pa was applied with the measurement of the airflow that can pass through the structure.

The tensile, tear, and punction tests were performed by using a dynamometer Mesdanlab Twistronic (Hansfield, Salfords, England). On the tensile test, a tension movement was applied under a constant rate and the maximum force and maximum elongation were measured, according to the standard NP EN 13934-2. The tear test was performed according to the standard ASTM D 2261–2017 and the average of the five highest strength values was determined. The puncture test measures the maximum force and energy absorbed during puncture with a piercing element according to the standards ISO 13996: 1999 and EN 863. The burst test follows the standard NP EN ISO 13938-1 and the burst resistance is measured under constant pressure.

The STF were impregnated on the auxetic knits by the pad-dry-cure process. In this process, the knit goes through a bath and then through two rollers, controlled by pressure and speed.

3. Results

3.1. Auxetic Behaviour

The knits produced on the seamless loom without STF were analyzed in order to see if they have auxetic behaviour. The samples with both patterns in the study produced with PA 78/68 × 2 with P0 knit stitch were evaluated. Accordingly, the measurement of the extension or contraction in the opposite direction in which the extension is executed was performed. The results are represented in Figure 2. Note that the wales and courses mentioned in the legend refer to the opposite direction in which the movement is imposed. The positive values represent the situation where the extension in one direction is followed by the extension in the opposite direction. Accordingly, the positive values represent the samples that have auxetic behaviour. Concerning the knits with the L pattern, it is observed that when the extension is imposed in the course direction, the opposite direction, i.e., the wale direction, always has a positive value. Thus, the L pattern has auxetic behaviour in the direction. In knits with the P pattern, the wales also have higher values; however, the final value of the extension is equal to zero. Thus it can generally be concluded that the L pattern has a greater prevalence of auxetic behaviour than the P pattern. This auxetics nature only happens when the extension happens in the course direction, being the wales direction the side that expands laterally. Hu et al. also concluded that, when the strain happens in the course direction, a higher auxetic behaviour is observed. The reason pointed out by the authors for this behaviour is the fact that, in the wale direction, the stripes are closer. Accordingly, when the course direction is extended, a higher transversion expansion effect occurs [12].

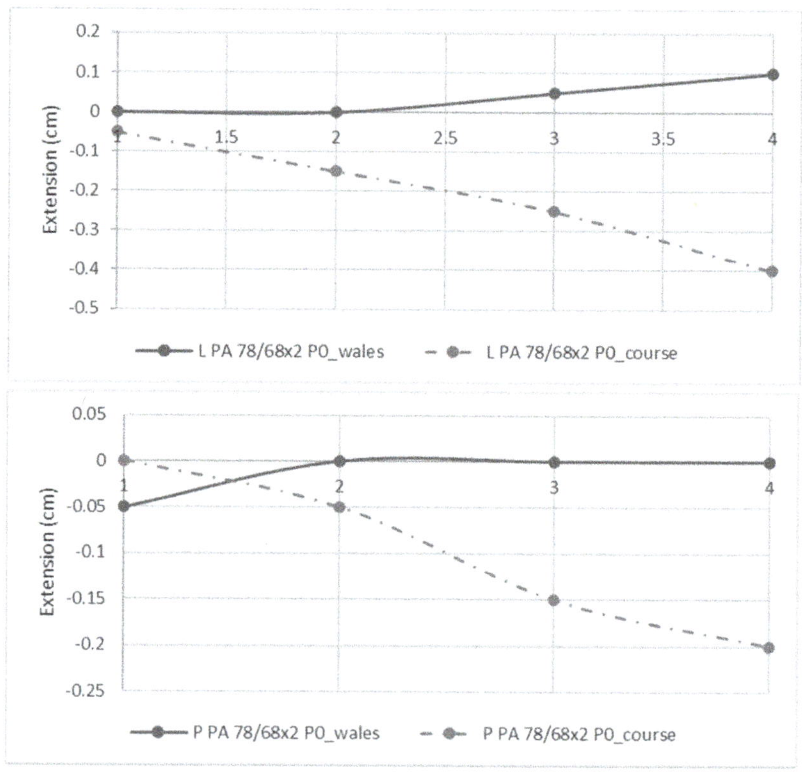

Figure 2. Extension of knits with L and P patterns in wales and course direction when imposed extension in the opposite direction. The *x*-axis represents the strain being applied.

3.2. Structural and Mechanical Characterization

The auxetic knits without STF were characterized and analysed to study their protective behaviour and their comfort parameters. The knits weight, specific air permeability (SAP), and the friction coefficient values (COF) of the L and P patterns produced in the several conditions in this study are shown in Figure 3. Regarding the weight values, it is observed that the L pattern has smaller values. Accordingly, during the production of the L knit, less raw material is required. It is also observed that when the knits are produced with a denser PA-PA 78/68 × 3-higher weight values are obtained. This result is logical since, in this situation, a higher amount of material is present in the same area. The change in the degree of tightness does not seem to promote changes in weight as significantly as the other parameters since no behavioural trend is observed. These results will directly influence the SAP values since denser structures limit the passage of air with greater impact due to the higher obstruction of pores [13]. The obtained results fall under this statement since the samples with denser PA have smaller SAP values. In COF values it is observed that the L pattern has higher values, regardless of production conditions. When a denser PA is applied it is also observed that the COF values are higher. However, the difference between the several values is small and in some samples is within the range of standard deviation values. Accordingly, it can be concluded that denser PA promotes the production of a more compact knit, which, in turn, will decrease the amount of air able to pass through the structure and will increase the friction intensity. The weight and air permeability parameters will affect the comfort parameters and they will influence properties like moisture management and thermoregulation [14]. Additionally, the friction will also affect the comfort parameters since when the skin contacts a surface, a touch perception occurs. Moreover, higher COF values may cause damage to the skin [15]. The obtained values fall under the ones referred by other authors [15,16].

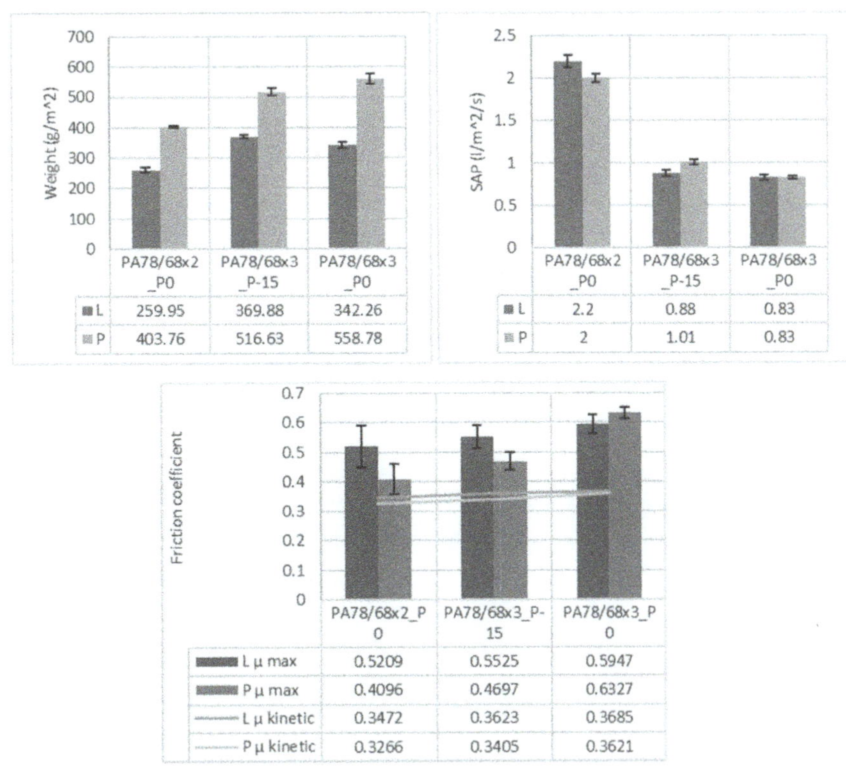

Figure 3. Weight, SAP and friction coefficient of knit samples.

The mechanical behaviour of the knits was also studied and were applied to different types of tests a tensile test, a tear test and a burst test. For each analysis, the corresponding maximum force achieved was analysed. The obtained results are represented in Figure 4. In the tensile test, regarding the L pattern, the maximum force values achieved are very similar in both knit directions—the wale and the course. On the other hand, in the P pattern differences were observed depending on the direction in which the tension is applied, obtaining higher values in the course direction. In Figure 1 it is possible to see that the L pattern has a very similar structure in both directions. However, the P pattern has different points depending on the observation direction. Due to this difference, the structure will have different behaviours depending on the direction in which the tension is applied. Yet, despite the P pattern having the highest force values in the course direction, it also has the smallest values in the wale direction. Accordingly, the L pattern may have much potential due to its intermediate values, not compromising its effectiveness in the wale direction. Concerning the types of PA, it is observed that the denser PA has the highest values of maximum force. Thus this compact arrangement will positively reinforce the structure, allowing it to achieve better mechanical performance. It is also important to note that the knit with the P pattern has a maximum force value in the course direction that is much higher than the other samples. This difference must be related to production problems instead of some standout mechanical properties of this sample, since this result is not observed in the other analysis and has no logical reason to happen. Accordingly, this result should be ignored in the analysis of this study. In the tear test, these differences imposed by the different directions were less noted; however, the same conclusions were observed, i.e., higher maximum forces are achieved when the PA 78/68 × 3 is applied. In the burst test, once again, the denser sample has the higher force values. Besides that, it was observed that the L pattern has much higher force values. In this final test, both knit directions are evaluated at the same time when executing one test since the assay is performed using a round sample. Accordingly, the differences previously observed in both directions of the P pattern will be conjugated in the final result. Due to this, the burst results with knits with the P pattern had much smaller force values than knits with the L pattern. Generally, concerning the mechanical parameters of the knits, it can be concluded that denser materials positively impact the achievement of higher force values. Samples with a tighter stitch seem to generally have higher force values, however, with a low impact.

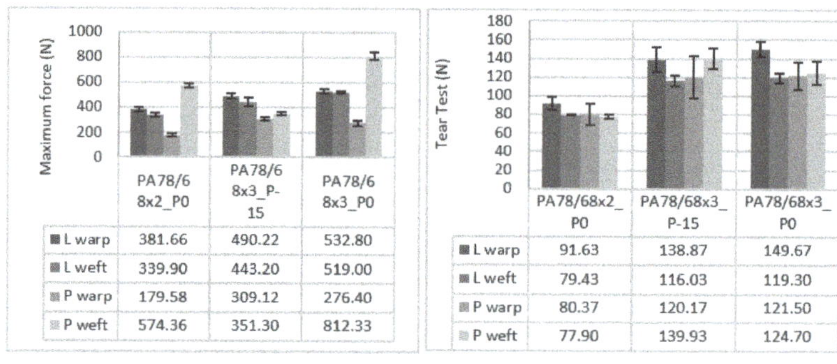

	PA78/68x2_P0	PA78/68x3_P-15	PA78/68x3_P0
L warp	381.66	490.22	532.80
L weft	339.90	443.20	519.00
P warp	179.58	309.12	276.40
P weft	574.36	351.30	812.33

	PA78/68x2_P0	PA78/68x3_P-15	PA78/68x3_P0
L warp	91.63	138.87	149.67
L weft	79.43	116.03	119.30
P warp	80.37	120.17	121.50
P weft	77.90	139.93	124.70

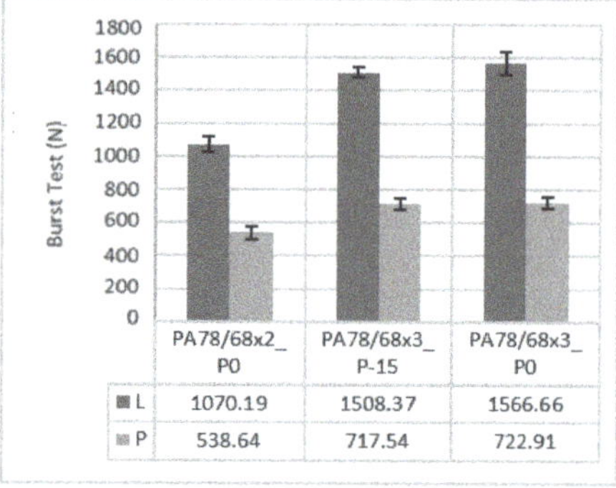

	PA78/68x2_P0	PA78/68x3_P-15	PA78/68x3_P0
L	1070.19	1508.37	1566.66
P	538.64	717.54	722.91

Figure 4. Maximum forces obtained in the tensile test, tear test and burst test.

3.3. Combination of Auxetic Knits with STF

The auxetic knits were combined with the STF to increase their protection capability. Other authors also studied the combination of these fluids with other materials to boost their protective properties. Santos et al. proved that the addition of STF to Kevlar fabrics improves the performance of these substrates under impact tests [17]. Lu et al. and Santos et al. also demonstrated that the combination of STF with materials like wale-knitted spacer fabrics and Kevlar woven samples, respectively, can be applied as a material for personal protection [18,19]. Within the present study, STF were combined with the auxetic knits by the pad-dry-cure process. The impregnation rates changed from 60 to 70%. The knits with L and P patterns, before and after STF impregnation, are shown in Figure 5.

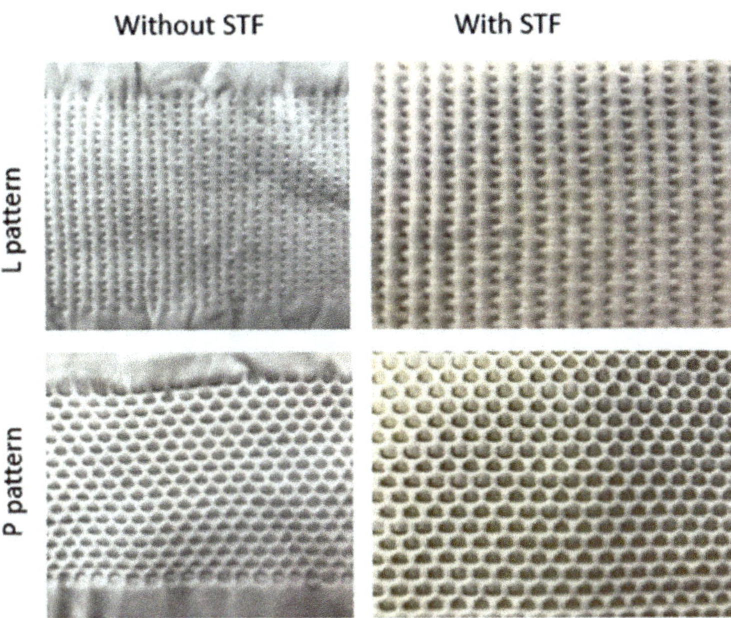

Figure 5. Knits with L and P patterns, with and without STF.

Once more, tensile tests were performed to see the mechanical behaviour of the samples. However, in this turn, the absorbed energy was also calculated to see the potential of the obtained materials to be applied in protective applications. The materials' ability to absorb energy is important in protection products due to the increase in the distribution of loads throughout the hierarchical architecture [20]. The graphical results obtained by knits with the L and P patterns are represented in Figures 6 and 7, respectively. When performing the tensile tests, it was observed that the knits impregnated with STF could stretch more. This higher stretch capability boosts the achievement of higher absorbed energy values. It was observed that when knits were combined with STF, higher maximum force and absorbed energy values were obtained. The only exception was the P pattern in the course direction. However, the incongruity of this sample was already mentioned in the analysis of the mechanical behaviour of the knits without STF. When comparing the L and P pattern, it is observed that the P pattern has higher energy absorbed values, making the knits stand out where denser PA was applied. This result follows the ones mentioned before, where it was concluded that applying higher amounts of mass in the same area unit increases the capability of the samples to support higher forces in the tensile movement. Moreover, as a consequence of this property, the samples with denser material will have the ability to absorb higher amounts of energy and much more potential to be applied as a protective material/product. Yan et al. also studied the influence of material density on protective applications' performance. In the mentioned study it was observed that when materials have higher densities they can reach higher forces when burst and puncture tests are performed. The higher amount of fibers and bonding points per unit of the area of materials with higher densities are the reasons given in the study for achieving higher mechanical behaviours [21].

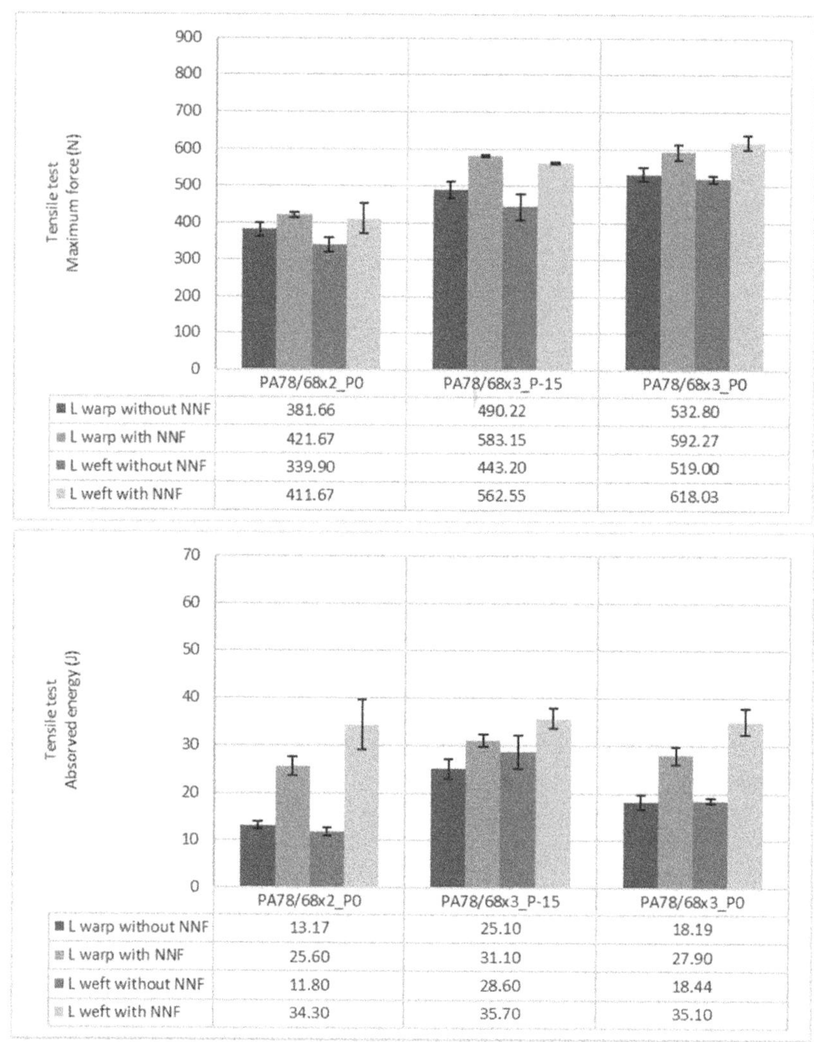

	PA78/68x2_P0	PA78/68x3_P-15	PA78/68x3_P0
L warp without NNF	381.66	490.22	532.80
L warp with NNF	421.67	583.15	592.27
L weft without NNF	339.90	443.20	519.00
L weft with NNF	411.67	562.55	618.03

	PA78/68x2_P0	PA78/68x3_P-15	PA78/68x3_P0
L warp without NNF	13.17	25.10	18.19
L warp with NNF	25.60	31.10	27.90
L weft without NNF	11.80	28.60	18.44
L weft with NNF	34.30	35.70	35.10

Figure 6. Maximum force and absorbed energy of knits with L pattern obtained on tensile tests.

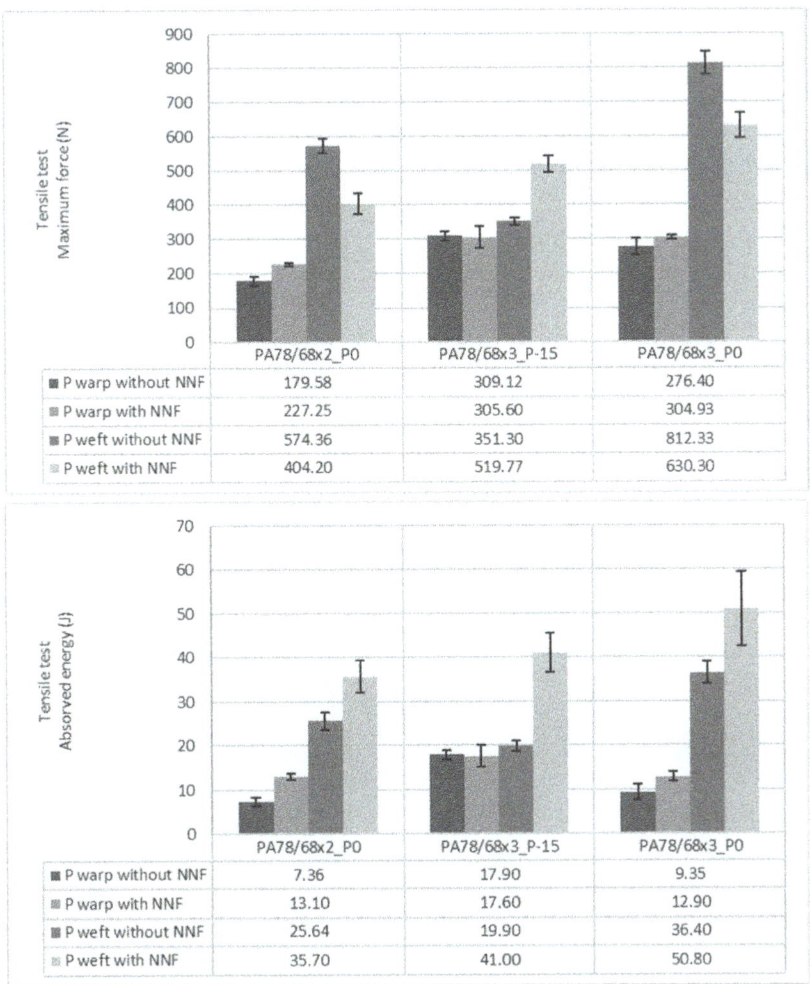

Figure 7. Maximum force and absorbed energy of knits with P pattern obtained on tensile tests.

To simulate a real low-velocity impact, puncture tests were performed with the calculation of maximum force and absorbed energy. The obtained results of knits with the L and P patterns—with and without STF—are represented in Figure 8. Once more it was observed that the addition of STF will increase the mechanical and protective behaviour of the samples in this study and that the P patterns have a higher performance in terms of mechanical behaviour. The samples with denser PA and impregnated with STF generally had higher values of energy absorbed, especially in the situation in which the stitch is tighter. These two combinations are the perfect match to correspond to the points mentioned by Yan et al., i.e., higher amount of fibers and bonding points per unit of area [21]. This combination will increase the protective effect of the samples.

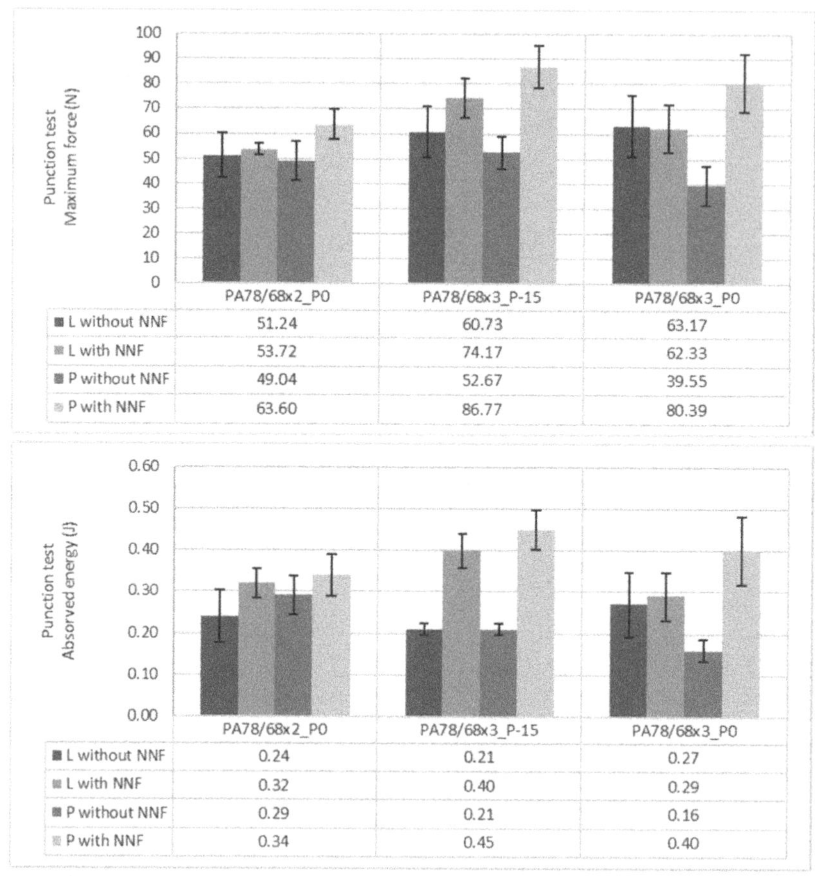

Figure 8. Maximum force and absorbed energy of knits with L and P pattern obtained on punction tests.

4. Conclusions

In this research, a combination of auxetic knits with STF was studied. First, the auxetic knits were characterized, and it was concluded that denser materials increase the mechanical performance of the final structure. When both technologies are combined through the pad-dry-cure process, a higher increase in mechanical behaviour happens. Moreover, puncture tests were performed to simulate the low-velocity impact and it was concluded that the absorbed energy increases with the presence of STF, being more visible in the knit with the P pattern. The increase of materials density, as well as the tightness stitch, will increase the protective behaviour mentioned. The product resulting from the combination of the technologies in this study is very light when compared to the most protective products on the market. Accordingly, this is a promising approach to be applied in personal protection applications.

Author Contributions: Conceptualization, V.P.; Investigation, V.P., H.D. and M.H.D.; Methodology, V.P., P.S., H.D. and M.H.D.; Project administration, P.S. and J.B.; Resources, H.D. and M.H.D.; Supervision, F.C. and R.F.; Writing—original draft, V.P.; Writing—review & editing, V.P. All authors have read and agreed to the published version of the manuscript.

Funding: This research was funded by Compete 2020 grant number 45211.

Institutional Review Board Statement: Not applicable.

Informed Consent Statement: Not applicable.

Data Availability Statement: Not applicable.

Conflicts of Interest: The authors declare no conflict of interest.

References

1. Safri, S.N.A.; Sultan, M.T.H.; Yidris, N.; Mustapha, F. Low velocity and high velocity impact test on composite materials—A review. *Int. J. Eng. Sci. (IJES)* **2014**, *3*, 50–60. [CrossRef]
2. Alcock, B.; Cabrera, N.O.; Barkoula, N.M.; Peijs, T. Low velocity impact performance of recyclable all-polypropylene composites. *Compos. Sci. Technol.* **2006**, *66*, 1724–1737. [CrossRef]
3. Duncan, O.; Shepherd, T.; Moroney, C.; Foster, L.; Venkatraman, P.D.; Winwood, K.; Allen, T.; Alderson, A. Review of auxetic materials for sports applications: Expanding options in comfort and protection. *Appl. Sci.* **2018**, *8*, 941. [CrossRef]
4. Zhou, L.; Zeng, J.; Jiang, L.; Hu, H. Low-velocity impact properties of 3D auxetic textile composite. *J. Mater. Sci.* **2017**, *53*, 3899–3914. [CrossRef]
5. Chang, Y.; Ma, P.; Jiang, G. Energy absorption property of warp-knitted spacer fabrics with negative Possion's ratio under low velocity impact. *Compos. Struct.* **2017**, *182*, 471–477. [CrossRef]
6. Xu, W.; Yan, B.; Hu, D.; Ma, P. Preparation of auxetic warp-knitted spacer fabric impregnated with shear thickening fluid for low-velocity impact resistance. *J. Ind. Text.* **2020**, 1528083720927013. [CrossRef]
7. Al-Rifaie, H.; Studziński, R.; Gajewski, T.; Malendowski, M.; Sumelka, W.; Sielicki, P.W. A new blast absorbing sandwich panel with unconnected corrugated layers—numerical study. *Energies* **2021**, *14*, 214. [CrossRef]
8. Sun, D.; Zhu, F.; Stylios, G.K. Investigation of composite fabric impregnated with non-Newtonian fluid for protective textiles. *J. Compos. Mater.* **2019**, 1–9. [CrossRef]
9. Santos, T.F.; Santos, C.M.; Aquino, M.S.; Oliveira, F.R.; Medeiros, J.I. Statistical study of performance properties to impact of Kevlar® woven impregnated with Non-Newtonian Fluid (NNF). *J. Mater. Res. Technol.* **2020**, *9*, 3330–3339. [CrossRef]
10. de Goede, T.C.; de Bruin, K.G.; Bonn, D. High-velocity impact of solid objects on Non-Newtonian Fluids. *Nature* **2019**, *9*, 1250. [CrossRef]
11. Lima, M.; Silva, L.F.; Vasconcelos, R.; Martins, J.; Hes, L. FRICTORQ, Tribómetro para Avaliação Objectiva de Superfícies Têxteis. *Communication* **2005**, 1–11.
12. Hu, H.; Zulifqar, A. Auxetic Textile Materials—A review. *J. Text. Eng. Fash. Technol.* **2017**, *1*, 00002. [CrossRef]
13. Debnath, S.; Madhusoothanan, M. Thermal insulation, compression and air permeability of polyester needle-punched nonwoven. *Indian J. Fibre Text. Res.* **2010**, *35*, 38–44.
14. Ogulata, R.T.; Mezarcioz, S.M. Optimization of air permeability of knitted fabrics with the Taguchi approach. *J. Text. Inst.* **2011**, *102*, 395–404. [CrossRef]
15. Vilhena, L.; Ramalho, A. Friction of Human Skin against Different Fabrics for Medical Use. *Lubricants* **2016**, *4*, 1–10. [CrossRef]
16. Čiukas, R.; Abramavičiūtė, J. Investigation of the Air Permeability of Socks Knitted from Yarns with Peculiar Properties. *Fibres Text. East. Eur.* **2010**, *18*, 84–88.
17. Santos, T.F.; Santos, C.M.; Fonseca, R.T.; Melo, K.M.; Aquino, M.S.; Oliveira, F.R.; Medeiros, J.I. Experimental analysis of the impact protection properties for Kevlar® fabrics under different orientation layers and non-Newtonian fluid compositions. *J. Compos. Mater.* **2020**, *54*, 3515–3526. [CrossRef]
18. Santos, T.F.; Santos, C.M.S.; Aquino, M.S.; Ionesi, D.; Medeiros, J.I. Influence of silane coupling agent on shear thickening fluids (STF) for personal protection. *J. Mater. Res. Technol.* **2019**, *8*, 4032–4039. [CrossRef]
19. Lu, Z.; Jing, X.; Sun, B.; Gu, B. Compressive behaviors of warp-knitted spacer fabrics impregnated with shear thickening fluid. *Compos. Sci. Technol.* **2013**, *88*, 184–189. [CrossRef]
20. Mohsenizadeh, M.; Gasbarri, F.; Munther, M.; Beheshti, A.; Davami, K. Additively-manufactured lightweight Metamaterials for energy absorption. *Mater. Des.* **2018**, *139*, 521–530. [CrossRef]
21. Yan, R.; Wang, R.; Lou, C.W.; Lin, J.H. Low-velocity impact and static behaviors of high-resilience thermal-bonding inter/intra-ply hybrid composites. *Compos. Part B Eng.* **2015**, *69*, 58–68. [CrossRef]

polymers

MDPI

Article

Mechanical Performance of Knitted Hollow Composites from Recycled Cotton and Glass Fibers for Packaging Applications

Hafsa Jamshaid [1], Rajesh Mishra [2,*], Muhammad Zeeshan [1], Bilal Zahid [3], Sikandar Abbas Basra [1], Martin Tichy [2] and Miroslav Muller [2]

[1] Faculty of Textile Engineering, National Textile University, Faisalabad 37610, Pakistan; hafsa@ntu.edu.pk (H.J.); z.muhammad25@yahoo.com (M.Z.); basra.ntu@gmail.com (S.A.B.)
[2] Department of Material Science and Manufacturing Technology, Faculty of Engineering, Czech University of Life Sciences Prague, Kamycka 129, 6-Suchdol, 165-00 Prague, Czech Republic; martintichy@tf.czu.cz (M.T.); muller@tf.czu.cz (M.M.)
[3] Textile Engineering Department, NED University of Engineering and Technology, Karachi 75270, Pakistan; drbilalzahid@neduet.edu.pk
* Correspondence: mishrar@tf.czu.cz

Citation: Jamshaid, H.; Mishra, R.; Zeeshan, M.; Zahid, B.; Basra, S.A.; Tichy, M.; Muller, M. Mechanical Performance of Knitted Hollow Composites from Recycled Cotton and Glass Fibers for Packaging Applications. *Polymers* **2021**, *13*, 2381. https://doi.org/10.3390/polym13142381

Academic Editor: Fabrizio Sarasini

Received: 27 June 2021
Accepted: 19 July 2021
Published: 20 July 2021

Publisher's Note: MDPI stays neutral with regard to jurisdictional claims in published maps and institutional affiliations.

Abstract: This research deals with the development of knitted hollow composites from recycled cotton fibers (RCF) and glass fibers (GF). These knitted hollow composites can be used for packaging of heavy weight products and components in aircrafts, marine crafts, automobiles, civil infrastructure, etc. They can also be used in medical prosthesis or in sports equipment. Glass fiber-based hollow composites can be used as an alternative to steel or wooden construction materials for interior applications. Developed composite samples were subjected to hardness, compression, flexural, and impact testing. Recycled cotton fiber, which is a waste material from industrial processes, was chosen as an ecofriendly alternative to cardboard-based packaging material. The desired mechanical performance of knitted hollow composites was achieved by changing the tube diameter and/or thickness. Glass fiber-reinforced knitted hollow composites were compared with RC fiber composites. They exhibited substantially higher compression strength as compared to cotton fiber-reinforced composites based on the fiber tensile strength. However, RC fiber-reinforced hollow composites showed higher compression modulus as compared to glass fiber-based composites due to much lower deformation during compression loading. Compression strength of both RCF- and GF-reinforced hollow composites decreases with increasing tube diameter. The RCF-based hollow composites were further compared with double-layered cardboard packaging material of similar thickness. It was observed that cotton-fiber-reinforced composites show higher compression strength, as well as compression modulus, as compared to the cardboard material of similar thickness. No brittle failure was observed during the flexural test, and samples with smaller tube diameter exhibited higher stiffness. The flexural properties of glass fiber-reinforced composites were compared with RCF composites. It was observed that GF composites exhibit superior flexural properties as compared to the cotton fiber-based samples. Flexural strength of RC fiber-reinforced hollow composites was also compared to that of cardboard packaging material. The composites from recycled cotton fibers showed substantially higher flexural stiffness as compared to double-layered cardboard material. Impact energy absorption was measured for GF and RCF composites, as well as cardboard material. All GF-reinforced composites exhibited higher absorption of impact energy as compared to RCF-based samples. Significant increase in absorption of impact energy was achieved by the specimens with higher tube thickness in the case of both types of reinforcing fibers. By comparing the impact performance of cotton fiber-based composites with cardboard packaging material, it was observed that the RC fiber-based hollow composites absorb much higher impact energy as compared to the cardboard-based packaging material. The current paper summarizes a comparative analysis of mechanical performance in the case of glass fiber-reinforced hollow composites vis-à-vis recycled cotton fiber-reinforced hollow composites. The use of recycled fibers is a positive step in the direction of ecofriendly materials and waste utilization. Their performance is compared with commercial packaging material for a possible replacement and reducing burden on the environment.

Keywords: recycled cotton; glass; hollow knitted composite; compression; flexural modulus; impact energy

1. Introduction

Textile structural composites are becoming more and more dominant as alternative materials to replace conventional load bearing materials, e.g., metal, wood, or concrete, due to their high performance to weight ratio. These materials are multifunctional in nature due to outstanding mechanical and physical properties which can be specially designed and engineered to meet the specific performance requirements of a particular application area. Textile-reinforced composite materials can exhibit excellent resistance to corrosion, wear, and even resistance to degradation at high temperatures [1]. These modern materials have a very wide range of applications in modern life, including sports equipment, automobiles, aeronautical components, buildings, infrastructure, etc. A composite material consists of two or more constituent materials which are mixed or bonded on a macroscopic level while the interfaces are in microscopic scale. Normally, a load bearing composite consists of a reinforcement which can be fibers, particles, flakes, or fillers. The reinforcement is impregnated in a matrix that can be polymers, metals, or ceramics, depending on the application area. When manufactured properly, the new combined material exhibits properties superior to the constituent materials [2].

Reinforcement used for composites can be natural, as well as synthetic, fibrous materials. The advantage of synthetic fiber-reinforced composite is its high strength and mechanical properties, which are more suitable to use in structural applications. The main problems associated with synthetic fiber-reinforced composites are the environmental aspects, e.g., production process, application, and afterlife disposal. It is harmful to the environment because it is not biodegradable and is made from nonrenewable resources [3]. Due to environmental concerns, a large amount of research is directed towards natural fibers. The prime reason for selection of natural fibers in new product development is their minimal contribution to the greenhouse effect. The most important concern is to protect our environment from pollution, and it can be achieved without compromising the performance and quality of the product. The solution is to use biodegradable materials which are obtained from natural and renewable sources. Due to environmental concerns, plant fiber-reinforced composites are receiving greater attention of researchers and industrialists because they are biodegradable, combustible, and lightweight [4]. Recently, natural fiber-reinforced composites have received great attention from researchers and industrialists as a replacement of synthetic fiber-reinforced composite. They have relatively good mechanical and physical characteristics that can be used in various applications. Natural fibers are bio-degradable, nonabrasive, nonhazardous, lightweight, and renewable materials [5]. Researchers are focusing on development of products from recyclable materials due to increasing environmental concern. In this context, recycled natural fiber-based textile waste can be used as a sustainable material in composite reinforcement. Several researchers have successfully used recycled cotton/polyester material to produce composites with significant mechanical and acoustic performance level [6,7]. Value added composite samples were also developed by using recycled cotton fiber waste from discarded denim fabrics. These materials exhibit sufficient thermal, acoustic, and mechanical performance [8,9].

Researchers have prepared thermoplastic composites using natural flax and thermoplastic polypropylene yarn on flat bed knitting machine by developing plain and rib structures and compared their mechanical properties. Three-dimensional knitted fabric reinforcement was developed on flat knitting machine for thermoplastic composites using glass and PP filaments. Results showed that mechanical properties of composites are affected by knitting structure and direction of inlay yarn [10,11].

Textile-reinforced hollow composites have numerous applications, such as sports equipment, pipes, drive shafts, printing rollers, landing gears for helicopters, rocket structure, structural building components, etc. Hollow composite tubes can be prepared by several methods, e.g., knitting, spacer weaving, braiding, stitching, etc. Researchers have investigated the deformation and fracture behavior of glass-epoxy braided circular tubes for different loading cases, like compression, torsion, combined tension-torsion, compression-torsion, etc., both experimentally and theoretically [12,13]. Other researchers proposed models for simulating the crushing behavior and predicting the energy absorption characteristics of triaxially braided carbon fiber/epoxy-vinyl ester composite tubes with both circular, as well as square, cross-sections [14]. Braided hollow textile preforms were used for development of composites by several researchers [15,16]. Many others have investigated the hybridization of glass woven fabric with a natural fiber mat for applications in the piping industry for commercial applications. Researchers have investigated the flexural stiffness of thick composite tubes which were manufactured by using an automated fiber placement machine [17].

Several researchers used finite element modeling for analysis of bending strength of cylindrical composites. They developed an enhanced version of finite element models (FEM) for elastic and non-linear plastic analysis of tubes [18]. In literature, models for both straight and curved composite tubes have been presented [19]. A beam element for analysis of straight and tapered composite tubes under general loading was investigated [20]. A model for four-point bending of a thermoplastic composite using a three-dimensional solid element was reported [21]. The lateral planar crushing and bending responses of carbon fiber-reinforced plastic (CFRP) square tube filled with aluminum honeycomb was investigated. The square tube was developed from plain weave of carbon fiber. Good agreement was obtained between numerically predicted results by FEM and experimentally measured results [22].

Knitting is well established as a technique to produce hollow and tubular structures. However, there is relatively very small amount of research conducted to use such structures for development of hollow/tubular composites. Despite the easy shaping capability of knitted fabrics, published literature is generally focused on the reinforcement of composites with knitted plates or flat structures. An advantage of knitted fabric-based hollow composites is the possibility of multiple layering by intermeshing of the loop structure in-between the different fabric layers. The structural maneuverability and compressional resilience in knitted hollow composites offers several research opportunities and product development possibilities. Such a research gap in the literature provided motivation for experimental investigation of hollow composites produced by knitting technology.

There is limited understanding of the mechanical properties and long-term durability of hollow knitted composites for application as structural components. In the current research, the mechanical properties of glass (GF)-reinforced hollow composites are investigated by varying the tube diameters for use in secondary structural elements, such as wall panels or door systems. Its advantages, such as being light weight, non-corrosive, and low maintenance cost, make such composites suitable alternative for steel, wood, and concrete materials.

A further objective of the present study is to develop knitted hollow composite preforms by using recycled cotton (RC) yarn, which has not been reported in literature. These composites are aimed at applications relating to packaging of heavy components. For packaging of heavy/bulkier goods, cardboards are normally used because of their cushioning properties. Cardboard is developed from paper which needs wood as a raw material; thus, it is not an environmentally-friendly option. A typical packaging is made of three layers of heavy paper, two flat layers with a wavy/corrugated one in the middle. Although it is hard enough not to break or tear, it cannot be reused. Moreover, it is not easily recyclable and dumped as landfill. Even the decomposition of cardboard materials generates methane, which is a major greenhouse gas with a global warming capacity 21 times more powerful than carbon dioxide. Its recycling also contributes to environmental pollution by using different sources of energy. During recycling, it needs almost 75% of the energy needed to make new one, while the quality is much inferior. In view of the existing issues, hollow knitting is a promising technology to prepare composite preforms which can replace conventional packaging materials based on hard paper. Use of recycled cotton obtained from industrial waste is a sustainable approach towards minimizing environment pollution.

Different mechanical tests, e.g., hardness, compression, flexural, and impact measurement, were carried out for the developed composite samples. The mechanical properties of knitted hollow composites developed from glass fiber were compared with hollow composites developed from recycled cotton fibers. Further, the cotton fiber-based samples were compared with cardboard packaging material of similar thickness. The findings provide new opportunities for an ecofriendly alternative material with superior cushioning, as well as protective performance.

2. Materials and Methods

2.1. Materials

Knitted hollow fabrics were developed by using recycled cotton (RC) fiber and glass (G) fiber yarn of 1800 denier on a V-bed flat knitting machine using gauge 7E. During formation of plain courses, both front and back needle beds remain operational simultaneously. However, during tube formation, both needle beds knit separately, as shown in Figure 1.

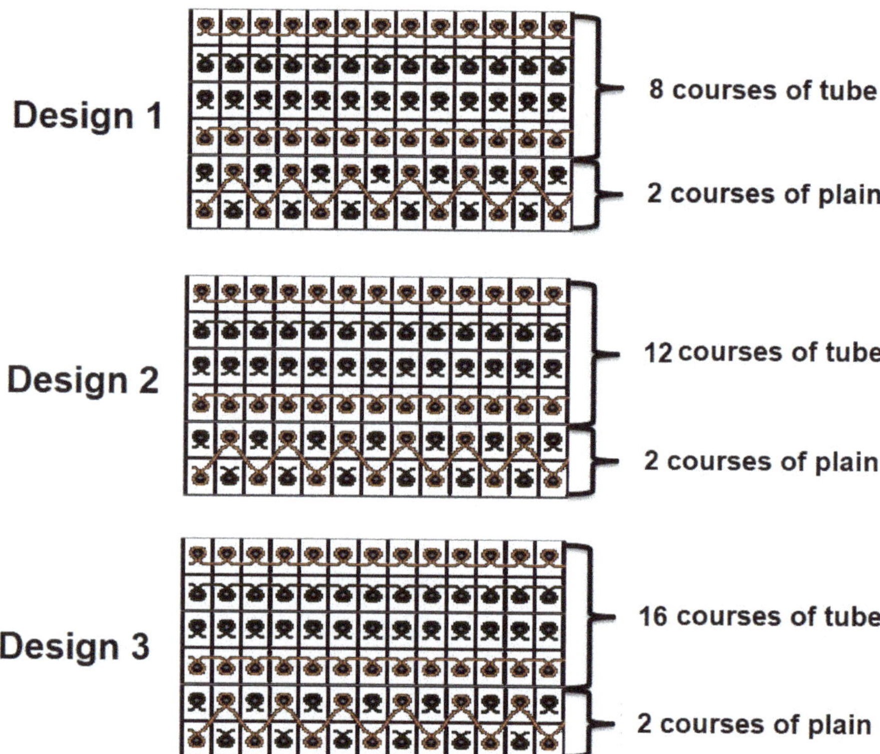

Figure 1. Knitting design (short representation).

All knitted tubular sample designs were developed on the SDS-ONE APEX platform [23]. Tubes with different thickness were developed by using different numbers of courses (8, 12, and 16). In total, six types of knitted preforms were developed/manufactured. Details of the fabric manufacture are given in Table 1. These specifications were selected in order to achieve the composite sample thickness similar to the thickness of commercially available double-layered carboard packaging material.

Table 1. Knitted fabric parameters.

Sample ID	Yarn Type	Linear Density of Yarn (Denier)	Plain Courses	Tube Courses	Wales (cm^{-1})	Courses (cm^{-1})	Stitch Length (cm)
C1	Recycled Cotton (RC)	1800	2	8	11	18	0.71
C2	Recycled Cotton (RC)	1800	2	12	11	18	0.71
C3	Recycled Cotton (RC)	1800	2	16	11	18	0.71
G1	Glass (G)	1800	2	8	12	16	0.69
G2	Glass (G)	1800	2	12	12	16	0.69
G3	Glass (G)	1800	2	16	12	16	0.70

2.2. Methods

2.2.1. Production of Composite Samples

Unsaturated polyester resin (UPR) (KZN Resins, Durban, South Africa), Malikens GP 555-04, which is dissolved in a 35 + 2% of organic liquid solvent "styrene", was used for preparation of composite samples. This is a thermoset polymer which, when oxidized, starts to convert from liquid to a gel, and later, to a hard solid form. Cobalt octoate (KZN Resins, Durban, South Africa), which is a metal salt of carboxylic acid, was used as an accelerator in the curing process of the polyester resin. MEKP 50 (Methyl ethyl ketone peroxide) (KZN Resins, Durban, South Africa), which is an organic substance, was used as hardener in the process. Composite samples were developed by inserting hexagonal metal pipes into the hollow fabrics. The metal pipes were first covered with plastic release film, as shown in Figure 2a, for easy removal of sample after composite manufacturing. After successful insertion of metal rods into the hollow fabric samples, as shown in Figure 2b, impregnation with resin was carried out. The impregnated samples were dried for 24 h and cured for 4 h at 100 °C. After the curing operation, the inserted pipes were removed, and the composite samples were obtained, as shown in Figure 2c. The internal diameters of the manufactured hollow composite samples were 8 ± 0.02 mm, 10 ± 0.02 mm, and 14 ± 0.02 mm, respectively, for both glass fiber- and recycled cotton fiber-based samples. Outer diameter was equivalent to that of double-layered cardboard sample. The test samples were cut from fabricated composites according to different standard requirements. The images of RC fiber knitted hollow composite, glass fiber knitted hollow composite, and conventional cardboard packaging material are shown in Figure 2d–f, respectively.

Mechanical behavior of commercially available double-layered cardboard (W) packaging material was also investigated in order to compare with recycled cotton fiber-based sample (C2), which is of most similar thickness (outer diameter).

Scanning electron microscopy (SEM) (NIST, Gaithersburg, MD, USA) images of recycled cotton fiber (RCF)- and glass fiber (GF)-based composite sample cross-sections are shown in Figure 3a,b, respectively.

These images indicate that the samples are free from any major voids, and the fibers are properly impregnated with the resin. The uniformity of impregnation also ensures a strong interface between the fibrous phase and the matrix phase.

2.2.2. Characterization

The density of the natural fiber-based composite materials using a polymeric matrix can be determined according to the standard ASTM D 792, using an analytical balance equipped with a stationary support for the immersion vessel, as shown in Figure 4a. In this method, a solvent, such as water or propanol, can be used as immersion liquid, depending on the density of the polymer.

In this investigation, the density was measured by using water as the immersion liquid. The samples absorb some water through the micro pores. In fact, the textile structures and their composites are almost always composed from micro pores, which can absorb water.

Initially, the dry mass of the composite samples was measured. Then, the hollow composite sample was immersed in the water column of known volume. The overall increase in water level (or volume) indicates the volume occupied by the solid (non-porous) portion of the composite. That means the sample occupies a space equivalent to its solid volume (not including its porosity). About 1 h of time was allowed for complete immersion and penetration of water through all possible pores. Then, the final water level was noted. The sample was taken out of the water column. Further, the mass of wet composite sample was measured, which is slightly higher than the dry mass of the same sample. The difference of mass indicates the mass of water absorbed through the micro pores. The volume corresponding to this mass difference is calculated and added to the solid volume of the composite sample in order to obtain the overall volume of the composite. It should, however, be noted that the water absorption capacity of developed composite samples was relatively much smaller as compared to their primary knitted structures and was only found to be around 1–2%. Density of composite samples was calculated from the dry mass and the overall volume, including micro pores. All the results of measurements are presented in Table 2.

(a) Steel rods with release film

(b) Insertion of rods in hollow knitted structures

(c) Composites after curing

(d) Recycled cotton fiber based hollow composite

(e) Glass fiber based hollow composite

(f) Double layered card-board packaging material

Figure 2. Photographs of developed composite samples versus double-layered cardboard.

Specimens of suitable dimensions were cut as per specified standards for physical and mechanical testing. Fiber volume fraction (Vf) of recycled cotton fiber-based hollow composite was maintained to be approximately 40%, and, for glass fiber-based composites, it was approximately 50%. The surface hardness of all the samples developed was measured by Barcol Hardness tester of Zwick/Roel, Brno, Czech Republic, as shown in Figure 4b, according to standard ASTM D2583. The test specimen is placed under the indenter of hardness tester, and uniform pressure of 1 bar was applied as per standard. Twenty measurements were conducted, and the mean value was calculated.

(a) RCF composite

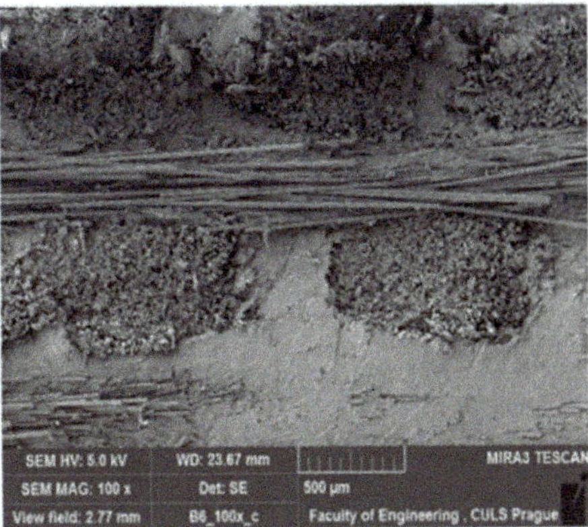

(b) GF composite

Figure 3. SEM images of cross-sections: (**a**) RCF sample, (**b**) glass fiber sample.

(a) Density measurement

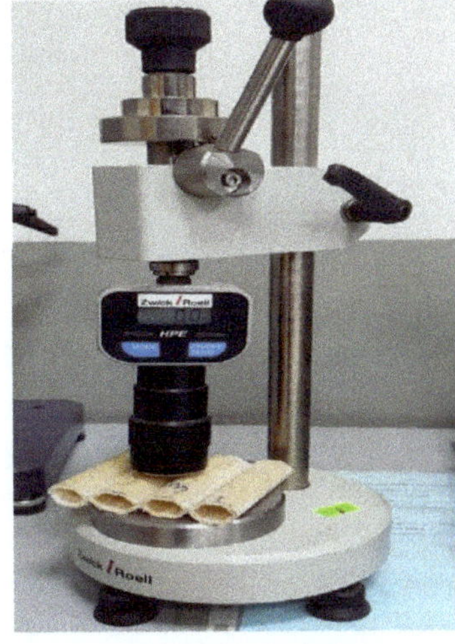

(b) Hardness test

(c) Compression test

(d) Flexural test

(e) Impact test

Figure 4. Physical and mechanical characterization of manufactured hollow composite samples.

Mechanical characterization, e.g., compression, flexural test, and impact measurements, were carried out for all the developed hollow composite samples as they are the most essential performance requirements for packing applications.

Table 2. Physical parameters of the developed composite samples.

Sample ID	Fiber Volume Fraction (Vf %)	Density (g/cm³)	Hardness (Barcol) Scale of (0–100)	Diameter of Tube	
				Inner Dia (mm)	Outer Dia (mm)
C1	40 ± 2	1.23 ± 0.02	82.7 ± 0.2	8.1 ± 0.1	11.8 ± 0.1
C2	40 ± 2	1.21 ± 0.02	81.9 ± 0.2	10.2 ± 0.1	14.8 ± 0.1
C3	40 ± 2	1.21 ± 0.02	80.1 ± 0.2	14.3 ± 0.1	18.5 ± 0.1
G1	50 ± 2	1.41 ± 0.02	88.0 ± 0.2	8.2 ± 0.1	10.5 ± 0.1
G2	50 ± 2	1.42 ± 0.02	87.7 ± 0.2	10.2 ± 0.1	12.5 ± 0.1
G3	50 ± 2	1.39 ± 0.02	86.6 ± 0.2	14.2 ± 0.1	16.5 ± 0.1
W	-	-	11.5 ± 0.2	-	15.0 ± 0.1

Compression strength indicates the resistance of a material to deformation under pressure. The compression strength of all the hollow composite specimens, as well as cardboard material, was determined by using Universal Testing Machine (Z100-100 KN) manufactured by Zwick/Roell, Brno, Czech Republic, as shown in Figure 4c, according to the standard ASTM D2412-11, at a crosshead speed of 1.3 mm per minute. This test determines the compression load-deflection characteristics of hollow composite samples subjected to loading between two parallel steel plates, as shown in Figure 5a. The measurements were repeated 20 times, and the mean value was calculated. The compression strength and strain were obtained using Equations (1) and (2), respectively. It was assumed that the tube will become elliptical during the load application [24].

$$\text{Compression strength} = \frac{F}{\Delta_y}\left(1 + \frac{\Delta_y}{2d}\right)^3, \tag{1}$$

$$\text{Compressive strain} = \frac{\Delta_y}{d} \times 100\%, \tag{2}$$

where: F is the applied load, d is the outside diameter, and Δ_y is the change in the outside diameter of the specimen in the load direction. Equation (3) was used to calculate the compression modulus.

$$\text{Compression modulus} = 0.149\ r^3 \times \text{Compression strength.} \tag{3}$$

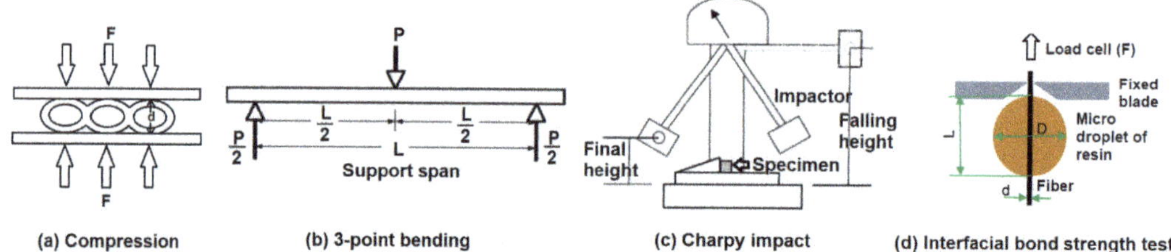

Figure 5. Principles of (**a**) compression test, (**b**) 3-point bending test, (**c**) Charpy impact test, and (**d**) fiber-matrix interfacial bond strength test.

The flexural behavior of the composite samples was evaluated by using the 3-point bending test by using Universal Testing Machine (Z100-100 KN), by Zwick/Roell, Brno, Czech Republic as shown in Figure 4d, according to the test method of ASTM D-7264. The same method was also used for evaluation of the double-layered cardboard packaging material. A specimen of rectangular shape having dimensions 120 mm × 13 mm was supported at the ends and deflected at the center point. As force was applied on the specimen, and it started deflecting from the center, its deflection and force were measured and recorded until the failure occurred or the maximum force reduced to 40%. The principle of 3-point bending is shown in Figure 5b. The gauge length/support span of 80 mm, deformation rate of 1 mm/min, and load of 5 kN was maintained. Twenty measurements were conducted, and the mean value was calculated. The flexural strength was calculated using Equation (4) [25].

$$\sigma = 3PL/2bh^2. \tag{4}$$

Flexural modulus was calculated using Equation (5)

$$E = PL^3/4ybh^3, \tag{5}$$

where P represents Load, L represents gauge length, b represents width, h represents thickness, and y represents deflection or strain during bending.

The Charpy impact test was performed by following the ISO-179 standard testing procedure. An impact testing machine (Model HIT50P) manufactured by Zwick/Roell, Germany, as shown in Figure 4e, was used for the test. A swinging hammer/pendulum with 21 J energy and velocity of 3.8 m/s was used to test the specimens for impact energy. The hollow composite specimens, as well as cardboard material, were tested without a notch. Samples were cut into size 80 mm $\times$ 10 mm for testing. The thickness and width of the samples were measured by Vernier caliper before the test. Specimens were placed on the specific slot, and the pendulum was allowed to impact in order to hit and break the specimen, as shown in Figure 5c. The measurements were repeated 20 times, and the mean value was calculated [26]. Impact energy was calculated as:

$$E = \text{Mass of impactor (m)} \times \text{acceleration due to gravity (g)} \times \text{falling height (h)}, \tag{6}$$

$$\text{Impact energy absorbed} =$$

$$\text{Energy of striking impactor} - \text{Residual energy of rebounding impactor.} \tag{7}$$

In order to determine the fiber-matrix interfacial bond strength, the single fiber pull out using the microdroplet test was conducted [2,27]. The single fibers of recycled cotton, as well as glass, were treated with microdroplets of the resin. Then, the fibers were cured under similar conditions as the composite samples. The impregnated fiber samples were dried for 24 h and cured for 4 h at 100 °C. The diameter of the microdroplets of resin was around 50 μm. The principle of microdroplet test to determine interfacial bond strength is shown in Figure 5d. The interfacial bond strength was calculated using Equation (8).

$$\text{Interfacial bond strength} = F/\pi\,d\,L, \tag{8}$$

where F is the maximum load, d is the average fiber diameter, and L is the length of fiber embedded in the droplet of resin. Ten measurements were carried out for cotton, as well as glass fibers, and the mean was calculated. The interfacial bond strength for cotton fiber was found to be 28.52 $\pm$ 0.2 MPa, and that for glass fiber was 19.35 $\pm$ 0.2 MPa. The stronger interface of cotton fiber with the unsaturated polyester resin can be attributed to the relatively rough fiber surface as compared to a smoother surface of the glass fiber.

3. Results and Discussion

3.1. Surface Hardness

It was observed that the surface hardness of glass fiber-based hollow composites and recycled cotton fiber-based samples are almost similar. In fact, the surface is mostly composed of the resin, and the fibers are embedded deeper inside. Therefore, the hardness of the surface is mainly dominated by the cured-resin hardness. All the hollow composites exhibit almost 7–8 times higher hardness as compared to double-layered cardboard material. This is an indication of longer service life in case the of the knitted hollow composites as compared to paper-based conventional packaging material. Further, it was observed that the hardness slightly decreases as the diameter of the tube increases. It can be attributed to decreasing curvature, which reduces the stiffness. It is well known that rigidity is inversely proportional to the radius of curvature. However, it must be noted that the change of surface hardness in this case is only marginal.

3.2. Compression Properties

The compression strength is one of the major properties required in composites used in packaging applications. It is generally accepted that fiber strength is the most important parameter responsible for composite strength. During mechanical testing, fiber fracture happens when the force exceeds the limiting strength of the fiber and interfacial bonding with the resin.

As glass is a relatively stronger fiber, its composites also exhibit substantially higher compression strength as compared to RC fiber composites. The trend is clearly visible from Figure 6.

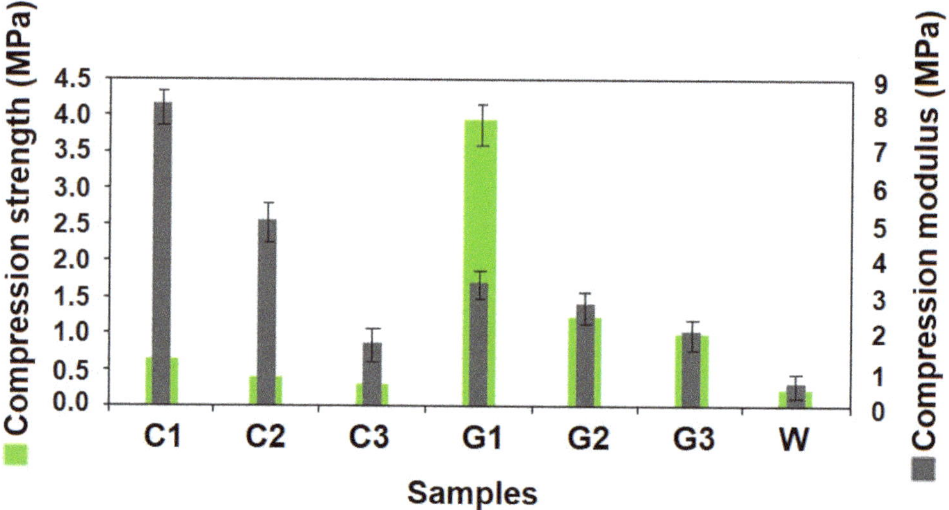

Figure 6. Compression strength and modulus of RC and glass fiber-reinforced hollow composites versus double-layered cardboard.

It is well known that glass fibers are more crystalline and more rigid as compared to natural origin cellulosic fibers as cotton. Therefore, a glass fiber-based composite offers higher stiffness and compressional strength as compared to recycled cotton fiber-based composites. It should be noted that the fineness/linear density of both glass and cotton yarns are the same. Moreover, the GF-based composite samples are developed with higher fiber volume fraction (Vf = 50%) as compared to RCF-based samples (40%). Thus, the fiber mechanical properties and the fiber volume fraction have significant influence on the overall composite mechanical performance. Such observations are also validated by the rule of mixture and the Halpin-Tsai equations shown below [28].

$$K_c = K_m \left[\frac{1 + \xi \zeta V_f}{1 - \eta V_f} \right], \tag{9}$$

$$\text{With, } \eta = \left[\frac{(K_f/K_m) - 1}{(K_f/K_m) + \zeta} \right], \tag{10}$$

where K_c represents the effective compressional (mechanical) property of the composite, while K_f and K_m are the corresponding fiber and matrix compressional (mechanical) properties, V_f denotes the fiber volume fraction, and ζ is a geometrical parameter, which represents the reinforcement geometry, packing geometry, and loading conditions. In the present analysis, the geometry is defined by the knitting pattern, and yarn fineness, which is same for both types of materials.

Compression strength of both RC- and glass fiber-reinforced composites show an inverse trend with increasing tube diameter. This fact is governed by basic relations in bending/compressional deformation. During compression, the hollow segment undergoes ovalization, and the tubes undergo bending deformation. Bending rigidity is always higher for a lower radius of curvature. Therefore, the smaller tube diameter results in higher stiffness, as well as compression strength. Sample C1, having the lowest diameter of 11.8 mm, shows 60.2% and 20.9% higher compression strength as compared to C2 and C3, respectively. Similarly, G1 show 966% and 292% higher compression strength as compared to G2 and G3, respectively. During the compression test, buckling is the main phenomenon responsible for failure of fiber reinforced hollow composites [29]. As the composite structure undergoes compression, the assembled fibers and yarns tend to spread and become misaligned. Higher diameter of the tube reduces curvature of the fibers and yarns on the surface. Thus, they are susceptible to deform to a higher extent during compression. The samples of hollow composites having higher diameter tend to offer more severe buckling phenomena and relatively lower resistance to compression load. The outcome is lower compression strength. These observations are also supported by previously reported literature [25,26]. All the developed hollow composite samples exhibit higher compression strength as compared to double-layered cardboard packaging material.

It is interesting to note that RC fiber-reinforced hollow composites show higher compression modulus as compared to glass fiber-reinforced composites. This observation is in contrast with the findings about compression strength. As is well known, modulus is a derived parameter which depends on both compressive stress and compressive strain. The shorter fiber length in recycled cotton enables much lower deformation compared to relatively much longer glass fibers during the compression test. In addition, it must be noted here that the tubes made from RC-based materials have a wall thickness almost twice that of glass fiber-based composite tubes. Higher thickness in this case is also obvious due to the higher thickness of cotton yarns, pertaining to lower density as compared to glass. Therefore, even with much lower compression strength, RC fiber-reinforced hollow composites exhibit significantly higher compression modulus. The knitted prepregs of RC also prove to be strongly bonded with the resin as per results of microdroplet test. Further, in the SEM image presented in Figure 3, a more uniform and deeper impregnation is observed in RCF-based samples as compared to GF-based samples. The hollow composite sample (C2) exhibits higher compression strength and significantly higher compression modulus as compared to the double-layered cardboard (W) packaging material of similar thickness. Thus, they can be easily used as replacement of cardboard-based packaging material with much superior compressional properties.

The differences in the stress-strain behavior during compression test for RC fiber composites and glass fiber composites are shown graphically in Figure 7a,b, respectively.

The compression stress-strain curves for the glass fiber-based hollow composites indicate much higher compression strength or peak compressive stress level. The inherent mechanical properties of glass, which are undoubtedly much higher than the recycled cotton fiber, are responsible for such behavior. It is also visible that the curves for RCF-based samples show higher slope values as compared to GF-based samples. This is indicative of the higher compression modulus in RCF-based hollow composites. This behavior is attributed to lower level of deformation before peak compression load. Shorter fiber length in recycled cotton and deeper impregnation of resin are the factors responsible for higher compression modulus observed in RCF-based hollow composites. Sample C1 offers higher stress compensation as compared to C2 and C3. It exhibits a permanent deformation after 8% compression, as shown in Figure 7a. Sample C1 shows the highest maximum compressive stress and lowest compressive strain due to the smallest tube diameter among the recycled cotton fiber-based samples. Smaller tube diameter, along with shorter RC fibers, enables the structure to bear higher compressive stress. This is due to the fact that both the upper and lower arms are connected by a shorter fiber column, which provides higher resistance to the applied compression load. Further the short fibers on the upper and lower surface can effectively align themselves and absorb the stresses. In the case of higher tube diameter, and longer fibers, there is more flattening and a higher chance of fiber slippage. These observations are also validated by reported literature [24,25,27,28].

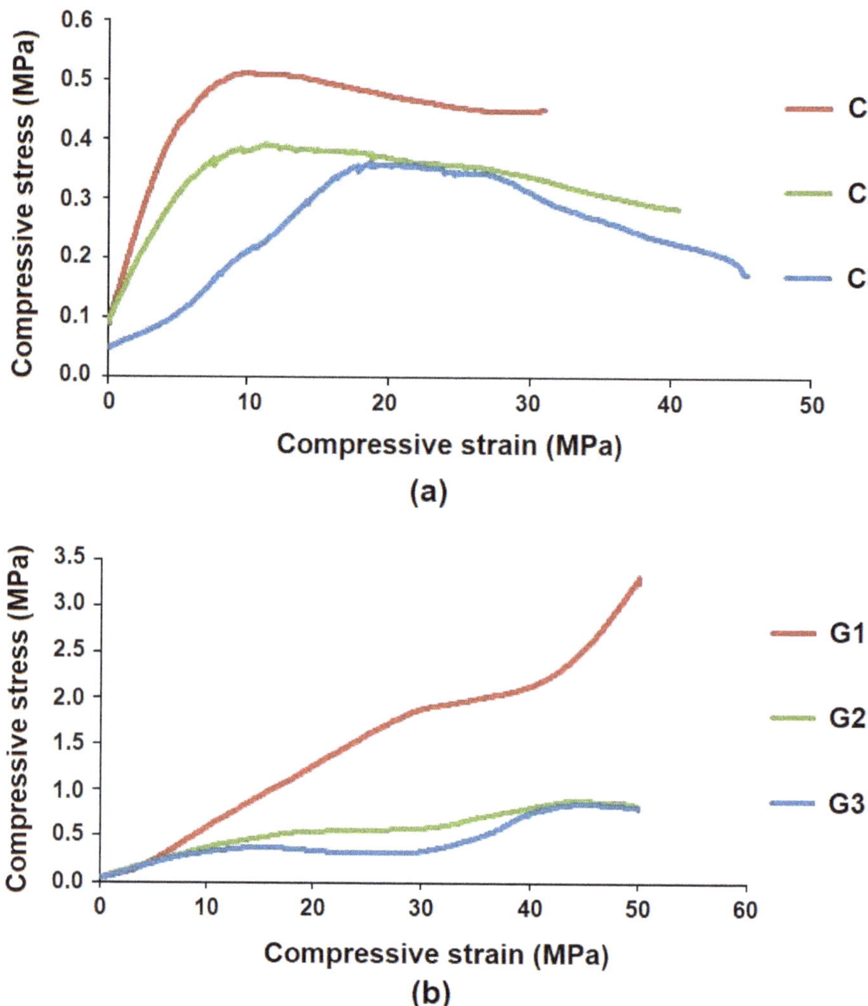

Figure 7. Compressive stress versus strain for (**a**) recycled cotton fiber-based hollow composites and (**b**) glass fiber-based hollow composites.

Similarly, among the glass fiber-reinforced composites, G1 shows the highest strength and peak stress level due to its smallest tube diameter, as shown in Figure 7b. The compressional strength increases with lower radius of curvature. The performance is governed by the geometry, which is defined by curvature, tube diameter, wall thickness, etc., as defined in the Halpin-Tsai equations [28].

Furthermore, all the glass fiber-based hollow composites exhibit two peaks in their compression curve, as shown in Figure 7b. The stress-strain behavior concerning the decrease and increase of the stiffness in glass fiber-reinforced hollow knitted composites are not completely unexpected. There are several research studies reported in literature where the compression behavior of knitted structures and their composites are described [30–36]. The initial part of the compression curve denotes the elastic stage, which corresponds to flattening and ovalization of hollow channels. The middle part showing a slight decrease is known as the plateau stage, which corresponds to deformation at the joining points. The third part, which again shows an increasing trend, denotes to the densification of fibers and load transfer to matrix. The dual peaks are more distinct in the sample (G3), which has the maximum tube diameter and thickness among all GF-based hollow composites. The first peak, which is observed at around 15% compressive strain, corresponds to the maximum elastic limit of the tubular structure. During this phase, the reinforcing fibers tend to spread and absorb the compressive stress. The first peak corresponds to the jamming state, which is the maximum limit before load transfer to the matrix phase. Subsequently, there is stress concentration at the weakest links in the hollow composite. These points are located at the joints and contact area between adjacent tubes. Stress concentration at these points results in deformation at the joining points of upper and lower half of the tubes, which results in a second peak at approximately 40% deformation level. These double peaks during compression test can be justified by the shape change in the tubes. The shape change (ovalization) effect is visible by flattening of the circular shape of the tube. This, in turn, decreases the stiffness of the composite structures. Such observations are also supported by reported literature [30–36]. In samples G1 and G2, the peaks are not as distinct as in G3 due to lower tube thickness/diameter. There is less flattening in samples of lower diameter of the hollow tubes. They offer more resistance to compressional deformation by virtue of the stiffness resulting from lower radius of curvature. Since the cotton fibers are relatively weaker, they fail/break before the second phase of compression occurs. Thus, RCF-based samples show only one distinct peak in the compression curve. Among the three RCF-based hollow composites, only sample C3 shows a slightly visible second peak. This is attributed to maximum diameter and flattening of the hollow tubes, which enables absorption of a small compressive stress in the second phase, though the overall peak stress is the minimum.

The compression behavior of the commercially available double-layered cardboard-based sample W was also tested and compared with the RCF-reinforced hollow composite sample (C2) having the nearest thickness. The comparison of the stress-strain curves is shown in Figure 8.

The sample C2 performs three times better than paperboard under compression load. This can be attributed to the strong interface between recycled cotton fibers and the polymer resin, as observed from SEM image in Figure 3. Further, the interfacial bond strength between recycled cotton fiber and the resin has also been measured by microdroplet test and found to be significantly high. As a result, the hollow columns of cotton fiber-based composites provide much better protection against relatively larger compressive stresses. The curved tubular hollow channels in the developed composite samples offer higher level of resistance to compressional deformation. Moreover, the reinforcing fibers can absorb the compressive stresses more effectively. On the other hand, the cardboard-based material is weaker and offers minimal resistance. Based on the rule of mixture and the Halpin-Tsai models, it can be predicted that fiber-reinforced composites can offer higher mechanical performance as compared to the constituent elements [25–28]. The cardboard material is relatively weaker, and there is absence of strong inter-polymer linkage as in fibers and polymeric resins.

The dual peak behavior is visible in double-layered cardboard as in the case of GF-based samples. The first peak is result of flattening and stress absorption. After the flattening of the cardboard paper material, the stress is accumulated at the joints of the cells/hollow tubes. The paper walls tend to buckle and bend. This buckling action enables further absorption of compressive stress. Thus, the second peak of stress is slightly higher than the first peak. Overall, paper-based hollow packaging material proves to be weaker and less resistant to compressive deformation.

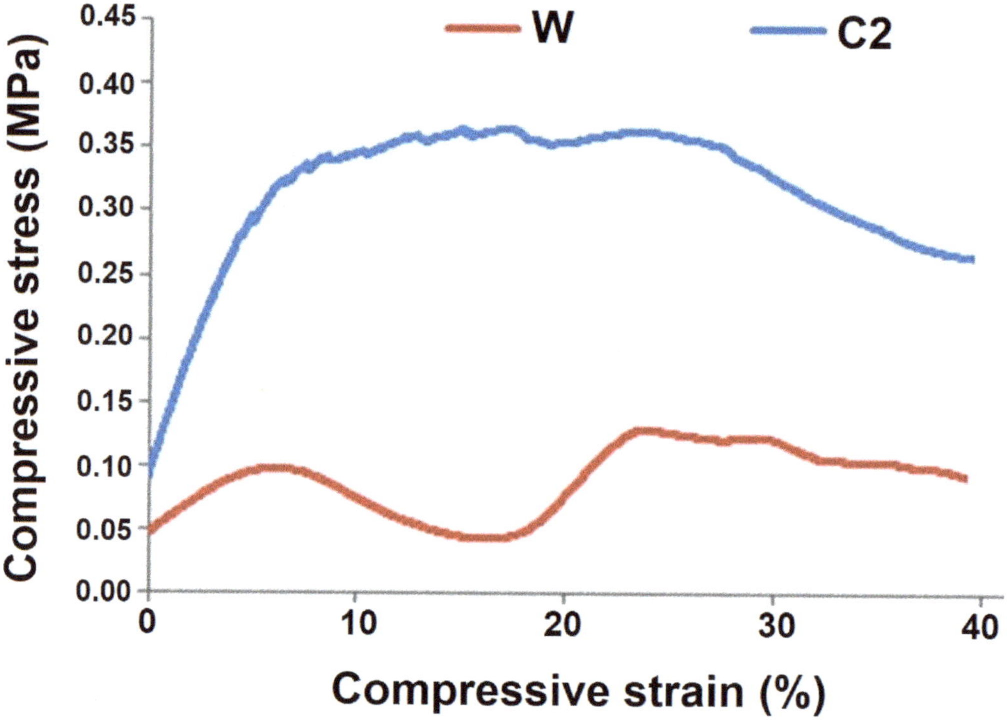

Figure 8. Compression behavior of RC fiber composite (C2) and cardboard material (W).

3.3. Flexural Properties

The failure of composites under flexural loading involves a combination of tensile failure, compression failure, shear, and/or delamination at different levels [25,26]. When a sample is subjected to bending deformation, the outer surface experiences tensile stress, while the inner surface experiences longitudinal compression. In the case of multiple layers of fibers, the tensile stress propagates inwards and causes delamination, which ultimately reduces the flexural resistance/strength. A comparative account of flexural behavior for all RCF- and GF-reinforced hollow composite samples, along with double-layered cardboard material, is shown in Figure 9.

Flexural strength also follows a similar decreasing trend as that of compression strength with the increase of tube diameter. The bending deformation translates into partial compression; thus, the trend is similar. Under 3-point bending mode, the support span undergoes a deflection as the load is applied. Initially, the load is taken up by the reinforcing fibers, which experience an extension on the outer layer and inward compression on the inner layer. Thus, the macroscale bending behavior is a cumulative response of the fiber tensile/elastic modulus and moment of inertia. Such behavior is reported by several other researchers in the available literature [24–27].

The flexural strength of GF-reinforced samples is much higher as compared to RCF-based samples. This is traced back to stronger glass fibers and relatively higher fiber volume fraction in glass fiber-reinforced hollow composites in the current study. The rule of mixture and the Halpin-Tsai model are validated for bending performance, as well [28].

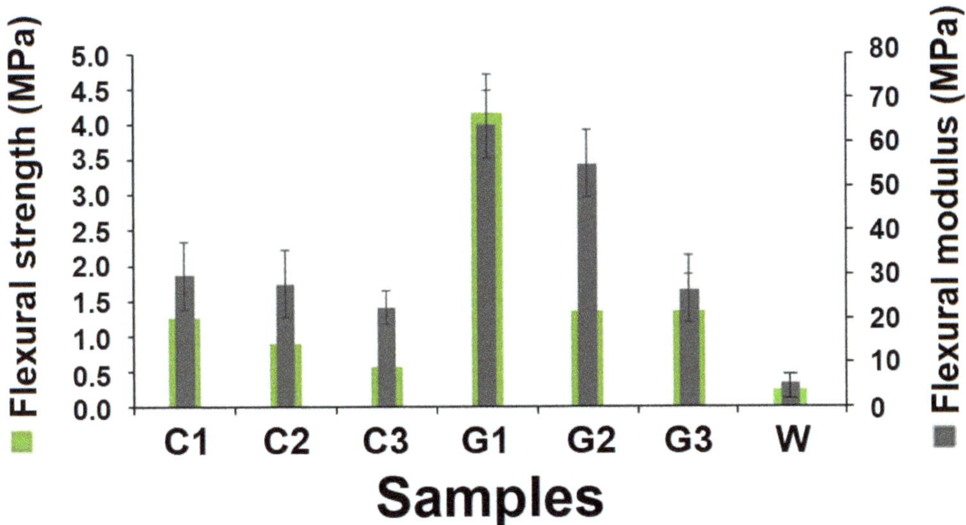

Figure 9. Flexural strength and modulus of RC and GF-reinforced hollow composites versus double-layered cardboard material.

From Figure 9, it is clearly visible that the flexural strength of C1 (1.24 MPa) is higher than C2 (0.89 MPa) and C3 (0.53 MPa). Similarly, flexural strength of G1 (4.16 MPa) is higher than that of G2 (1.36 MPa) and G3 (1.35 MPa), respectively. The smaller tube diameter enables the sample to withstand higher level of bending stress. According to fundamentals of bending, the bending rigidity is inversely proportional to the fourth power of the tube. Thus, the higher curvature or lower radius of curvature is more beneficial to attain higher stiffness. Further, the glass fiber-reinforced hollow composites are substantially stiffer as compared to the recycled cotton fiber-based samples. Based on the Halpin-Tsai model, a higher bending rigidity of fibers and higher fiber volume fraction results in higher overall stiffness in the composite [26–28]. The glass fibers are much stiffer as compared to cotton fibers due to higher level of crystallinity. Thus, the bending performance of the constituent fibers is reflected in the flexural performance of the knitted hollow composites.

The flexural stress-strain behavior of RCF-based samples and GF-based samples are shown in Figure 10a,b, respectively. The figure shows that hollow composite samples C1 and G1 have relatively higher and steeper stress-strain curves as compared to other samples in their respective groups because of relatively smaller tube diameters. This is due to the fact that both the upper and lower arms are connected by a shorter continuous fiber column, which provides higher resistance to the applied bending load. The weak link in tubular channels are the inter-tube joints. With smaller diameters, the link is smaller and does not allow any fracture to initiate. Further, the fibers on the upper and lower surface can effectively align themselves and absorb the stresses. In the case of higher tube diameter, there is more flattening and a higher chance of fiber slippage. These observations are also validated by reported literature [24,25,27,28].

The maximum bending stress is absorbed by samples C1 and G1 with a deformation of 13% and 15%, respectively. On the other hand, C2, C3 and G2, G3 manage maximum stresses at a deformation of 24%, 40% and 16%, 17%, respectively, as shown in Figure 10. It can be observed that higher tube diameter enables higher flexural strain. This is indicative of the fiber slippage and flattening in samples C2, C3 and G2, G3. Further, the stress-strain curves show multiple steps in the case of higher diameters. Such behavior is observed for both RCF- and GF-based samples. This can be attributed to a step wise stick-slip behavior exhibited by the fibers in the hollow composites. These peaks also indicate matrix cracking, cracking of the supporting layer, and ovalization effect as discussed previously in the compression study. Such observations are also reported by other researchers [24,26,27].

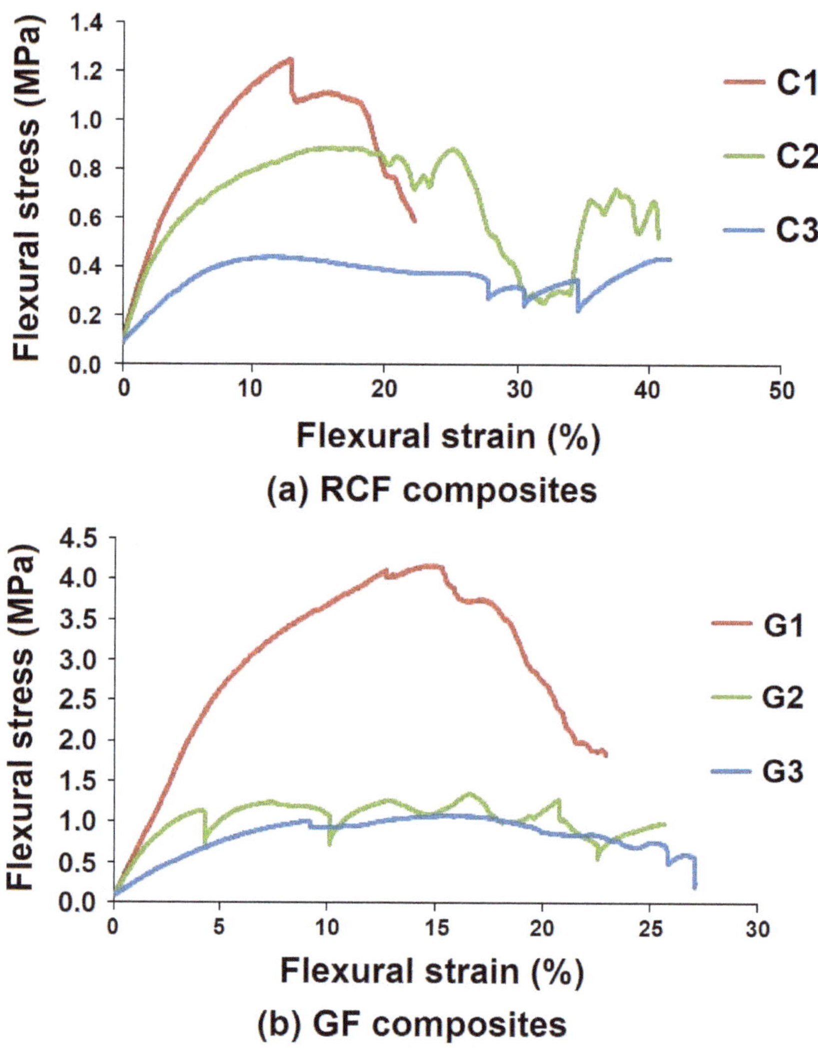

Figure 10. Flexural stress versus strain of (**a**) recycled cotton fiber-based hollow composites and (**b**) glass fiber-based hollow composites.

The slope of the flexural curves defines the flexural modulus, which also indicates stiffness. It can be observed that samples C1 and G2 show maximum stiffness in their respective groups of samples. Such observations are supported by theoretical models, as well as experimental studies, reported in the literature [28]. During the flexural test, no brittle failure was observed (especially in specimens with higher tube diameter). This is indicative of very good service performance of such hollow composites as a packaging material.

Since sample C2 has the nearest thickness as that of the cardboard-based material, their flexural performance is compared in Figure 11.

The flexural stress-strain curves of RC fiber-reinforced hollow composite sample (C2) and double-layered cardboard material (W) is shown in Figure 11. The cardboard material exhibits a rather flat stress-strain behavior. The material cannot survive higher flexural stresses. On the other hand, a fiber structure-based hollow composite shows substantially higher flexural rigidity, even with the use of recycled cotton material. Thus, it is deemed to be suitable to replace the cardboard in the packaging applications. Textile structural composites show geometry related performance and can be designed in special way to perform load bearing functions. The results are validated by theoretical models reported in the literature [13,21,28].

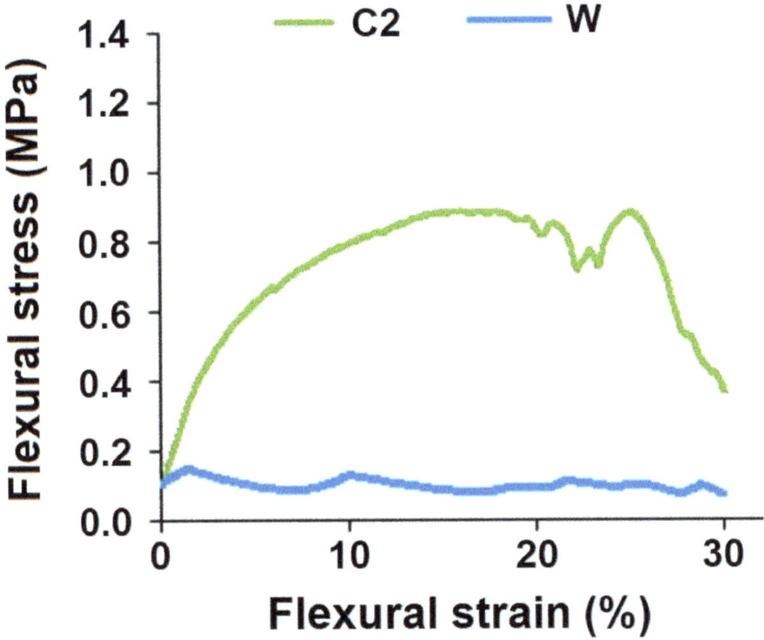

Figure 11. Flexural behavior of RC fiber-based hollow composite (C2) and cardboard material (W).

3.4. Impact Properties

Impact testing helps to understand the primary cause of failure or delamination of composites due to sudden impact of stone pellets, metal edges, rods, baggage loading, and dropping, or even during maintenance. Fiber mechanical properties have a very significant influence on impact properties of the textile-reinforced composites since these are the primary load bearing elements. These reinforcing elements absorb major portion of the energy during impact. The impact properties of the composite materials are directly related to their toughness [37–40]. Unnotched specimens of GF-reinforced hollow composites were not at all broken during the Charpy test in the present investigation. A comparative account of impact energy absorbed by the hollow composite specimens developed from RC fiber and glass fiber, as well as double-layered cardboard, is given in Figure 12.

Figure 12 indicates a significant increase in impact energy absorbed by the specimens with increase in tube diameter/thickness. This behavior is observed in the case of RCF-based, as well as GF-based, hollow composites. Such improvement in impact energy absorption with increasing tube diameter is in striking contrast to the results obtained in compression and bending tests. In general, impact energy absorbed by a sample depends upon its ability to deform/extend over a longer period of time and, thus, to absorb the impact energy for the total work done during this deformation. Among the recycled cotton fiber-based samples, impact energy absorption for C3 is 23.3% higher than C2 and 55.2% higher than C1, respectively. Similarly, among the glass fiber-reinforced hollow composites, impact energy absorbed increases with an increase in tube diameter. Sample G3 shows 25.3% higher impact energy absorption as compared to sample G2, and 43.1% higher absorption than G1. Such behavior can be explained in terms of the flattening and ovalization of hollow composites having higher diameter of the tubes. By such structural deformations, the hollow composites with higher thickness/diameter of tubes are able to absorb higher amount of energy exerted by the impactor. On the other hand, samples with smaller tube diameter proved to be stiffer and less capable of absorbing impact energy. The observations of compressional and flexural performance support such behavior. A sample which is relatively more compressible (lower resistance to compression) and relatively less stiff proves to be more efficient absorber of impact energy. The maximum absorption of the impact energies by samples C3 and G3 in their respective groups is because of higher tube diameters, which offer higher deformation/elongation before attaining peak stress. The higher impact energy is also attributed to the change in momentum of the hammer. Samples C3 and G3 having higher tube diameter provide longer impact time before peak force, hence absorbing the highest amount of energy. The behavior can be traced back to the hardness results given in Table 2. In order to absorb higher impact energy, the material needs to be softer and allow higher deformation under impact. However, the hardness is a surface property and depicts resistance to localized surface deformation only. A softer surface initiates the energy absorption, which is further enhanced with higher tube thickness/diameter. On comparing the results of the hardness tests with impact tests, a negative trend is obtained. Higher surface hardness is associated with lower absorption of impact energy. Similar results are also reported in the literature [27]. However, further deeper analysis will be needed in this aspect.

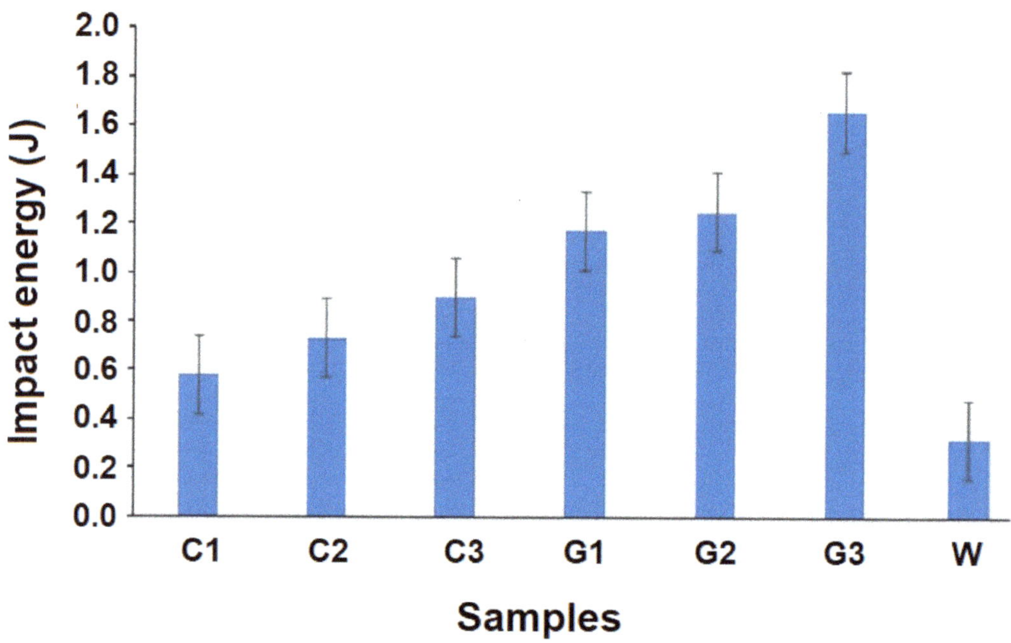

Figure 12. Impact energy absorbed by RC and glass fiber-based hollow composites versus double-layered cardboard.

In general, glass fiber-based hollow composites exhibit higher impact energy absorption as compared to the recycled cotton fiber-reinforced composite samples. This is mainly due to higher tensile strength and modulus of glass fibers, as well as the higher fiber volume fraction with respect to the recycled cotton fiber-based hollow composites. The Halpin-Tsai model based on the rule of mixture can be successfully used to predict such performance in fiber-reinforced composites [28].

All the knitted hollow composites developed exhibit substantially higher absorption of impact energy as compared to the cardboard-based packaging material. The cardboard-based packaging material is composed of weaker cellulosic material, which has relatively inferior mechanical performance. On the other hand, fiber-reinforced composite structures, and more specifically knitted hollow composites, are very efficient cushioning materials, which exhibit an efficient absorption mechanism [41,42]. It can be noted that the cardboard materials completely broke down and were destroyed after the impact test. Therefore, such conventional packaging materials cannot be reused, whereas the hollow composite samples did not break completely. While recycled cotton fiber-reinforced samples developed minor cracks in some cases, glass fiber-based samples showed absolutely no sign of any structural damage. They can probably be used in even higher risk packaging purposes as aircraft components, military equipment, automotive parts, sports equipment, etc. The use of recycled cotton fibers is a sustainable and ecofriendly approach, while reducing the environmental burden of cardboard-paper-based packaging materials.

The ovalization effect after the impact test in the case of RC fiber- and glass fiber-reinforced hollow composite is shown in Figure 13.

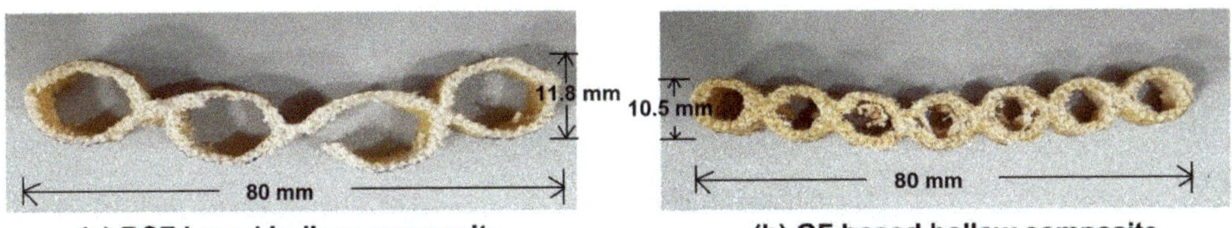

Figure 13. Ovalization after impact testing in (**a**) recycled cotton fiber-based hollow composites and (**b**) glass fiber-based hollow composites.

Though the RC fiber-based hollow composites shown in Figure 13a develop some cracking at a few joints, the glass fiber-reinforced composite sample in Figure 13b is completely undamaged.

4. Conclusions

In the current study, an attempt has been made to develop knitted hollow composites by using recycled cotton fibers and glass fibers. Composite samples developed were subjected to hardness test, compression, flexural, and impact loading. Glass fiber-reinforced hollow composites exhibit substantially higher compression strength as compared to RC fiber-based composites. However, RC fiber-reinforced hollow composites show higher compression modulus as compared to glass fiber-based samples due to shorter fiber length, which enables much lower deformation during compression loading.

Compression strength of both RC- and glass fiber-reinforced composites decreases with increasing tube diameter. The RCF-based hollow composites were compared with a commercial cardboard-based packaging material of equivalent thickness. Substantially higher compression strength, as well as compression modulus, was observed in RCF-based hollow composites as compared to the double-layered cardboard packaging material of similar thickness. No brittle failure was observed during the flexural test, and samples with smaller tube diameter exhibited higher stiffness. RC fiber-reinforced hollow composites show substantially higher flexural stiffness as compared to double-layered cardboard material. Significant increase in absorption of impact energy was achieved by the specimens with higher tube diameter. The ovalization and flattening effect enable higher absorption of impact energy. The RC fiber-based hollow composites absorb much higher impact energy as compared to the cardboard-based packaging material. The findings are in accordance with previous research and theoretical models based on the rule of mixture, as well as the Halpin-Tsai model.

RC fiber-based hollow composite is proven as a replacement for paperboard packaging material in order to utilize industrial waste and reduce the environmental pollution. Theses composites can be used as separation/packaging material for heavy goods. Recycled cotton is more appropriate and ecofriendly alternative for cardboard-based packaging. In addition, such composites can be used repeatedly as their performance and durability are much higher than cardboard packaging materials.

Glass fiber-based hollow composites can be used in relatively higher load bearing applications, e.g., separator for offices, kiosks, boats, and light weight shelters. Its sound insulation properties can also be improved by inserting porous sound absorbing material inside the tubes. These tubes can be filled with foam or any honeycomb structures for construction elements in body parts of lightweight electric vehicles, which can survive crash/impact to reasonably a higher extent. In composite structures, the honeycomb core plays a vital role in energy absorption properties. In the case of hollow structures, the fabric consists of one or more layers of triangular, trapezoidal, or hexagonal cross-sectional shapes, which are self-opening. These geometrical variations will further enhance the applicability of hollow composites in several engineering applications. This research opens new directions for further investigation on hollow composites with different core geometries and core fillings.

Author Contributions: H.J., R.M., M.Z., B.Z., S.A.B., M.T. and M.M. experiment design; H.J., R.M., M.Z., B.Z., S.A.B. and M.T. testing of mechanical properties and data analysis; H.J., R.M., M.Z., B.Z., S.A.B., M.T. and M.M. wrote and edited the paper; H.J., R.M. and M.M. language correction; R.M. communication with the editors; H.J., R.M., M.T. and M.M. administration and funding. All authors have read and agreed to the published version of the manuscript.

Funding: The work was supported by the internal grant agency of the Faculty of Engineering, Czech University of Life Sciences Prague (no. 2021:31140/1312/3108).

Institutional Review Board Statement: Not applicable.

Informed Consent Statement: Not applicable.

Data Availability Statement: Not applicable.

Conflicts of Interest: The authors declare no conflict of interest.

References

1. Ramakrishna, H.; Priya, S.; Rai, S. Effect of fly ash content on impact, compression, and water absorption properties of epoxy toughened with epoxy phenol cashew nutshell liquid-fly ash composites. *J. Reinf. Plast. Compos.* **2006**, *25*, 455–462. [CrossRef]
2. Jamshaid, H.; Mishra, R.; Pechociakova, M.; Noman, M.T. Mechanical, thermal and interfacial properties of green composites from basalt and hybrid woven fabrics. *Fibers Polym.* **2016**, *17*, 1675–1686. [CrossRef]
3. Sargianis, J.; Kim, H.; Andres, E.; Suhr, J. Sound and vibration damping characteristics in natural material based sandwich composites. *Compos. Struct.* **2013**, *96*, 538–544. [CrossRef]
4. Mohanty, A.; Misra, M.; Drzal, L. *Natural Fibers, Biopolymers and Biocomposites*; CRC Press: Boca Raton, FL, USA, 2005.
5. Joserph, P.; Joseph, K.; Thomas, S. Effect of processing variables on the mechanical properties of sisal-fiber-reinforced polypropylene composites. *Compos. Sci. Technol.* **1999**, *59*, 1625–1640. [CrossRef]
6. Sakthivel, S.; Kumar, S.; Melese, B.; Mekonnen, S.; Solomon, E.; Edae, A.; Abedom, F.; Gedilu, M. Development of nonwoven composites from recycled cotton/polyester apparel waste materials for sound absorbing and insulating properties. *Appl. Acoust.* **2021**, *180*, 108126. [CrossRef]
7. Mishra, R.; Behera, B.K. Recycling of textile waste into green composites: Performance characterization. *Polym. Compos.* **2014**, *35*, 1960–1967. [CrossRef]
8. Sezgin, H.; Kucukali-Ozturk, M.; Berkalp, O.B.; Yalcin-Enis, I. Design of composite insulation panels containing 100% recycled cotton fibers and polyethylene/polypropylene packaging wastes. *J. Clean. Prod.* **2021**, *304*, 127132. [CrossRef]
9. Mishra, R.; Huang, J.; Kale, B.; Zhu, G.; Wang, Y. The production, characterization and applications of nanoparticles in the textile industry. *Text. Prog.* **2014**, *46*, 133–226. [CrossRef]
10. Li, H.; Li, Z.; Liu, L. Flax/PP weft-knitted thermoplastic composites and its tensile properties. *J. Reinf. Plast. Compos.* **2010**, *29*, 1820–1825. [CrossRef]
11. Hoffmann, G.; Diestel, O.; Cherif, O. Thermoplastic composite from innovative flat knitted 3D multi-layer spacer fabric using hybrid yarn and the study of 2D mechanical properties. *Compos. Sci. Technol.* **2010**, *70*, 363–370. [CrossRef]
12. Harte, A.; Fleck, N. Deformation and failure mechanisms of braided composite tubes in compression and torsion. *Acta Mater.* **2000**, *48*, 1259–1271. [CrossRef]
13. Mishra, R.; Gupta, N.; Pachauri, R.; Behera, B.K. Modelling and simulation of earthquake resistant 3D woven textile structural concrete composites. *Compos. Part B Eng.* **2015**, *81*, 91–97. [CrossRef]
14. Beard, S.; Chang, F. Energy absorption of braided composite tubes. *Int. J. Crashworthiness* **2002**, *7*, 191–206. [CrossRef]

15. Ziegmann, G.; Dickert, M.; Cristaldi, G. Properties and performances of various hybrid glass/natural fiber composites for curved pipes. *Mater. Des.* **2009**, *30*, 2538–2542. [CrossRef]
16. Wu, X.; Zhang, Q.; Zhang, W. Axial compression deformation and damage of four-step 3-D circular braided composite tubes under various strain rates. *J. Text. Inst.* **2016**, *107*, 1584–1600. [CrossRef]
17. Hu, D.; Luo, M.; Yang, J. Experimental study on crushing characteristics of brittle fiber/epoxy hybrid composite tubes. *Int. J. Crashworthiness* **2010**, *15*, 401–412. [CrossRef]
18. Ahmed, M.; Hoa, S. Flexural stiffness of thick-walled composite tubes. *Compos. Struct.* **2016**, *149*, 125–133. [CrossRef]
19. Yan, A.; Jospin, R.; Nguyen, D. An enhanced pipe elbow element application in plastic limit analysis of pipe structures. *Int. J. Numer. Meth. Eng.* **1999**, *46*, 409–431. [CrossRef]
20. Bathe, K.; Almeida, C. A simple and effective pipe elbow element-linear analysis. *J. Appl. Mech.* **1980**, *47*, 93–100. [CrossRef]
21. Qi, X.; Jiang, S. Design and analysis of a filament wound composite tube under general loadings with assistance of computer. In Proceedings of the 2nd International Conference on Education Technology and Computer (ICETC), Shanghai, China, 22–24 June 2010.
22. Xu, D.; Derisi, B.; Hoa, S. Stress distributions of thermoplastic composite tubes subjected to four-point loading. In Proceedings of the 1st Joint Canadian-American International Conference, Delaware, DE, USA, 15–17 September 2009.
23. Available online: https://www.shimaseiki.com/product/design/virtual_sampling/ (accessed on 19 July 2021).
24. Almeida, J., Jr.; Ribeiro, M.L.; Tita, V.; Amico, S.C. Damage modeling for carbon fiber/epoxy filament wound composite tubes under radial compression. *Compos. Struct.* **2017**, *160*, 204–210. [CrossRef]
25. Mishra, R.; Behera, B.K.; Mukherjee, S.; Petru, M.; Muller, M. Axial and radial compression behavior of composite rocket launcher developed by robotized filament winding: Simulation and experimental validation. *Polymers* **2021**, *13*, 517. [CrossRef]
26. Hassan, T.; Jamshaid, H.; Mishra, R.; Khan, M.Q.; Petru, M.; Novak, J.; Choteborsky, R.; Hromasova, M. Acoustic, mechanical and thermal properties of green composites reinforced with natural fibers waste. *Polymers* **2020**, *12*, 654. [CrossRef] [PubMed]
27. Sarr, M.M.; Inoue, H.; Kosaka, T. Study on the improvement of interfacial strength between glass fiber and matrix resin by grafting cellulose nanofibers. *Compos. Sci. Technol.* **2021**, *211*, 108853. [CrossRef]
28. Halpin, A.J.C.; Kardos, J.L. The halpin-tsai equations: A review. *Polym. Eng. Sci.* **1976**, *16*, 344–352.
29. Liu, Q.; Xu, X.; Ma, J. Lateral crushing and bending responses of CFRP square tube filled with aluminum honeycomb. *Compos. Part B Eng.* **2017**, *118*, 104–115. [CrossRef]
30. Hamidon, M.; Sultan, M.; Ariffin, A. Investigation of Mechanical Testing on Hybrid Composite Materials. In *Failure Analysis in Biocomposites, Fiber-Reinforced Composites and Hybrid Composites*; Elsevier: Amsterdam, The Netherlands, 2019; pp. 133–156.
31. Lu, Z.; Jing, X.; Sun, B.; Gu, B. Compressive behaviors of warp-knitted spacer fabrics impregnated with shear thickening fluid. *Compos. Sci. Technol.* **2013**, *88*, 184–189. [CrossRef]
32. Asayesh, A.; Amini, M. The effect of fabric structure on the compression behavior of weft-knitted spacer fabrics for cushioning applications. *J. Text. Inst.* **2020**, 1–12, Online First. [CrossRef]
33. Liu, Y.; Hu, H.; Zhao, L.; Long, H. Compression behavior of warp-knitted spacer fabrics for cushioning applications. *Text. Res. J.* **2012**, *82*, 11–20. [CrossRef]
34. Arumugam, V.; Mishra, R.; Tunak, M. In-plane shear behavior of 3D knitted spacer fabrics. *J. Ind. Text.* **2016**, *46*, 868–886. [CrossRef]
35. Arumugam, V.; Mishra, R.; Tunak, M. In plane shear behavior of 3D warp-knitted spacer fabrics: Part-II: Effect of structural parameters. *J. Ind. Text.* **2018**, *48*, 772–801. [CrossRef]
36. Arumugam, V.; Mishra, R.; Salacova, J. Investigation on thermo-physiological and compression characteristics of weft knitted 3D spacer fabrics. *J. Text. Inst.* **2017**, *108*, 1095–1105. [CrossRef]
37. Ahmed, M.M.; Dhakal, H.N.; Zhang, Z.Y.; Barouni, A.; Zahari, R. Enhancement of impact toughness and damage behaviour of natural fibre reinforced composites and their hybrids through novel improvement techniques: A critical review. *Compos. Struct.* **2021**, *259*, 113496. [CrossRef]
38. Li, F.S.; Gao, Y.B.; Jiang, W. Design of high impact thermal plastic polymer composites with balanced toughness and rigidity: Toughening with one phase modifier. *Polymer* **2019**, *170*, 101–106. [CrossRef]
39. Liu, W.; Zhang, J.Q.; Hong, M.; Li, P.; Xue, Y.H.; Chen, Q.; Ji, X.L. Chain microstructure of two highly impact polypropylene resins with good balance between stiffness and toughness. *Polymer* **2020**, *188*, 122146. [CrossRef]
40. Han, S.; Zhang, T.; Guo, Y.; Li, C.; Wu, H.; Guo, S. Brittle-ductile transition behavior of the polypropylene/ultra-high molecular weight polyethylene/olefin block copolymers ternary blends: Dispersion and interface design. *Polymer* **2019**, *182*, 121819. [CrossRef]
41. Hajjari, M.; Nedoushan, R.J.; Dastan, T.; Sheikhzadeh, M.; Yu, W.R. Lightweight weft-knitted tubular lattice composite for energy absorption applications: An experimental and numerical study. *Int. J. Solid Struct.* **2021**, *213*, 77–92. [CrossRef]
42. Khondker, O.A.; Leong, K.H.; Herszberg, I.; Hamada, H. Impact and compression-after-impact performance of weft-knitted glass textile composites. *Compos. Part A Appl. Sci. Manuf.* **2005**, *36*, 638–648. [CrossRef]

 *polymers*

Article

Experimental Investigation of Wavy-Lap Bonds with Natural Cotton Fabric Reinforcement under Cyclic Loading

Viktor Kolář [1], Miroslav Müller [1], Martin Tichý [1], Rajesh Kumar Mishra [1,*], Petr Hrabě [1], Kristýna Hanušová [1] and Monika Hromasová [2]

[1] Department of Material Science and Manufacturing Technology, Faculty of Engineering, Czech University of Life Sciences Prague, Kamycka 129, Suchdol, 16500 Prague, Czech Republic; vkolar@tf.czu.cz (V.K.); muller@tf.czu.cz (M.M.); martintichy@tf.czu.cz (M.T.); hrabe@tf.czu.cz (P.H.); hanusovakristyna@tf.czu.cz (K.H.)

[2] Department of Electrical Engineering and Automation, Faculty of Engineering, Czech University of Life Sciences Prague, Kamycka 129, Suchdol, 16500 Prague, Czech Republic; hromasova@tf.czu.cz

* Correspondence: mishrar@tf.czu.cz

Abstract: This study is focused on the mechanical properties and service life (safety) evaluation of hybrid adhesive bonds with shaped overlapping geometry (wavy-lap) and 100% natural cotton fabric used as reinforcement under cyclic loading using various intensities. Cyclic loading were implemented between 5–50% (267–2674 N) and 5–70% (267–3743 N) from the maximum strength (5347 N) measured by static tensile test. The adhesive bonds were loaded by 1000 cycles. The test results demonstrated a positive influence of the used reinforcement on the mechanical properties, especially during the cyclic loading. The adhesive bonds Tera-Flat withstood the cyclic load intensity from 5–70% (267–3743 N). The shaped overlapping geometry (wavy-lap bond) did not have any positive influence on the mechanical performance, and only the composite adhesive bonds Erik-WH1 and Tera-WH1 withstood the complete 1000 cycles with cyclic loading values between 5–50% (267–2674 N). The SEM analysis results demonstrated a positive influence on the fabric surface by treatment with 10% NaOH aqueous solution. The unwanted compounds (lignin) were removed. Furthermore, a good wettability has been demonstrated by the bonded matrix material. The SEM analysis also demonstrated micro-cracks formation, with subsequent delamination of the matrix/reinforcement interface caused by cyclic loading. The experimental research was conducted for the analysis of hybrid adhesive bonds using curved/wavy overlapping during both static and cyclic loading.

Keywords: quasi-static test; cyclic fatigue; wavy-lap bond; natural cotton fabric; polymer composite; mechanical properties; service life; safety; SEM

Citation: Kolář, V.; Müller, M.; Tichý, M.; Mishra, R.K.; Hrabě, P.; Hanušová, K.; Hromasová, M. Experimental Investigation of Wavy-Lap Bonds with Natural Cotton Fabric Reinforcement under Cyclic Loading. *Polymers* **2021**, *13*, 2872. https://doi.org/10.3390/polym13172872

Academic Editor: Francesco Mollica

Received: 5 August 2021
Accepted: 25 August 2021
Published: 26 August 2021

Publisher's Note: MDPI stays neutral with regard to jurisdictional claims in published maps and institutional affiliations.

1. Introduction

Adhesive bonding technology represents one of the most promising methods of material bonding. This technology finds its application in automotive, aviation, and electrotechnical industries [1,2]. The dynamic development of adhesive bonding technology is demonstrated by the wider possibilities offered by this process as compared to the conventional technologies of bonding materials (welding, soldering, etc.). Significant advantages are observed as compared to conventional technologies in a wide spectrum of bonding materials, along with lower component costs and lower labor requirements [3]. Adhesive technology could also fulfill supporting roles, such as sealing, clamping, and securing [4,5]. Currently, there are plenty of research opportunities dealing with adhesive bonding. Their aim is to improve the efficiency of using such material under loading conditions. The majority of research deals with the static strength of adhesive bonds [6–8]. The mechanical properties of adhesive bonds could be influenced by physical and chemical factors (wettability, adhesion and cohesion, aging, and environmental degradation) [4,9,10], technological factors (roughness and structure of bonded surface and filler material) [11–13], and constructional factors (overlapping length and geometry and type of applied load) [14–18]. The resulting performance of the adhesive bond is governed by the synergy of these factors, i.e., the effect of their mutual interaction [19].

Shaped overlapping geometry is one of the factors that could positively influence the adhesive bond strength and the internal stress [20,21]. Another reason for using shaped overlapping geometry is to address a more complex constructional requirement of the adhesive systems when non-flat-lap bonds are used. Many researchers have dealt with shaped overlapping geometry. Zeng and Sun [22] came up with a solution of wavy-lap bonds and detected an increase in shear strength as compared to flat-lap bonds during a static test. Ávila and Bueno [23] conducted a similar experiment and detected an increase of 41% in shear strength under the static test. Müller [24] tested the influence of various adhesive types on the strength of wavy-lap bonds. A number of researchers are devoted to modification of various types of wavy-lap bonds. Jaiswal et al. [25] tested adhesive bonds with teeth of different depth created on the lap surface to increase the static tensile strength. Haghpanah et al. [26] tested adhesive bonds with different adherend geometry using positive and negative teeth. Razavi et al. [27] dealt with sinusoidal geometry of adherend lapping in their research.

The wettability of natural fiber reinforcement is a significant factor influencing its bonding properties. Deteriorated wettability (adhesion) of natural-fiber-based reinforcements, which usually decreases the shear strength of adhesive bonds, leads to significant disadvantages for their use in the polymer composites [28–32]. Deteriorated wettability of such natural fiber surfaces could be minimized by chemical treatment in aqueous solution of NaOH, plasma treatment of its surface, or by other methods. [33]. Alkali treatment with NaOH solution improves the surface structure of the reinforcement. The improvement is caused by removal of unwanted layers, e.g., lignin, oils, and fats, from the reinforcement fiber surface [34–36]. The surface treatment leads to improvement of interaction at the interphase boundary, i.e., on the interface of the natural reinforcement and matrix [37]. It leads to improvement of mechanical properties, especially the shear strength of adhesive bonds [19,35,38].

During application, the bonded materials are loaded, not only under static condition, but also by cyclic loading. A number of studies have dealt with cyclic loading of different intensities in the field of fiber–polymer composites [39,40]. With adhesive bonds, it cannot be expected that quality will be preserved throughout their service life. Operating conditions usually include the action of the cyclic loading, i.e., cyclic fatigue. Cyclic fatigue is characterized by propagation of cracks inside of the adherend and subsequent permanent damage to the adhesive bonds [41]. The process itself leads to relatively lower values of cyclic loading due to delamination between adherend and bonded material, which negatively influences the service life of the adhesive bonds [42]. The strength and fatigue service life of adhesive bonds are even lower at smaller numbers of repeating cycles. The tests of cyclic loading are essential for practical application of adhesive bonds [42–44]. The researchers demonstrated that the hybrid composite layer of adhesive bonding can positively influence the mechanical properties and extend their service life under cyclic loading [19,35,45].

The experimental research was mainly focused on the hybrid adhesive bonds with shaped overlapping geometry (wavy-lap) and 100% natural cotton fabric as reinforcement. Adhesive bonds were exposed to cyclic loading of various intensities, and the results of mechanical properties and service life (safety) were evaluated. Cyclic loading (cyclic fatigue) represents a common cause of failure in adhesive bonds due to delamination of reinforcement and the matrix. Based on previous research to achieve optimum results for mechanical properties and service life during cyclic loading, the bonding materials and procedure were chosen. That included selection of 100% cotton fabric as reinforcement [19]. The previous studies have focused on flat geometries, while some of the real applications are in the form of curved (wavy) shapes.

The aim of this study's research was to evaluate the influence of the shaped adherend geometry (wavy-lap) and reinforcing natural cotton fabric with modified surface. 10% aqueous solution of NaOH was used for pretreatment of the cotton fabrics. Mechanical properties (tensile strength, deformation-strain, modulus of elasticity) and related service life and safety of the hybrid adhesive bonds with composite layer of adhesive was evaluated by cyclic loading of various intensities. Selected mechanical parameters provide an overview of the behavior of adhesive bonds and cyclic loading to approximate realistic loading conditions and their subsequent application.

2. Materials and Methods

2.1. Materials

2.1.1. Bonded Material (Adherend)

Structural carbon steel S235J0 (Ferona a.s., Prague, Czech Republic) with dimensions 1 mm thickness, 100 ± 0.25 mm length, and 25 ± 0.25 mm width was used as an adherend. The adherend dimensions were established by the ČSN EN 1465 standard [46]. Basic mechanical properties and indicative chemical composition are listed in Tables 1 and 2.

Table 1. Basic mechanical properties of the S235J0 steel at 20 °C temperature [47].

Tensile Strength	340–470 MPa
Yield strength	225–235 MPa
Elastic modulus	212 GPa
Elongation	24%

Table 2. Indicative chemical composition of S235J0 steel.

C (%)	Mn (%)	P (%)	S (%)	Cu (%)	N (%)	Fe (%)
≤ 0.19	≤ 1.50	≤ 0.04	≤ 0.04	≤ 0.60	≤ 0.014	≤ 99.55

The shaping of the adherends (height h of wavy-lap bonds) was achieved by using a pressing form (Figure 1). The adherends were placed in a form, and by using 850 N (F) force, the shaped geometry with different wave heights $h_1 = 2.43 \pm 0.10$ mm (marked as WH1) and $h_2 = 4.82 \pm 0.13$ mm (marked as WH2) was obtained. The types of adherends and principle of measurement for the height (h) of a wave are shown in Figure 2.

Figure 1. Pressing form for the shaping of adherends.

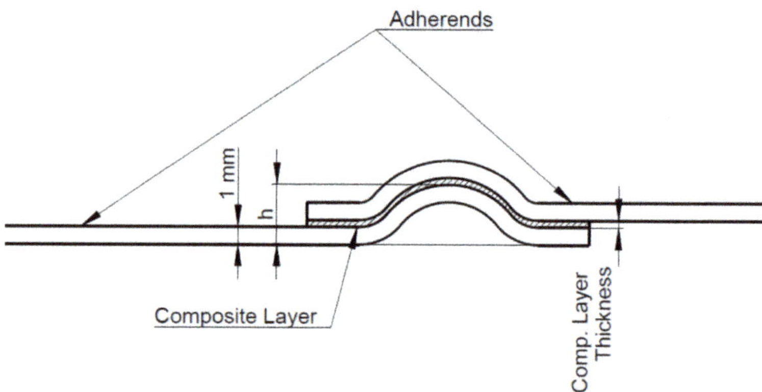

Figure 2. The types of shaped adherends and the measurement principle for height *(h)* of a wave.

The surface of the adherends was mechanically treated in a blasting cabin using abrasive Garnet MESH 80 and then chemically treated in an acetone bath just before bonding. These methods for surface treatments were proven as optimal in terms of mechanical properties of adhesive bonds by several studies [45,48]. The roughness of adherends' surfaces were measured using profilometer Mitutoyo Surftest 301 (Mitutoyo Europe GmbH, Neuss, Germany). Value Ra = 3.65 ± 0.12 μm and Rz = 11.19 ± 0.37 μm.

2.1.2. Matrix and Reinforcement

2 types of 100% natural cotton fabric were used as reinforcement. Their characteristics are listed in Table 3.

Table 3. Reinforced fabric characteristics [46].

Fabric	Geometry	Areal Density	Warp-Way Strength (200 × 50 mm)	Weft-Way Strength (200 × 50 mm)
		$g \times m^{-2}$	N	N
Tera	Plain	290	950	900
Erik	Plain	190	850	800

The surface of natural cotton fabric was alkali-treated before application of the adhesive layer. Treating the surface leads to improvement of wettability and thus improves the performance of the bond, mainly its strength [33,49,50]. The following steps were used for the surface treatment:

1. Soaking the fabrics in hot water (100 °C) for removal of starch;
2. Rinsing with cold water for removal of residual impurity;
3. Soaking the fabrics in 10% NaOH solution for 30 min. Distilled water was used to create the solution;
4. Repeated washing of the alkali-treated fabrics with cold water;
5. Drying the fabric in a laboratory oven at 105 °C temperature for 24 h [51].

Structural two-component epoxide resin CHS-Epoxy 324 (Epoxy 1200) (Havel Composites CZ s.r.o., Svésedlice, Czech Republic) with P11 hardener (Havel Composites CZ s.r.o., Svésedlice, Czech Republic) was used as a matrix (in weight ratio 100:7 according to the manufacturer's recommendation). According to the manufacturer, resin is suitable for metal bonding [52].

2.1.3. Preparation of Adhesive Bonds

The research was based on modified norm ČSN EN 1465. The norm ČSN EN 1465 establishes the lapping length to be 12.5 ± 0.25 mm. The length of the lapping was based on shaped adherend geometry and was identical for all of the adhesive bond types (29 ± 1.31 mm) so that the results could be compared. The bonds were loaded with 750 g (7.4 N) weights and left to be hardened at 21 ± 2 °C laboratory temperature and 45 ± 7% relative air humidity for 24 h. The adhesive layer thickness was measured using Gwyddion software (version 2.49, David Nečas and Petr Klapetek, VUT Brno, Brno) from scanning electron microscope (SEM) images. Type, shape, and adhesive layer thickness of the hybrid adhesive bonds are listed in Table 4.

Table 4. Types, shape, and adhesive layer thickness of the hybrid adhesive bonds.

Bond Type	Adherend Geometry (Shape)	Adhesive Layer Thickness (μm)	Characteristics
Resin	Flat	33 ± 3	Adhesive bonds with pure resin and flat shape, WH1 and WH2
	WH1		
	WH2		
Erik	Flat	432 ± 12	Adhesive bonds with composite layer with Erik fabric and flat shape, WH1 and WH2
	WH1		
	WH2		
Tera	Flat	614 ± 9	Adhesive bonds with composite layer with Tera fabric and flat shape, WH1 and WH2
	WH1		
	WH2		

2.2. Methods

The testing of mechanical properties was realized on a universal testing machine LABTest 5.50 ST (LABORTECH s.r.o., Opava, Czech Republic) with measuring unit AST KAF 50 kN (LABORTECH s.r.o., Opava, Czech Republic) and evaluation software Test & Motion (version 4.5.0.15, LABORTECH s.r.o., Opava, Czech Republic) at 21 ± 1 °C laboratory temperature and 44 ± 4% relative air humidity. The testing of mechanical properties during cyclic loading, i.e., tensile strength and extension upon rupture, was based on setting the standard value obtained during static tensile test (ČSN EN 1465) consisting of 7 adhesive bonds marked as Resin-Flat with testing speed 0.6 mm × min^{-1}. The testing speed during static test was chosen on the basis of the ČSN EN 1465 standard, which defines the test duration in the interval of 60 ± 2 s.

The maximum average load of 5347 ± 157 N (average value from 7 Resin-Flat adhesive bonds) was obtained. Cyclic loading (quasi-static test) consisted of 1000 cycles with testing speed 6 mm × min^{-1} within the limits of 5%, 50% and 70% of maximum load. The lower limit was 5% = 267 N and the upper limit was 50% and 70% from the maximum load, i.e., 50% = 2674 N and 70% = 3743 N. The testing speed during the cyclic test was chosen based on the characteristics of cyclic loading, which often results in sharp fluctuations in its intensity. For this reason, the test speed was higher than for static tests. The time delay between lower and upper limit was set for 0.5 s. When 1000 cycles were finished, a static tensile test automatically followed and ran until complete failure of adhesive bond with 0.6 mm × min^{-1} speed. Static test was only realized if 1000 cycles were finished. If they were not, the test was concluded. Every testing sequence consisted of 7 testing samples.

The analysis of variance was used to evaluate the executed experiments, i.e., ANOVA-F test in STATISTICA 12 (version 12, StatSoft CR, Prague, Czech Republic) program. The Resin-Flat was set as the reference. The statistical dependency of 0.05 limit (95% confidence interval) between average and each experiment variant was evaluated. The null hypothesis H_0 presents a statistically insignificant difference between measured data ($p > 0.05$). Alternative hypothesis H_1 rejects null hypothesis H_0 and presents statistically significant difference between measured data ($p < 0.05$).

Hybrid layer of adhesive bonds was evaluated using scanning electron microscope MIRA 3 TESCAN GMX SE (Tescan Brno s.r.o., Brno, Czech Republic). The interaction at interphase boundary between reinforcement/matrix and adherend/composite layer was evaluated. Samples were coated with gold using Quorum Q150R ES (Tescan Brno s.r.o., Brno, Czech Republic) device for the microscopy.

3. Results and Discussion

Strength of adhesive bonds depends on many factors. An important factor is the overlapping length, i.e., the area that conveys adhesion stress. It is not possible to apply a random amount of adhesive layer during practical application. This restriction is due to the increasing weight, constructional limitations, and shape complexity of the final product. This study focuses on surface modification before adhesive bonding using forming, specifically, adherend forming using a specific angle. Eventually, a wavy profile forming on the surface helped the wetting of adhesive become more efficient [24,53–58]. The wavy geometrical shape of a bonded surface usually had a positive effect on the tensile strength of adhesive bonds [24,53–58]. However, the results did not demonstrate a significant influence of geometry of adhesive bonded surface by using two types of bonded material. A significant improvement was observed by using reinforcing cotton fabric in the hybrid adhesive layer. This was demonstrated by an increase in service life of the adhesive bond during low-cycle fatigue, an essential aspect for adhesive bond application.

Adhesive bonds were initially evaluated by static tensile test. The mechanical properties of adhesive bonds (Resin, Erik, and Tera) and different lapping construction (Flat, WH1, and WH2) under static tests are listed in Table 5. The influence of the shape change along with the reinforcement fabrics on mechanical properties is described based on their dependency in Figures 3–5 where the data are compared to the result of Resin-Flat bonds.

Table 5. Results of static tensile tests of adhesive bonds and statistical evaluation of data (*p*-value).

Adhesive Bond		Static Test					
		Tensile Strength		Strain		Modulus of Elasticity	
	Shape	MPa	*p*-value	%	*p*-value	MPa	*p*-value
Resin	Flat	7.38 ± 0.22	-	14.30 ± 1.88	-	52.46 ± 6.65	-
	WH1	3.91 ± 0.23	0.01	4.51 ± 0.73	0.01	88.96 ± 16.02	0.01
	WH2	2.45 ± 0.13	0.01	3.92 ± 0.54	0.01	63.67 ± 8.14	0.01
Erik	Flat	6.53 ± 0.38	0.01	8.00 ± 1.59	0.01	83.68 ± 10.87	0.01
	WH1	5.31 ± 0.29	0.01	6.27 ± 0.65	0.01	85.18 ± 5.29	0.01
	WH2	2.99 ± 0.33	0.01	6.69 ± 3.12	0.01	52.48 ± 17.16	0.50
Tera	Flat	7.12 ± 0.74	0.22	12.03 ± 2.70	0.07	61.53 ± 10.43	0.06
	WH1	4.30 ± 0.83	0.01	4.92 ± 0.91	0.01	87.61 ± 7.91	0.01
	WH2	2.67 ± 0.43	0.01	7.16 ± 3.14	0.01	44.99 ± 19.14	0.19

The static tensile test results demonstrated a quite severe deformation, $14.3 \pm 1.88\%$, for the Resin-Flat adhesive bond, as shown in Table 5. The adhesive bond strength, however, was the highest, 7.38 ± 0.22 MPa, among all tested samples. The change of geometry from standard lapped bond Resin-Flat construction to shaped lapped bonds Resin-WH1 and Resin-WH2 did not have a positive influence on the tensile strength during static tests. As shown in Figure 3, the strength of Resin-WH1 decreased by 47% to 3.91 ± 0.23 MPa, and that of Resin-WH2 decreased by 67% to 2.45 ± 0.13 MPa.

The tensile strength of Erik-Flat decreased slightly by 10% to 6.53 ± 0.38 MPa compared to Resin-Flat. The strength in Erik-WH1 decreased by 28% to 5.31 ± 0.29 MPa. The drop in this case is not as big as by Resin-WH1. The strength in Erik-WH2 decreased by 60% to 2.99 ± 0.33 MPa. This drop was 7% lower in sample Resin-WH2. The results clearly demonstrate that the Erik fabric positively influenced the tensile strength in samples Erik-WH1 and WH2, as seen in Figure 3.

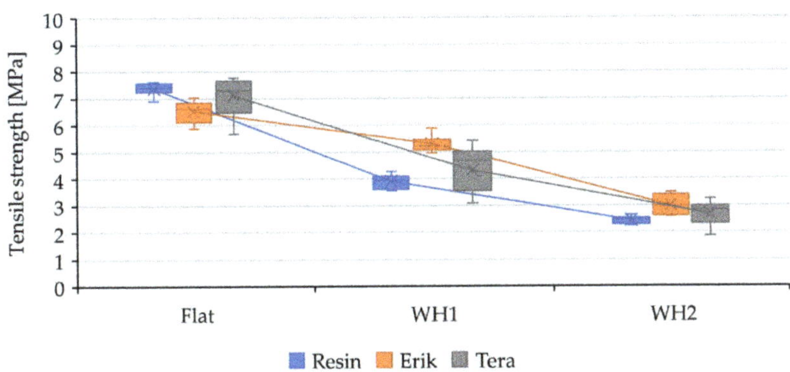

Figure 3. Evaluation of the tensile strength of adhesive bonds under static loading and their dependance on the bond shape.

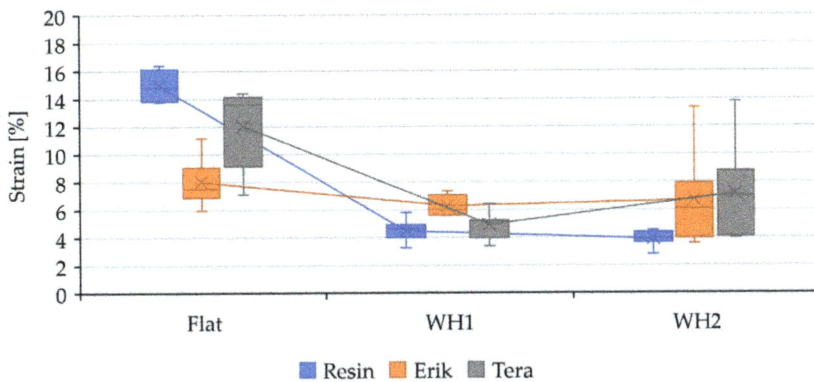

Figure 4. Evaluation of the strain in adhesive bonds under static loading and their dependance on the bond shape.

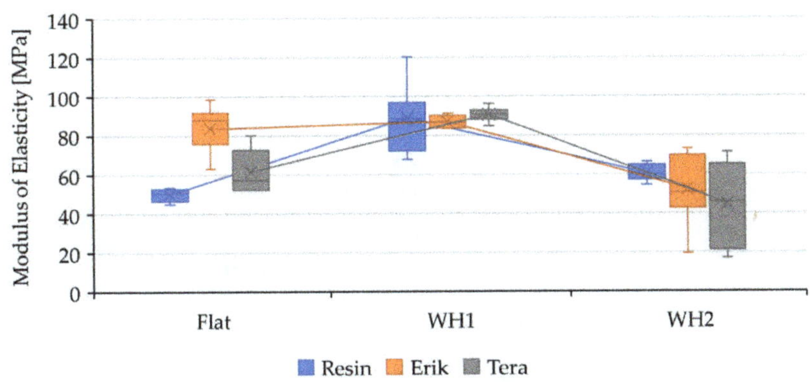

Figure 5. Evaluation of the modulus of elasticity of adhesive bonds under static loading and their dependance on the bond shape.

The Tera-Flat achieved 7.12 ± 0.74 MPa strength during static loading, which is 3% lower compared to Resin-Flat. It is, however, statistically insignificant. The strength in Tera-WH1 decreased by 42% to 4.30 ± 0.83 MPa. This drop was 5% lower than sample Resin-WH1. The strength in Tera-WH2 decreased by 64% to 2.67 ± 0.43 MPa, 3% lower than Resin-WH2. The results demonstrate that the Tera fabric slightly increased the tensile strength of Tera-WH1 and Tera-WH2, as seen in Figure 3.

The static tensile test results demonstrated quite severe deformation (strain), 14.3 ± 1.88%, in Resin-Flat, as seen in Figure 4 A severe deformation points to a rather low load-bearing capacity of the bond. Previous research also shows that an adhesive bond with such severe deformation cannot withstand cyclic loading [19]. This fact was proven by cyclic loading in 5–50% (267–2674 N) and 5–70% (267–3743 N) intervals, where the adhesive bond with pure resin did not withstand the load in any of the intervals. The fracture area showed adhesive-cohesive structure. The endurance of the resin bond was not influenced by WH1 and WH2 modification. In sample Resin-WH1, the deformation decreased to 4.51 ± 0.73%, as shown in Table 5. In Resin-WH2, the deformation again decreased to 3.92 ± 0.54%. Too-moderate deformation with moderate strength shows low endurance of the bond during cyclic loading [19]. That is why the adhesive joints did not withstand the cyclic loading.

Deformation in Erik-Flat positively decreased to 8 ± 1.59%. This drop defines an increase in the bond rigidity while maintaining strength and thus improved endurance during cyclic tests, as shown in Figures 3 and 4. The deformation in Erik-WH1 positively decreased to 6.27 ± 0.65%. Even though the construction/geometry of the bond was changed, the rigidity was preserved, resulting into endurance of the bond during cyclic loading in a 5–50% interval. Erik-WH2 showed a higher deformation, 6.69 ± 3.12%, associated with a lower strength, as shown in Figures 3 and 4. This demonstrates a lower endurance under cyclic loading.

Similar deformation occurred in the case of Tera-Flat. The observed deformation was 12.03 ± 2.70%, which is lower only by 2.3% (statistically insignificant, p-value 0.22). This small difference in deformation caused a sufficient increase in rigidity of the bond under cyclic loading, in both 5–50% and 5–70% intervals. Deformation in Tera-WH1 positively decreased to 4.92 ± 0.91%, maintaining optimal ratio between strength and deformation and thus the rigidity of adhesive bond, as shown in Figures 3 and 4. Tera-WH2 showed 7.16 ± 3.14% deformation. That is a rather huge deformation associated with a lower strength. That demonstrates a low endurance during cycling loading.

Figure 5 shows the modulus of elasticity in the bonded samples. The Resin-Flat bond exhibited a modulus of 52.46 ± 6.65 MPa. The modulus in the Erik-Flat bond increased to 83.68 ± 10.87 MPa. In the case of the Tera-Flat bond, the modulus increased to 61.53 ± 10.43 MPa. The Erik-Flat and Tera-Flat samples showed higher modulus of elasticity and thus improved performance under cyclic loading. The Resin-WH1 bond showed a modulus of 88.96 ± 16.02 MPa, while the Erik-WH1 bond showed 85.18 ± 5.29 MPa and Tera-WH1 showed 87.61 ± 7.91 MPa. In the case of wavy-shaped bond WH1, there was an increase in the modulus of elasticity. The Resin-WH2 bond exhibited a lower modulus of 63.67 ± 8.14 MPa. Erik-WH2 bond showed a modulus of 52.48 ± 17.16 MPa, and Tera-WH2 showed 44.99 ± 19.14 MPa. Wavy-shaped bond WH2 exhibited a lower modulus compared to WH1, which would affect endurance and fatigue properties under cyclic loading.

The results of cyclic mechanical tests of adhesive bonds with reinforcing fabrics Tera and Eric with different bond shapes are listed in Table 6. The results of the static tests showed that Resin-Flat, WH1, and WH2 did not withstand any intensity of cyclic loading. Wave-shaped geometries of Resin-WH1 and WH2 did not affect mechanical properties during static test positively enough to be able to resist the cyclic loading. As a result, neither of the shaped adhesive bonds with pure resin performed well during the cyclic loading in 5–50% (267–2674 N) and 5–70% (267–3743 N) intervals.

Erik-Flat withstood cyclic loading in the 5–50% interval with a moderate increase in strength to 7.13 ± 0.52 MPa, as shown in Figure 6. The deformation increased to 12.97 ± 4.06% at the same time, as shown in Figure 7. The increased deformation leads to endurance of the bond during cyclic loading. The bond did not reach the parameters high enough to withstand 5–70% load.

Table 6. Results of cyclic tensile tests of adhesive bonds in the load intervals 5–50% and 5–70%.

Adhesive Bond		Cyclic Test (5–50%)			Cyclic Test (5–70%)		
		Tensile Strength	Strain	Finished Test Samples (1000 cycles)	Tensile Strength	Strain	Finished Test Samples (1000 cycles)
	Shape	MPa	%		MPa	%	
Resin	Flat	-	-	0/7	-	-	0/7
	WH1	-	-	0/7	-	-	0/7
	WH2	-	-	0/7	-	-	0/7
Erik	Flat	7.13 ± 0.52	12.97 ± 4.06	7/7	-	-	3/7
	WH1	5.29 ± 0.32	5.99 ± 0.80	7/7	-	-	0/7
	WH2	-	-	0/7	-	-	0/7
Tera	Flat	7.45 ± 0.01	14.15 ± 2.82	7/7	7.49 ± 0.29	14.76 ± 2.41	7/7
	WH1	5.66 ± 0.40	6.79 ± 0.58	7/7	-	-	0/7
	WH2	-	-	0/7	-	-	0/7

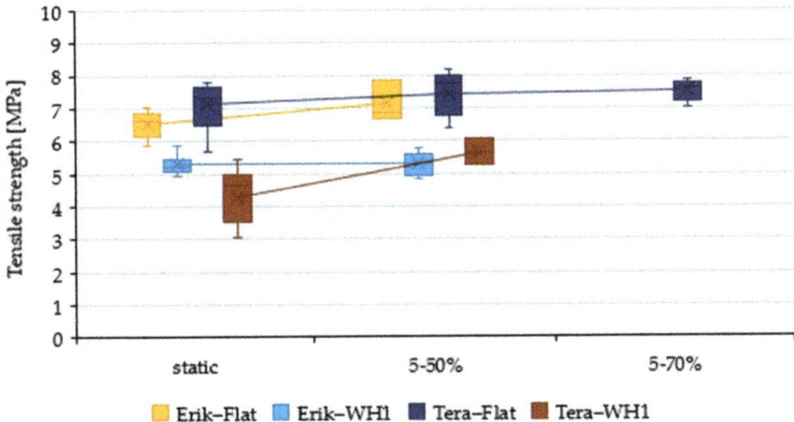

Figure 6. Evaluation of the tensile strength of the adhesive bonds under static loading and cyclic loading in the load intervals 5–50% and 5–70%.

Erik-WH1 showed strength (5.29 ± 0.32 MPa) and did not change significantly as compared to static tensile strength, as shown in Figure 6. Deformation also did not show significant changes (5.99 ± 0.80 MPa), as shown in Figure 7. Parameters were sufficient to withstand 5–50% load, but they were not sufficient to withstand 5–70% load. Due to lower strength and higher deformation, the adhesive bonds did not pass the cyclic loading. Both Tera-Flat and Tera-WH1 showed enhanced tensile strength during cyclic loading, along with reduced deformation. This shows a self-reinforcing effect, as shown in Figures 6 and 7.

Tera-Flat exhibited 7.45 ± 0.01 MPa strength and 14.15 ± 2.82% deformation during 5–50% cyclic loading. It was even higher during 5–70% cyclic loading. The strength of 7.49 ± 0.29 MPa together with 14.76 ± 2.41% deformation was observed. Strength of sample Tera-WH1 increased to 5.66 ± 0.40 MPa together with the deformation of 6.79 ± 0.58%. The bond did not withstand 5–70% cyclic loading. Tera-WH2 did not withstand any cyclic loading.

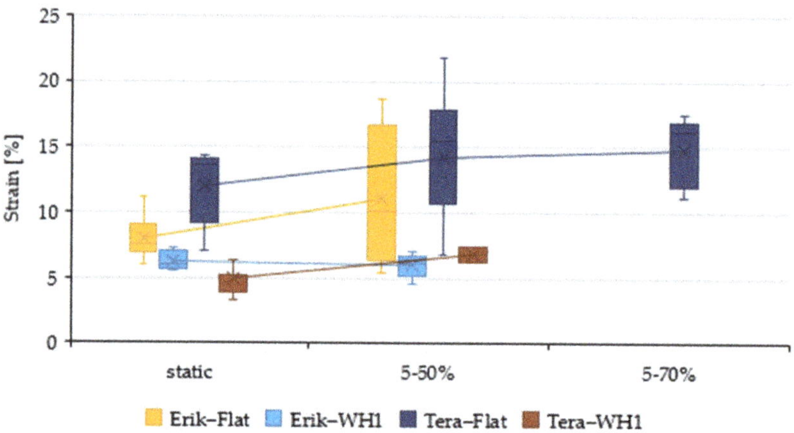

Figure 7. Evaluation of the strain of the adhesive bonds under static loading and cyclic loading in the load intervals 5–50% and 5–70%.

It is evident from Table 6 that the Erik and Tera reinforcements positively influenced the service life and therefore the safety of the adhesive bonds, especially for the bonds marked as Flat and WH1, which correspond with the modulus of elasticity results. Similar results, showing an increase in the service life and safety of the adhesive bonds under cyclic loading by the formation of a composite adhesive layer, have been found by other studies [35,59,60].

Figure 8 demonstrates viscoelastic behavior (creep) of Erik-Flat and Erik-WH1 during the 5–50% cyclic loading. It clearly shows continuous extension during cyclic loading corresponding to continuous bond fatigue. The longer the extension, the sooner the bond breaks and does not withstand the given number of cycles (1000 cycles). Figure 8 also shows that Erik-WH1 suffered longer extension, which results in lower endurance of the bond. Figure 9 demonstrates cyclic loading of Tera-Flat and Tera-WH1. The behavior is similar to Erik-WH1 (Figure 8). Tera-Flat undergoes lower extension during the cyclic loading, resulting in enhanced capacity for subsequent maximum load. Thanks to this characteristic, Tera-Flat withstood the 5–70% cyclic load.

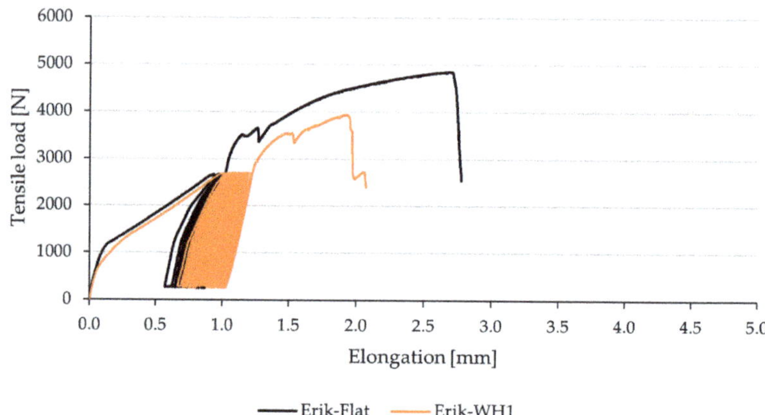

Figure 8. Viscoelastic behavior of adhesive bonds with reinforcing fabric Erik and different shapes during 5–50% cyclic load.

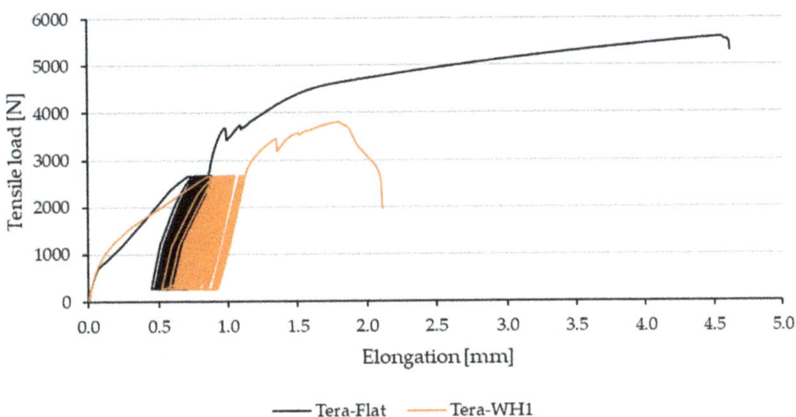

Figure 9. Viscoelastic behavior of adhesive bonds with reinforcing fabric Tera and different shapes during 5–50% cyclic load.

Figure 10A shows a microscopic view of the Erik cotton fabric that was used as reinforcement in an adhesive bond. Figure 10B,C show apparent details in microstructures of Erik fabric before and after alkali treatments, respectively. By analyzing the scanning electron microscopy (SEM) images, it was proved that alkali treatment dissolves surface layers of lignin from the cotton fibers in the fabric. The Figure 10C also shows no disintegration of the fiber bundles caused by the NaOH solution treatment. Disintegration of the fibers due to alkali treatment is negative [36,61] and may have a significant negative impact on the mechanical properties of the fibers in the fabric [62–65].

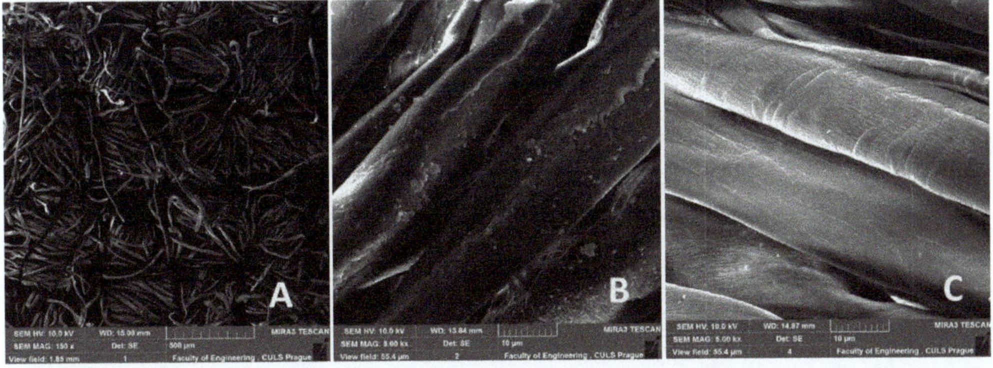

Figure 10. SEM images: (**A**): Cotton fabric Erik (MAG 150×); (**B**): Detailed look at the fabric—warp (cotton fiber) without alkali treatment (MAG 5000×); (**C**): Detailed look at the fabric—warp (cotton fiber) with alkali treatment in 10% NaOH solution for 30 min (MAG 5000×).

The cross-section of adhesive bonds presented in Figure 11 clearly shows the difference between each tested variant of adhesive bonds. It also shows the arrangement of adherent and adhesive layers in the bonded material. Adhesive layer in Figure 11A,B is composite, consisting of reinforcing cotton fabric Erik/Tera and resin (structural two-component epoxide resin). Furthermore, it demonstrates that every variant of the experiment had a different thickness of the adhesive layer, listed in Table 4. The cross-sections (Figure 11A,C) show integrity of adhesive layer, which was not exposed to the cyclic loading, and Figure 11B shows adhesive layer exposed to 1000 cycles in 5–50% intervals (267–2674 N).

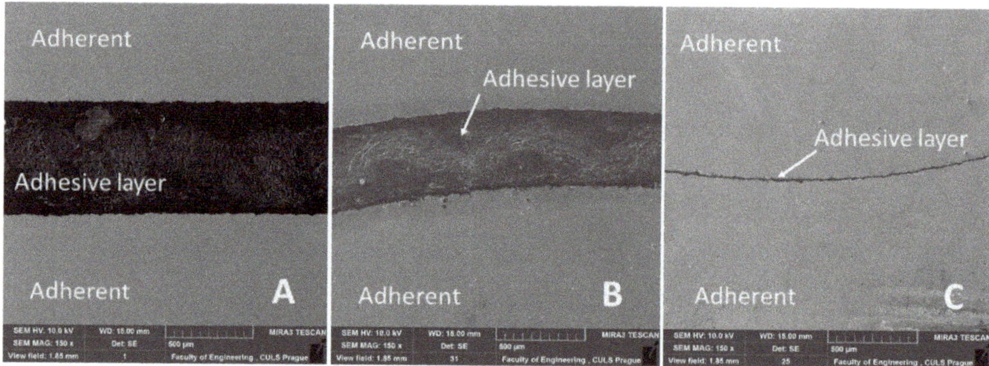

Figure 11. SEM images of samples cut through the adhesive bond: (**A**): cut through Tera-Flat, 0 cycles (MAG 150×), (**B**): cut through Erik-WH1, 1000 cycles in interval 5–50% (267–2674 N) (MAG 150×), (**C**): cut through Resin-WH2, 0 cycles (MAG 150×).

From the cross-section of the adhesive bond presented in Figure 12A, the warp and weft of Tera fabric bonded with the resin is visible. It shows the intimate interaction of resin and reinforcing fabric consisting of cotton fibers along warp and weft. Figure 12A–C and Figure 13A show good wettability of bonded material (adherent) with resin. Wettability defines the basic assumption of quality in adhesive bonds [66–68]. A detailed look at Figure 12C reveals a slight delamination at the adhesive layer and adherend boundary in identical adhesive bonds due to cyclic loading. Figure 12B shows obvious delamination due to cyclic loading in adhesive layer. Not only is the damage to adhesive layer visible, but it also shows damage to the bonded material. Delamination in any part of an adhesive bond leads to the possibility of rupture and thus damaging the integrity of the adhesive bond, leading to failure [69]. Research results reveal that Tera-Flat after treatment with 10% NaOH solution for 30 min demonstrates improved service life of adhesive bonds through 1000 cycles in intervals 5–50% (267–2674 N) and also in 5–70% (267–3743 N), as shown in Table 5.

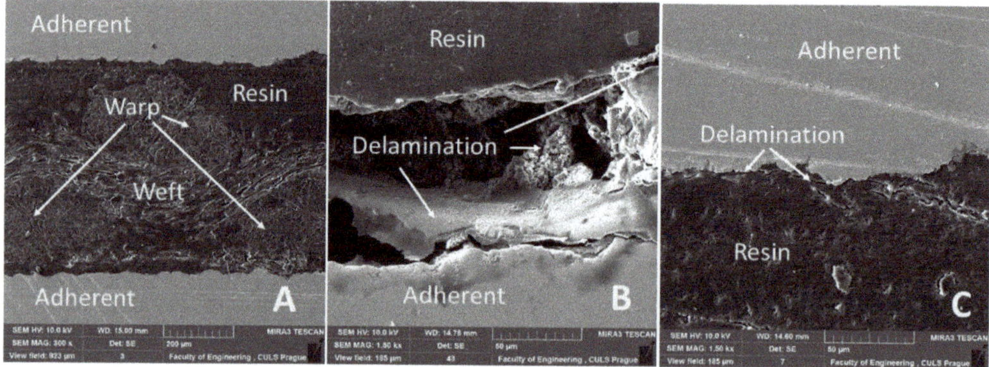

Figure 12. SEM images of cross-section through adhesive bonds: (**A**): cut through Tera-Flat reinforced bond, 1000 cycles in interval 5–70% (267–3743 N) (MAG 300×); (**B**): cross-section of Erik-WH1, 1000 cycles in interval 5–50% (267–2674 N) (MAG 1500×); (**C**): cross-section of Tera-Flat, 1000 cycles in interval 5–70% (267–3743 N) (MAG 1500×).

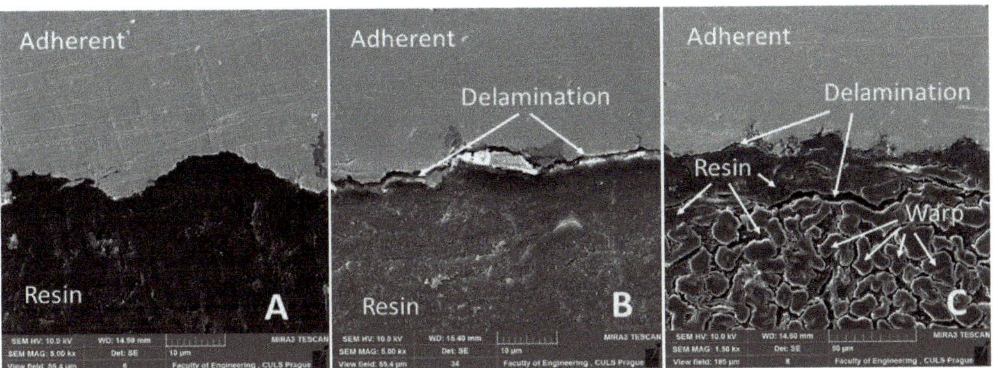

Figure 13. SEM images of cross-section of tested adhesive bond: (**A**): cross-section of Resin-Flat (MAG 5000×), (**B**): cross-section of Erik-WH1, 1000 cycles in interval 5–50% (267–2674 N) (MAG 5000×), (**C**): cross-section of Tera-Flat, 1000 cycles in interval 5–70% (267–3743 N) (MAG 5000×).

Results of the SEM cross-section analysis focused on evaluating adhesive bonds exposed to dynamic loading during cyclic tests. The images demonstrated initiation of micro-cracks in adhesive bond, which leads to delamination. Small cracks appeared inside the adhesive layer (Figure 13C), as well as on the boundary between the adhesive layer and bonded material (Figure 13B). Adhesive bonds that were not exposed to cyclic loading showed no micro-cracks after SEM analysis.

The research involved cyclic tensile testing of wavy-shaped adhesive bonds to understand the fatigue and service behavior. A number of investigations have been carried out regarding adhesive bonds with modified adherend shapes under static loading [26,70]. However, in practical applications, the cyclic loading of non-flat geometries is more relevant. The practical solutions involve several curved elements which undergo cyclic loading and deformation. Sometimes, there is the necessity to create shaped bonds, which reduces the strength substantially. Extensive research needs to be carried out with several other shapes to find practical solutions that suit the design requirements while exhibiting good mechanical performance and service life.

4. Conclusions

Experimental results of wavy-lap bonds with natural cotton fabric reinforcement under cyclic loading proved that:

- Wave-shaped bonds WH1 and WH2 reduced the overall strength of the resin under static tests. For Resin-WH1, the strength decreased by 47% to 3.91 ± 0.23 MPa. For Resin-WH2, the strength decreased by 67% to 2.45 ± 0.13 MPa. Resin-Flat, WH1, and WH2 failed the cyclic tests.
- The reinforcing fabric has a positive effect on the mechanical performance of the adhesive bonds. The reinforcing fabrics Erik and Tera did not increase the overall strength of the bond but positively reduced the deformation of the bond and thus increased the elastic modulus and service life of the adhesive bonds under cyclic loading. Erik-Flat and Erik-WH1 passed the 5–50% (267–2674 N) cyclic tests. Tera-Flat and WH1 also passed the 5–50% (267–2674 N) cyclic test. Tera-Flat further passed the 5–70% (267–3743 N) cyclic test.
- SEM analysis showed a positive effect of alkali treatment (10% aqueous NaOH solution) on the fabric surface. The unwanted layers of lignin, oils, and fats were removed. The SEM analysis showed improved wettability of the reinforcing fabrics Erik and Tera due to the alkali treatment with 10% NaOH solution. The SEM analysis also showed the formation of micro-cracks with subsequent delamination due to cyclic loading at the adhesive/adherend interface and at the matrix/reinforcement interface.

- The results of this research demonstrate the ability of natural fabrics to act as reinforcement to increase the service life and safety of hybrid adhesive bonds under cyclic loading. Hybrid adhesive bonds create an interesting alternative in the design of adhesive bonding technology. The use of shaped design for the overlapped bonds is an interesting area that needs to be studied further.

Author Contributions: V.K., M.T., M.M., R.K.M. and P.H.—experiment design; V.K., M.M., M.T. and P.H.—methodology; P.H., V.K., M.T., M.M. and R.K.M.—testing of mechanical properties and data analysis; V.K., M.T., M.M., K.H. and R.K.M. wrote and edited the paper; M.T., M.M., V.K. and R.K.M.—project administration; K.H. and R.K.M.—language correction; V.K., M.M. and M.T.—resources; M.T. and R.K.M.—communication with the editors; M.M., V.K. and M.H. performed the SEM analysis; V.K., M.M., R.K.M. and M.T.—supervision. All authors have read and agreed to the published version of the manuscript.

Funding: This paper is supported by the internal grant agency of Faculty of Engineering, grant no. 2021:31140/1312/3108: "Experimental research of hybrid adhesive bonds with multilayer sandwich construction, Czech University of Life Sciences Prague".

Institutional Review Board Statement: Not applicable.

Informed Consent Statement: Not applicable.

Data Availability Statement: Not applicable.

Conflicts of Interest: The authors declare no conflict of interest.

References

1. Barnes, T.; Pashby, I. Joining techniques for aluminium spaceframes used in automobiles: Part II—Adhesive bonding and mechanical fasteners. *J. Mater. Process. Technol.* **2000**, *99*, 72–79. [CrossRef]
2. Preu, H.; Mengel, M. Experimental and theoretical study of a fast curing adhesive. *Int. J. Adhes. Adhes.* **2007**, *27*, 330–337. [CrossRef]
3. Banea, M.D.; Da Silva, L.F.M. Adhesively bonded joints in composite materials: An overview. *Proc. Inst. Mech. Eng. Part L J. Mater. Des. Appl.* **2009**, *223*, 1–18. [CrossRef]
4. Adams, R.D. *Adhesive Bonding: Science, Technology and Applications*; Woodhead Publishing: Abington Cambridge, UK, 2005, ISBN 9781855737419.
5. Pizzi, A.; Mittal, K.L. *Handbook of Adhesive Technology*; CRS Press Taylor & Francis Group: Boca Raton, FL, USA, 2003, ISBN 0824709861.
6. Müller, M.; Valášek, P. Composite adhesive bonds reinforced with microparticle filler based on egg shell waste. *J. Phys. Conf. Ser.* **2018**, *1016*, 12002. [CrossRef]
7. Bahrami, B.; Ayatollahi, M.; Beigrezaee, M.; da Silva, L. Strength improvement in single lap adhesive joints by notching the adherends. *Int. J. Adhes. Adhes.* **2019**, *95*, 102401. [CrossRef]
8. Stoeckel, F.; Konnerth, J.; Gindl-Altmutter, W. Mechanical properties of adhesives for bonding wood—A review. *Int. J. Adhes. Adhes.* **2013**, *45*, 32–41. [CrossRef]
9. Kolář, V.; Tichý, M.; Müller, M.; Valášek, P.; Rudawska, A. Research on influence of cyclic degradation process on changes of structural adhesive bonds mechanical properties. *Agron. Res.* **2019**, *17*, 1062–1070. [CrossRef]
10. Han, X.; Crocombe, A.; Anwar, S.N.R.; Hu, P. The strength prediction of adhesive single lap joints exposed to long term loading in a hostile environment. *Int. J. Adhes. Adhes.* **2014**, *55*, 1–11. [CrossRef]
11. Krolczyk, G.; Raos, P.; Legutko, S. Experimental analysis of surface roughness and surface texture of machined and fused deposition modelled parts. *Teh. Vjesn.* **2014**, *21*, 217–221.
12. Nieslony, P.; Krolczyk, G.M.; Wojciechowski, S.; Chudy, R.; Zak, K.; Maruda, R.W. Surface quality and topographic inspection of variable compliance part after precise turning. *Appl. Surf. Sci.* **2018**, *434*, 91–101. [CrossRef]
13. Rudawska, A. Selected aspects of the effect of mechanical treatment on surface roughness and adhesive joint strength of steel sheets. *Int. J. Adhes. Adhes.* **2014**, *50*, 235–243. [CrossRef]
14. Bresson, G.; Jumel, J.; Shanahan, M.E.; Serin, P. Strength of adhesively bonded joints under mixed axial and shear loading. *Int. J. Adhes. Adhes.* **2012**, *35*, 27–35. [CrossRef]
15. Mishra, R.; Militky, J.; Gupta, N.; Pachauri, R.; Behera, B. Modelling and simulation of earthquake resistant 3D woven textile structural concrete composites. *Compos. Part B Eng.* **2015**, *81*, 91–97. [CrossRef]
16. Broughton, W.R.; Mera, R.; Hinopoulos, G. *Cyclic Fatigue Testing of Adhesive Joints Test Method Assessment*; NPL Report; National Physical Laboratory: Teddington, UK, 1999.
17. Akrami, R.; Anjum, S.; Fotouhi, S.; Boaretto, J.; De Camargo, F.V.; Fotouhi, M. Investigating the effect of interface morphology in adhesively bonded composite wavy-lap joints. *J. Compos. Sci.* **2021**, *5*, 32. [CrossRef]

18. Behera, B.K.; Pattanayak, A.K.; Mishra, R. Prediction of fabric drape behaviour using finite element method. *J. Text. Eng.* **2008**, *54*, 103–110. [CrossRef]

19. Tichý, M.; Kolář, V.; Müller, M.; Mishra, R.K.; Šleger, V.; Hromasová, M. Quasi-static shear test of hybrid adhesive bonds based on treated cotton-epoxy resin layer. *Polymers* **2020**, *12*, 2945. [CrossRef]

20. Boss, J.; Ganesh, V.; Lim, C.T. Modulus grading versus geometrical grading of composite adherends in single-lap bonded joints. *Compos. Struct.* **2003**, *62*, 113–121. [CrossRef]

21. Da Silva, L.; Adams, R.D. Techniques to reduce the peel stresses in adhesive joints with composites. *Int. J. Adhes. Adhes.* **2007**, *27*, 227–235. [CrossRef]

22. Zeng, Q.-G.; Sun, C.T. Novel design of a bonded lap joint. *AIAA J.* **2001**, *39*, 1991–1996. [CrossRef]

23. Ávila, A.F.; Bueno, P.d.O. Stress analysis on a wavy-lap bonded joint for composites. *Int. J. Adhes. Adhes.* **2004**, *24*, 407–414. [CrossRef]

24. Müller, M. Research on constructional shape of bond at connecting galvanized sheet of metal. *Manuf. Technol.* **2015**, *15*, 392–396. [CrossRef]

25. Jaiswal, P.; Hirulkar, N.; Papadakis, L.; Jaiswal, R.R.; Joshi, N.B. Parametric study of non flat interface adhesively bonded joint. *Mater. Today Proc.* **2018**, *5*, 17654–17663. [CrossRef]

26. Haghpanah, B.; Chiu, S.; Vaziri, A. Adhesively bonded lap joints with extreme interface geometry. *Int. J. Adhes. Adhes.* **2014**, *48*, 130–138. [CrossRef]

27. Razavi, S.; Berto, F.; Peron, M.; Torgersen, J. Parametric study of adhesive joints with non-flat sinusoid interfaces. *Theor. Appl. Fract. Mech.* **2017**, *93*, 44–55. [CrossRef]

28. Herrera-Franco, P.; Valadez-González, A. A study of the mechanical properties of short natural-fiber reinforced composites. *Compos. Part B Eng.* **2005**, *36*, 597–608. [CrossRef]

29. Alkbir, M.F.; Sapuan, S.M.; Nuraini, A.A.; Ishak, M.R. Fibre properties and crashworthiness parameters of natural fibre-reinforced composite structure: A literature review. *Compos. Struct.* **2016**, *148*, 59–73. [CrossRef]

30. Aziz, S.H.; Ansell, M.P. The effect of alkalization and fibre alignment on the mechanical and thermal properties of kenaf and hemp bast fibre composites: Part 1—Polyester resin matrix. *Compos. Sci. Technol.* **2004**, *64*, 1219–1230. [CrossRef]

31. Dalmay, P.; Smith, A.; Chotard, T.; Sahay-Turner, P.; Gloaguen, V.; Krausz, P. Properties of cellulosic fibre reinforced plaster: Influence of hemp or flax fibres on the properties of set gypsum. *J. Mater. Sci.* **2010**, *45*, 793–803. [CrossRef]

32. Mishra, R.; Militky, J.; Baheti, V.; Huang, J.; Kale, B.; Venkataraman, M.; Bele, V.; Arumugam, V.; Zhu, G.; Wang, Y. The production, characterization and applications of nanoparticles in the textile industry. *Text. Prog.* **2014**, *46*, 133–226. [CrossRef]

33. Petrásek, S.; Müller, M. Mechanical qualities of adhesive bonds reinforced with biological fabric treated by plasma. *Agron. Res.* **2017**, *15*, 1170–1181.

34. Valadez-Gonzalez, A.; Cervantes-Uc, J.; Olayo, R.; Herrera-Franco, P. Effect of fiber surface treatment on the fiber–matrix bond strength of natural fiber reinforced composites. *Compos. Part B Eng.* **1999**, *30*, 309–320. [CrossRef]

35. Müller, M.; Valášek, P.; Kolář, V.; Šleger, V.; Gürdil, G.A.K.; Hromasová, M.; Hloch, S.; Moravec, J.; Pexa, M. Material utilization of cotton post-harvest line residues in polymeric composites. *Polymers* **2019**, *11*, 1106. [CrossRef]

36. Valášek, P.; Müller, M.; Šleger, V.; Kolář, V.; Hromasová, M.; D'Amato, R.; Ruggiero, A. Influence of alkali treatment on the microstructure and mechanical properties of coir and abaca fibers. *Materials* **2021**, *14*, 2636. [CrossRef] [PubMed]

37. Nam, T.H.; Ogihara, S.; Tung, N.H.; Kobayashi, S. Effect of alkali treatment on interfacial and mechanical properties of coir fiber reinforced poly (butylene succinate) biodegradable composites. *Compos. Part B Eng.* **2011**, *42*, 1648–1656. [CrossRef]

38. Fan, T.; Hu, R.; Zhao, Z.; Liu, Y.; Lu, M. Surface micro-dissolve method of imparting self-cleaning property to cotton fabrics in NaOH/urea aqueous solution. *Appl. Surf. Sci.* **2017**, *400*, 524–529. [CrossRef]

39. Ferdous, W.; Manalo, A.; Yu, P.; Salih, C.; Abousnina, R.; Heyer, T.; Schubel, P. Tensile fatigue behavior of polyester and vinyl ester based gfrp laminates—A comparative evaluation. *Polymers* **2021**, *13*, 386. [CrossRef]

40. Ferdous, W.; Manalo, A.; Peauril, J.; Salih, C.; Reddy, K.R.; Yu, P.; Schubel, P.; Heyer, T. Testing and modelling the fatigue behaviour of GFRP composites—Effect of stress level, stress concentration and frequency. *Eng. Sci. Technol. Int. J.* **2020**, *23*, 1223–1232. [CrossRef]

41. Saraç, I.; Adin, H.; Temiz, Ş. Experimental determination of the static and fatigue strength of the adhesive joints bonded by epoxy adhesive including different particles. *Compos. Part B Eng.* **2018**, *155*, 92–103. [CrossRef]

42. Hafiz, T.; Wahab, M.A.; Crocombe, A.; Smith, P. Mixed-mode fracture of adhesively bonded metallic joints under quasi-static loading. *Eng. Fract. Mech.* **2010**, *77*, 3434–3445. [CrossRef]

43. Kelly, G. Quasi-static strength and fatigue life of hybrid (bonded/bolted) composite single-lap joints. *Compos. Struct.* **2006**, *72*, 119–129. [CrossRef]

44. Šleger, V.; Müller, M. Quasi static tests of adhesive bonds of alloy AlCu4Mg. *Manuf. Technol.* **2015**, *15*, 694–698. [CrossRef]

45. Kolář, V.; Müller, M.; Mishra, R.; Rudawska, A.; Šleger, V.; Tichý, M.; Hromasová, M.; Valášek, P. Quasi-static tests of hybrid adhesive bonds based on biological reinforcement in the form of eggshell microparticles. *Polymers* **2020**, *12*, 1391. [CrossRef]

46. International Organization for Standardization. *ČSN EN 1465—Adhesives—Determination of Tensile Lap-Shear Strength of Bonded Assemblies*; Czech Standardization Institute: Prague, Czech Republic, 2009.

47. DIN 17120 Grade St 37-3—Low Carbon Steel—Matmatch. Available online: https://matmatch.com/materials/minfm31305-din-17120-grade-st-37-3 (accessed on 13 May 2021).

48. Rudawska, A.; Zaleski, K.; Miturska, I.; Skoczylas, A. Effect of the application of different surface treatment methods on the strength of titanium alloy sheet adhesive lap joints. *Materials* **2019**, *12*, 4173. [CrossRef] [PubMed]

49. Svitap.cz. Available online: https://www.tkaniny-svitap.cz/kcfinder/upload/file/vzorkovnik.pdf (accessed on 19 May 2021).

50. Boopathi, L.; Sampath, P.; Mylsamy, K. Investigation of physical, chemical and mechanical properties of raw and alkali treated Borassus fruit fiber. *Compos. Part B Eng.* **2012**, *43*, 3044–3052. [CrossRef]

51. Bunsell, A.R. *Handbook of Tensile Properties of Textile and Technical Fibres*, 2nd ed.; Woodhead Publishing: Sawston, UK, 2018.

52. CHS-EPOXY 324 (EPOXY 1200) and Hardener P11. Available online: https://havel-composites.com/uploads/files/products/335/2bf05fdbe434cf27e38cd1554c43a9a4792aa4db.pdf (accessed on 19 May 2021).

53. Kolář, V.; Müller, M.; Tichý, M.; Rudawska, A.; Hromasová, M. Influence of preformed adherent angle and reinforcing glass fibre on tensile strength of hybrid adhesive bond. *Manuf. Technol.* **2019**, *19*, 786–791. [CrossRef]

54. You, M.; Li, Z.; Zheng, X.-L.; Yu, S.; Li, G.-Y.; Sun, D.-X. A numerical and experimental study of preformed angle in the lap zone on adhesively bonded steel single lap joint. *Int. J. Adhes. Adhes.* **2009**, *29*, 280–285. [CrossRef]

55. Ávila, A.F.; Bueno, P.D.O. An experimental and numerical study on adhesive joints for composites. *Compos. Struct.* **2004**, *64*, 531–537. [CrossRef]

56. Campilho, R.; de Moura, M.; Domingues, J. Numerical prediction on the tensile residual strength of repaired CFRP under different geometric changes. *Int. J. Adhes. Adhes.* **2009**, *29*, 195–205. [CrossRef]

57. Müller, M.; Herák, D. Dimensioning of the bonded lap joint. *Res. Agric. Eng.* **2010**, *56*, 59–68. [CrossRef]

58. Grant, L.; Adams, R.; da Silva, L.F. Experimental and numerical analysis of single-lap joints for the automotive industry. *Int. J. Adhes. Adhes.* **2009**, *29*, 405–413. [CrossRef]

59. Tichý, M.; Kolář, V.; Muller, M.; Valasek, P. Quasi-static tests on polyurethane adhesive bonds reinforced by rubber powder. In *Engineering for Rural Development*; Latvia University of Life Sciences and Technologies: Jelgava, Latvia, 2019; Volume 18, pp. 1035–1041.

60. Zavrtálek, J.; Müller, M.; Šléger, V. Low-cyclic fatigue test of adhesive bond reinforced with glass fibre fabric. *Agron. Res.* **2016**, *14*, 1138–1146.

61. Symington, M.C.; Banks, W.M.; West, O.D.; Pethrick, R.A. Tensile testing of cellulose based natural fibers for structural composite applications. *J. Compos. Mater.* **2009**, *43*, 1083–1108. [CrossRef]

62. Kalia, S.; Kaith, B.; Kaur, I. Pretreatments of natural fibers and their application as reinforcing material in polymer composites-A review. *Polym. Eng. Sci.* **2009**, *49*, 1253–1272. [CrossRef]

63. Li, X.; Tabil, L.G.; Panigrahi, S. Chemical treatments of natural fiber for use in natural fiber-reinforced composites: A review. *J. Polym. Environ.* **2007**, *15*, 25–33. [CrossRef]

64. Valasek, P.; Müller, M.; Šleger, V. Influence of plasma treatment on mechanical properties of cellulose-based fibres and their interfacial interaction in composite systems. *BioResources* **2017**, *12*, 5449–5461. [CrossRef]

65. Kabir, M.M.; Wang, H.; Lau, K.T.; Cardona, F. Chemical treatments on plant-based natural fibre reinforced polymer composites: An overview. *Compos. Part B Eng.* **2012**, *43*, 2883–2892. [CrossRef]

66. Miturska, I.; Rudawska, A.; Müller, M.; Hromasová, M. The Influence of mixing methods of epoxy composition ingredients on selected mechanical properties of modified epoxy construction materials. *Materials* **2021**, *14*, 411. [CrossRef]

67. Miturska, I.; Rudawska, A.; Müller, M.; Valášek, P. The influence of modification with natural fillers on the mechanical properties of epoxy adhesive compositions after storage time. *Materials* **2020**, *13*, 291. [CrossRef]

68. Rudawska, A. The effect of the salt water aging on the mechanical properties of epoxy adhesives compounds. *Polymers* **2020**, *12*, 843. [CrossRef] [PubMed]

69. Ayatollahi, M.R.; Samari, M.; Razavi, S.M.J.; da Silva, L. Fatigue performance of adhesively bonded single lap joints with non-flat sinusoid interfaces. *Fatigue Fract. Eng. Mater. Struct.* **2017**, *40*, 1355–1363. [CrossRef]

70. Da Silva, L.F.; Carbas, R.; Critchlow, G.; Figueiredo, M.; Brown, K. Effect of material, geometry, surface treatment and environment on the shear strength of single lap joints. *Int. J. Adhes. Adhes.* **2009**, *29*, 621–632. [CrossRef]

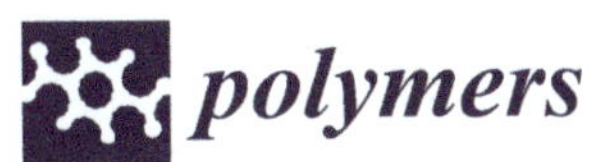

Article

Characterization of Hybrid Composites with Polyester Waste Fibers, Olive Root Fibers and Coir Pith Micro-Particles Using Mixture Design Analysis for Structural Applications

Muhammad Rizwan Tufail [1], Hafsa Jamshaid [1], Rajesh Mishra [2,*], Uzair Hussain [1], Martin Tichy [2] and Miroslav Muller [2]

[1] Faculty of Textile Engineering, National Textile University, Faisalabad 37610, Pakistan; formanite819@gmail.com (M.R.T.); hafsa@ntu.edu.pk (H.J.); uzair@ntu.edu.pk (U.H.)
[2] Department of Material Science and Manufacturing Technology, Faculty of Engineering, Czech University of Life Sciences Prague, Kamycka 129, 165 00 Prague 6-Suchdol, Czech Republic; martintichy@tf.czu.cz (M.T.); muller@tf.czu.cz (M.M.)
* Correspondence: mishrar@tf.czu.cz

Citation: Tufail, M.R.; Jamshaid, H.; Mishra, R.; Hussain, U.; Tichy, M.; Muller, M. Characterization of Hybrid Composites with Polyester Waste Fibers, Olive Root Fibers and Coir Pith Micro-Particles Using Mixture Design Analysis for Structural Applications. *Polymers* **2021**, *13*, 2291. https://doi.org/10.3390/polym13142291

Academic Editor: Fabrizio Sarasini

Received: 10 June 2021
Accepted: 12 July 2021
Published: 13 July 2021

Abstract: In the present work, hybrid composites were developed by using polyester waste fibers along with natural origin materials: olive root fibers and coir pitch filler. Such composite panels can be used as a potential alternative for fiber glass sunshade panels and room dividers in buildings. Hybrid composites were fabricated by mixing polyester waste fibers and olive root fibers in different ratios (0:100, 33:67, 67:33 and 100:0). Coir pith micro-particles with an average size of 312 d.nm were used as filler in the polyester matrix at three different levels (0%, 5%, and 10%) of the overall matrix weight. Mechanical properties, e.g., tensile strength, flexural strength and impact strength, thermal properties, e.g., coefficient of linear thermal expansion, thermo-gravimetric analysis (TGA) and environmental properties, e.g., water absorption, loss of density after exposure to weathering were characterized. For comparison purposes, a commercially available fiber glass sunshades sample was also investigated. Mixture design analysis was used to optimize the ratio of all components in the composite. Graphical comparison of experimental results using regression models showed a high degree of correlation. An optimized formulation of composite with an objective of maximization of tensile strength, flexural strength, impact strength and minimization of water absorption, density loss, as well as coefficient of linear thermal expansion, was determined at 70.83 wt%, 15.15 wt%, and 14.01 wt% of polyester waste fibers, olive root fibers and coir pith micro-fillers, respectively. Overall, it can be concluded that the developed hybrid composites from waste fibrous materials can be used as a promising alternative and a value-added application in buildings and construction purposes.

Keywords: hybrid composite; polyester waste fiber; olive root fiber; coir pith filler; building materials; mixture design analysis

1. Introduction

Composite material performance and properties depend on the properties of individual components, mixture proportions and their inter-facial compatibility. The composite materials primarily consist of two constituents, one of which is the reinforcement material that could be treated with chemicals for surface modification [1] in order to improve binding and handling properties. The second component is the matrix which serves to protect the reinforcement material from environmental and external damage by transfer of the load [2]. In addition, they may contain a third component called "fillers" which are mixed with polymeric matrix in order to improve its mechanical and thermal properties. Fillers are used for reducing the overall weight and cost of the composite while enhancing its performance [3,4]. Natural fiber-based composites are environmentally superior in specific fields of application [5]. The global interest in natural fiber-based composites is growing because they are alternatives for synthetic or man-made fibers such as glass or carbon [6]. Natural fiber reinforcements are a renewable resource, and their production requires little energy. They have very low carbon dioxide emissions and are not dependent on petroleum-based precursors. Sisal, bamboo, jute, and coir are widely used as engineering materials in different industries [7,8].

A proper material design helps in achieving a balance in performance and cost for application in numerous fields ranging from the building industry to the automobile industry [9,10]. Recently, different industries are looking into the replacement of polymer or glass fiber composites. The natural fiber-based composites have relatively lower mechanical performance and higher water absorption properties compared to the glass or carbon fiber-based composites. Therefore, a hybrid fabrication method is used to produce composite products with improved mechanical, thermal, water resistance, and water absorption properties. In this technique, composite materials are manufactured by using two or more reinforcing materials in a single matrix or two polymers blended matrices with single natural fiber reinforcement. The reinforcing materials can be made of natural or synthetic fibers depending on the required performance level [11,12]. Polyester fibers are consumed in different textile applications. Their origin is from petroleum-based materials which are not sustainable, and their biodegradability is another issue. Polyester waste fibers can be recycled in order to reduce environmental pollution by not putting them into landfills. That would prevent the chemicals from leaching into the soil. Reusing is another option that is more energy-efficient than recycling, especially when it can be used for value-added products. Franciszczak et al. revealed that the injection-molded composite of short PET fibers reinforced to PETG (Polyethylene terephthalate glycol) matrix display good impact strength on the same level of polypropylene reinforced by glass fibers [13]. Wu et al. determined that the self-reinforced PET composite of high tenacity PET as reinforcement and copolymerized PET as matrix show excellent flexural and impact behavior [14]. Ahmad et al. studied the composite of hybrid fabric by the interlacement of warp (hemp yarn) and weft (PET yarn) produced by vacuum infusion process with epoxy resin. Such composites display an enhancement in tensile strength [15]. Composite samples were produced from six layers of woven kenaf fabrics arranged at symmetric angles (0°, 45°, 90°) and impregnated with PET in POM (Polyoxymethylene) matrix. The hybrid composite samples showed increases in the modulus and strength [16]. Wu et al. studied that self-reinforced PET composite of recycled PET homopolymer filaments serving as the reinforcements and copolymerized PET (mPET) filaments as the matrix exhibit improved resistance to creep deformation [17]. Composites prepared by hand layup and compression molding of coir pith/nylon/epoxy showed that chemical treatment improves the water resistance and offers optimum retention of impact strength in composites [18]. Essabir et al. developed polypropylene hybrid composites by using coir fibers and coir shell micro-particles under injection molding. The findings indicate that coir shell particles have low thermal degradation compared to coir fiber [19]. Islam et al. concluded that hand layup of coir mat/polyester composite with coir powder as filler with varying content of (10, 20, 30, 40, 50, and 100%) shows an improvement in mechanical properties up to 30% and further addition of filler results in deterioration of mechanical properties [20]. Hybrid composite with coir pith/nylon fabric/epoxy made by hand lay-up showed that NaOH treatment of coir pith increases the composite's impact strength and water resistance [21,22]. A hybrid composite of coir pith, rice husk, and groundnut shell was prepared with epoxy resin. The results revealed that the hybridization of micro-particles greatly influenced the tensile and flexural properties of the composites [23,24]. Composites based on recycled PET fibers with an average length of (2 to 20 mm) in the polyester matrix exhibited good dispersion, interfacial adhesion, and high affinity of PET fibers with matrix leading to improved mechanical performance [25]. Abdulla et al. developed the hybrid composite of kenaf and PET fibers reinforced POM using compression molding technique and exposed the samples to ultraviolet penetration, moisture, and water spray in the weathering chamber. The increasing PET contents indicated better retention in mechanical properties when exposed to UV and moisture as compared to natural fiber-rich composites. Thus, hybrid composites are more suitable for outdoor applications [26]. The hybrid composite of kenaf/PET fiber in the POM (polyoxymethylene) matrix was investigated to study the mechanical as well as moisture absorption properties. The impact strength of composites using longer fibers was found to be higher than composites reinforced with short fibers [27]. Studies on composites reinforced with glass fibers, waste polyethylene terephthalate fibers and kenaf fibers developed by compression molding concluded that the tensile strength of glass fiber-reinforced PET waste and kenaf hybrid composites was higher than the pure PET waste composites. They can be used for construction materials like wall and partition materials [28]. The injection-molded polypropylene composites were prepared with distinct mixtures of the wood flour and the olive mill sludge (OMS). With increasing olive mill sludge, the water-resistance of composites increases. However, the flexural characteristics deteriorate with the increase in OMS flour content [29].

Mixture design analysis is used where the response depends on mixture proportions of various components and not on the absolute amount of incorporating materials. For example in an alloy, the mechanical properties may depend on variable proportions of the various metals but not on the absolute amount of each component. The mixture design experiments are widely used nowadays in order to determine the optimum formulation of various components and blending ratios in multicomponent composite materials. The specific objective is to determine, analyze and optimize the most preferred mixture proportion in a composite product at the lowest price [30,31].

The main aim of this research is to analyze the effect of polyester waste fibers, olive root fibers as reinforcement and coir pith as filler on the mechanical properties (tensile strength, flexural strength and impact strength), weathering effects (water absorption and density loss) and the thermal properties (coefficient of linear thermal expansion and thermogravimetric analysis) of hybrid composites based on a polyester matrix. After modeling the mixture design, the aim is to determine the optimized component proportion for the experimental responses through maximizing and minimizing each component. The mixture design analysis was applied through Minitab statistical software in order to model the relationship between input variables and the experimental responses.

2. Materials and Methods

2.1. Materials

Polyester waste fibers of varying length in the range of 2–5 mm was provided by Diamond export industries Pvt. Ltd., Jaranwala Road, Khurrianwala, Faisalabad, Pakistan. This waste was removed in the rinsing process during the manufacturing of fabric. Olive root sticks were procured from a local market in Faisalabad, Pakistan. The olive root fibers were obtained by beating/pressing the sticks and extracting the fibers manually. The fibers were cut into a staple of 2–5 mm. Coconut husk was collected from local vendors and was dried under the sun to remove the moisture from them. The raw materials used are shown in Figure 1.

The coir pith was extracted manually and cleaned from dust and dirt by washing with distilled water. Fibers were chopped into 2 mm length. Selected samples were treated with 2% NaOH solution for 4 h at room temperature. Further, the alkali-treated samples were thoroughly washed with distilled water, so as to leave no trace of alkali. The alkali treatment removes pectins, fats, lignin and hemicellulose from the fiber, thereby increasing the percentage of cellulose content. This also improves the adhesivity of coir surface with the resin in a composite. In order to obtain the required particle size, alkali-treated coir pith was subjected through the ball mill process as shown in Figure 2. Z-average size of ball-milled coir pith micro-particles was measured to be 312 d.nm as per stander protocol on Malvern-zeta sizer.

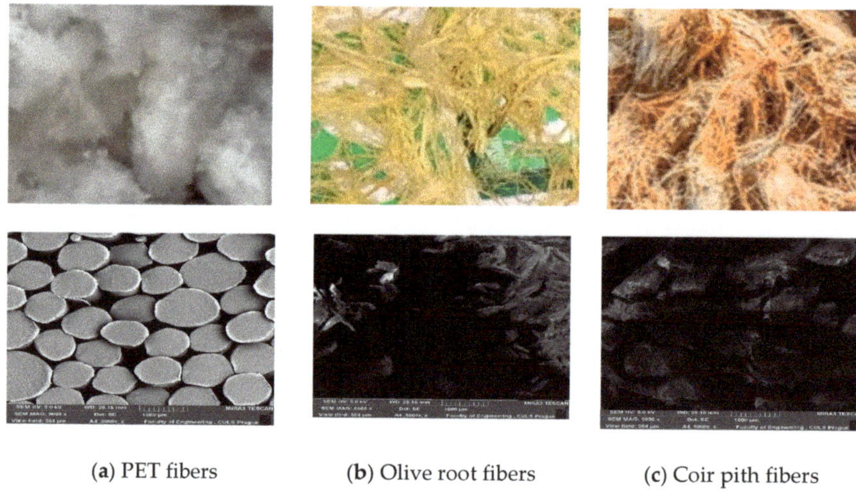

(**a**) PET fibers   (**b**) Olive root fibers   (**c**) Coir pith fibers

Figure 1. SEM images of fibers used.

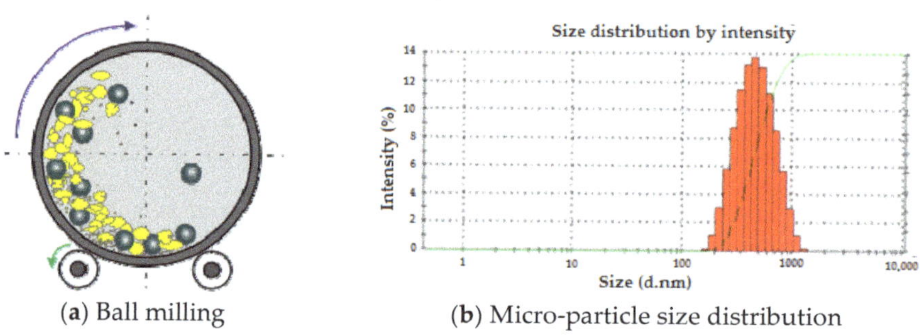

(**a**) Ball milling   (**b**) Micro-particle size distribution

Figure 2. Ball milling to prepare coir pith micro-particles.

The commercially available unsaturated polyester resin was obtained from Changzhou Rixin Resin Co., Ltd., China. Unsaturated polyester resin is a thermosetting polymer, which is chemically similar to epoxy resin and vinyl ester resins. It has excellent and long-term durability to water. It offers adequate resistance to different chemicals, to a large range of substances, from vegetable oil to sulfuric acid. The list of some chemicals' resistance along their resistive level is shown in Table 1.

Table 1. Resistance against different chemicals.

Chemicals	Resistance Level
Dilute acid	+ + +
Dilute alkalis	+ + +
Oil and greases	+ +
Aliphatic hydrocarbons	+
Aromatic hydrocarbons	+
Halogenated hydrocarbons	+
Alcohols	+ + +

Resistance levels: + + + + very good, + + + good, + + moderate, + poor.

Unsaturated polyester resin also exhibits adequate resistance against extreme weathering conditions like rain, ice, snow, strong winds, high temperature, and UV rays. The physical, mechanical and thermal properties of polyester resin are given in Table 2.

Table 2. Properties of unsaturated polyester resin.

Properties	Value
Physical State	Liquid at 25 °C
Color	Yellowish
Density, g/cm^3	1.12 at 20 °C
Viscosity, mPoise	300–600 at 25 °C
Storage Temperature	15–25 °C
Barcol Hardness	40
Tensile Strength, MPa	12
Tensile Modulus, GPa	2.1
Elongation at break, %	2.1
Flexural Strength, MPa	25
Flexural Modulus, GPa	0.9
Heat deflection temperature (HDT,1.80 MPa), °C	80
Glass transition temperature (Tg), °C	115–120

COBALT was used as an initiator and methyl ethyl ketone peroxide (MEKP) as a curing agent were procured from local market.

2.2. Methods

Twelve samples of hybrid composites were prepared with different compositions of reinforcement materials by weight percentage as given in Table 3 of the experiment design. The mold of size (340.8 mm × 340.8 mm × 3 mm) was used, and mold-releasing wax was applied to it before developing the samples. Polyester waste fibers and olive root fibers were blended in different ratios (0:100, 33:67, 67:33 and 100:0). The reinforcing fibers were loaded as per the required blend ratio and were arranged in the form of a matt inside the mold. The unsaturated polyester resin was mixed with 1 wt% COBALT as an accelerator and 0.5 wt% ethyl methyl ketone peroxide (EMKP) as a matrix curing agent. Coir pith micro-particles were used as filler in the polyester matrix at three different levels (0%, 5%, and 10%) of the overall matrix weight. The matrix with micro-particles was poured and applied by the hand lay-up method. The total aggregate of the loaded mixture components of polyester waste fibers (Xp), olive root fibers (Xo), and coir pith (Xc) is equal to 100 wt% as shown in Table 3. After impregnation, the specimens were subjected to compression molding, under a pressure of 10 kg/cm^2 at a temperature of 100 °C for 15 min. All samples were stored post-curing at room temperature for 24 h, and then cut into required testing specimens. For all the samples, a thickness of 3 ± 0.1 mm was maintained. The commercial fiber glass sample which was used for comparison purposes also has the same thickness.

Table 3. Design of experiment.

Samples	Xp	Xo	Xc
C1	100.00	0.00	0.00
C2	67.00	33.00	0.00
C3	33.00	67.00	0.00
C4	0.00	100.00	0.00
C5	87.33	0.00	12.67
C6	58.52	28.81	12.67
C7	28.81	58.52	12.67
C8	0.00	87.33	12.67
C9	77.52	0.00	22.48

Table 3. *Cont.*

Samples	*Xp*	*Xo*	*Xc*
C10	51.94	25.58	22.48
C11	25.58	51.94	22.48
C12	0.00	77.52	22.48

2.3. Characterization

2.3.1. Mechanical Testing

Tensile, flexural and impact tests of the hybrid composite were carried out according to ASTM D3039, ASTM D7264 and ASTM D 790-02 respectively. The specimen testing was performed under controlled environmental conditions at 23 °C and relative humidity at 65%. The tensile test of the specimens of size (25 mm × 203 mm) was carried out on a universal testing machine (model UMT Z 100 Allroundline, Zwick–Germany). The flexural tests (3-point bending) of specimens with size (13 mm × 120 mm) were carried out using a thickness to span ratio of 24. The impact test was conducted on a pendulum impact tester (model HIT5.5P, ZWICK–Germany).

2.3.2. Environmental Degradation/Weathering Related Properties

The environmental degradation or the impact of weathering on the hybrid composites for outdoor exposure was tested according to ASTM D2565-99. The test was conducted on xenon arc weather-Odometer (Atlas Ci 4000) in two cycles. In the first cycle of 18 h, the specimen was exposed to light for 102 min, and water was sprayed for 18 min. Black panel temperature was maintained at 40 °C, having a radiant flux of 41.5 W·m^{-2} in the range of 300 to 400 nm. In the second cycle, the specimen was exposed in the dark with a relative humidity of 95% with no water spray for 6 h at 38 °C black panel temperature. In order to analyze the impact of weathering, the density of composite samples was measured before and after the weathering condition. Further, the percentage decrease in the total density of composite samples was calculated. Surface images were analyzed in order to visually examine the change in the surface of hybrid composite samples both before and after the weathering conditions. The water absorption behavior of hybrid composites was studied according to ASTM D570. The sample was cut into a circular disk of 5 cm diameter. Firstly, the specimen samples were dried at 40 °C for 24 h in an oven. All the specimens were then cooled at room temperature for 30 min. The conditioned specimens were then immersed in distilled water at 23 °C for 24 h. Before and after immersion, the specimens were weighed to the nearest value of 0.001 g. The water absorption capacity was calculated by weight difference, and using the following equation.

Water absorption capacity,

$$w\% = \left(\frac{w_t - w_0}{w_0} \right) \times 100 \tag{1}$$

w_0 = oven-dry weight, w_t = weight of specimen after immersion.

2.3.3. Thermal Properties

The thermal properties (coefficient of linear thermal expansion (CTE) and thermogravimetric analysis TGA) of the hybrid composite were determined as per standard procedures. The effect of temperature on thermal expansion of the specimens was studied according to ASTM E831 of thermal expansion on DIL801L Dilatometer–TA Instrument with the temperature ranging from room temperature to 150 °C. Thermogravimetric analysis (TGA) of the samples was carried out according to ASTM E1131 on NETZSCH TG 209F1 Libra. A 5 mg sample size was used to determine the TGA parameters. Thermogravimetric analysis was carried out under an oxygen environment from room temperature to 600 °C with a rate of increase of 10 °C/min. The effect of increasing temperature on the degradation profile was studied and peak degradation temperatures were recorded.

3. Results and Discussion

In this research, Mixture Design Analysis was used to forecast corresponding responses, i.e., mechanical properties (tensile strength, flexural strength, impact strength), environmental effects (loss of density after exposure to weathering chamber and water absorption) as well as coefficient of thermal expansion by the input of component proportion, e.g., polyester waste fibers (Xp), olive root fibers (Xo) and coir pith filler (Xc). The effect of polyester waste fibers and olive root fiber as reinforcement and coir pith as incorporating filler into polyester resin on its mechanical, environmental, and thermal properties are evaluated. The responses are the functions of properties of different components in the mixture. The obtained experimental results for mechanical properties, environmental degradation properties, and thermal properties are shown in Table 4. At 95% ($\alpha = 0.05$) confidence level ANOVA tables were generated to determine the results. According to the p-value (probability value), the significance level of each term was determined. If the p-value will be more than 95% ($\alpha \leq 0.05$) each input component will have a significant effect on response and in case if probability value will be less than 95% ($\alpha \geq 0.05$) then the influence could not be considered significant and should be rejected from the final analysis because they do not have a significant effect. Moreover, the null hypothesis will be eliminated.

Table 4. Properties of samples investigated.

Sample Code	Tensile Strength (MPa)	Tensile Moduli (GPa)	Flexural Strength (MPa)	Flexural Moduli (GPa)	Impact Strength (KJ/m^2)	Water Absorbtion (%)	Initial Density (g/cm^3)	Weathered Density (g/cm^3)	CTE (10^{-6}/K)
C1	16.93	3.25	36.31	1.14	6.6886	3.24	0.984	0.968	−16.56
C2	20.74	3.71	48.32	1.61	5.9985	4.13	0.964	0.942	−11.86
C3	16.58	3.97	35.83	1.84	4.5893	4.94	1.182	1.146	−9.1
C4	15.46	5.21	34.84	2.02	4.0293	7.64	1.156	1.111	−8.14
C5	25.22	2.85	79.02	2.12	7.0858	4.41	1.199	1.15	−9.86
C6	27.02	3.29	86.90	4.11	6.8967	5.78	1.135	1.083	−8.25
C7	24.64	3.80	73.42	4.46	6.2576	6.33	0.968	0.908	−6.37
C8	22.32	3.86	55.22	5.44	5.6868	8.92	0.983	0.896	−4.72
C9	17.09	2.31	34.74	1.25	9.1548	6.17	1.162	1.051	11.64
C10	19.53	2.62	44.56	1.93	9.0445	7.03	1.009	0.887	10.72
C11	16.76	2.81	33.22	3.35	8.5619	9.48	1.087	0.946	8.76
C12	14.86	3.23	31.90	3.87	7.3260	10.9	1.132	0.964	7.31
G	88.83	10.66	158.66	4.99	64.61	2.58	1.148	1.136	3.4

G-Fiber glass commercial sample.

The experimental results reflect the influence of fiber properties, fiber volume fraction and interfacial strength on the mechanical, thermal and environmental properties of the hybrid composites. The constituents of the hybrid composites are PET fibers, olive root fibers, coir pith micro-particles and polyester resin. The individual component properties and their corresponding proportion play a vital role in deciding the performance of the hybrid composites. A blend proportion of 67:33 for PET:Olive root fibers results in the highest tensile strength in the composite samples. The PET fibers are stronger than olive root fibers and thus as the olive root fiber proportion increases, the tensile strength decreases. The addition of 12.67% coir pith micro-particles, leads to an increase in tensile strength. As the proportion of coir pith micro-particles further increases, there is a decrease in the overall fiber component in order to maintain the total content of reinforcement constant. Based on the Halpin–Tsai model, the tensile strength decreases [32,33]. With respect to the tensile modulus of the hybrid composite samples, they are inversely proportional to the tensile deformation of individual components. The olive root fibers are less extensible as compared to PET fibers. Thus, the increasing proportion of olive root fibers, leads to an increase in the tensile modulus. An increase in coir pith particles in the mixture results in a decrease in fibrous components. Therefore, a decrease in tensile modulus is observed.

The flexural properties are very much similar to tensile properties. The bending deformation in a composite sheet, results in tensile deformation on the outer surface and longitudinal compression on the inner surface. The observed behavior is following the Halpin–Tsai model equations [32,33]. A blend proportion of 58.52% (PET), 28.81% (Olive root fiber) and 12.67% (Coir pith particles) results in the highest tensile and flexural strength.

Impact performance is a multiaxial deformation, unlike tensile or bending. It depends on the mechanical performance of components as well as their interaction in the radial and thickness direction. PET fibers being stronger than other reinforcing components in the mixture, are very much dominant in deciding the impact energy absorption. Further, a highly extensible fiber enables better absorption of the momentum of impact. Thus, the higher proportion of PET, results in higher impact strength. The addition of coir pith micro-particles results in an increase in impact strength, despite the corresponding decrease in the fibrous component. This is quite different from the observations regarding tensile and bending performance. This behavior can be attributed to the interfacial bonds provided by the micro-particles with the resin component and thus improvement in the multiaxial resistance to impact. This behavior is not so prominent under uniaxial deformations, e.g., tensile or bending.

Water absorption is dependent on the hygroscopic component proportion. Both olive root fibers, as well as coir pith, are water-absorbing components, unlike PET fibers. Thus, a decreasing proportion of PET fibers and an increasing proportion of olive root fiber and coir pith particles result in an increase in water absorption capacity. In comparison to the other two components, the coir pith particles are more porous and thus are able to absorb a higher amount of liquid water. An increasing proportion of coir pith micro-particles results in increase in water absorption capacity, The Halpin–Tsai model is validated by these observations [33].

Due to the weathering treatment, there is a degradation of some components which results in loss of weight and an increase in the volume of the composites. The decrease in density is dependent on the proportion of degradable components in the mixture. Since olive root and coir pith are biodegradable components, a higher proportion of such materials leads to more decrease in density.

The coefficient of thermal expansion (CTE) is governed by several models, e.g., the Voigt model [34] and the Reuss model [35]. The overall CTE of hybrid composites is governed by the % fraction, elastic modulus and CTE of individual components. As PET is a thermoplastic fiber, it might deform and melt during heating and thus causing shrinkage of the composite. As the portion of PET fibers is decreased, there is a decrease in the level of thermal deformation. The increase in cellulosic components decreases the level of thermal shrinkage or expansion. Thus, the thermal stability of hybrid composites is improved. The addition of 12.67% coir pith particles reduces the thermal shrinkage in hybrid composites. However, with an increasing proportion of coir particles and a corresponding decrease in fibrous components, there is a significant increase in the thermal expansion coefficient. Unlike fibrous reinforcements, coir pith micro-particles are discontinuous elements. They might not restrict the thermal deformations in the resin phase as efficiently as the fibers. This might be the reason behind higher CTE in coir particle-rich samples. It is consistently observed that decreasing proportion of PET fibers and increasing proportion of olive root fibers results in a decrease in thermal expansion coefficient. Such observations are also supported by reported literature from other researchers [34,36].

3.1. Scanning Electron Microscopy (SEM) of Composite Samples

The hybrid composite samples were tested for the mechanical, thermal and environmental/weathering properties. SEM images of tensile-tested hybrid composite samples are shown in Figure 3.

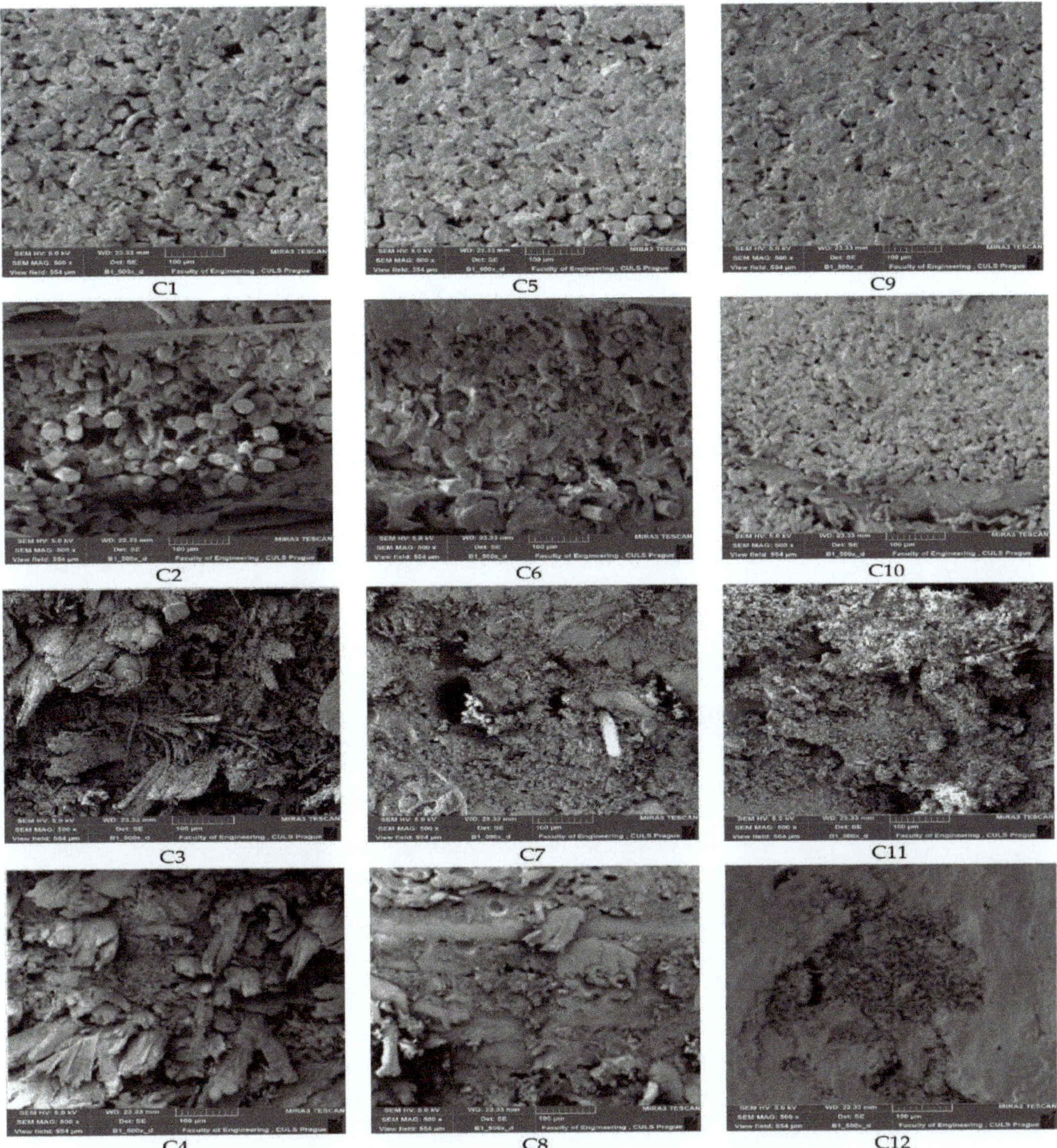

Figure 3. SEM images of tensile tested hybrid composite samples (C1–C12 as in Table 3).

It is visible from the SEM images that the hybrid composites and composites are free from any major defects. The fibrous reinforcement is uniformly impregnated by the matrix phase. When the composite samples contain only one type of fiber, i.e., pure polyester or pure olive root fiber, there is more uniformity. While mixing the polyester and olive root fibers, there are a few minor voids in between the dissimilar fibers. This might be because the diameter, as well as surface morphology of polyester and olive root fibers, is quite different from each other. The pure polyester fiber-based composites show much better impregnation with unsaturated polyester resin. In this case, both the reinforcing fibers as well as resin are based on similar polymers and thus result in better impregnation.

On the other hand, olive root fibers are mainly composed of cellulose. Therefore, there are a few minor voids at the fiber–matrix interface. Such voids however are compensated by the hybridization effect in several samples. The addition of coir pith-based cellulosic micro-particles helps in a much better interface with the resin. The uniformly dispersed micro-particles fill all the minor voids around the fibrous phase. Such effect is visible in all the pure as well as hybrid composite samples prepared during this investigation.

3.2. Modeling and Statistical Analysis of Experimental Data

Significance models were chosen for all responses using linear, quadratic and special cubic functions of Scheffe's polynomial equations as given below [37,38].

$$Y = b1X1 + b2X2 + b3X3 \tag{2}$$

$$Y = b1X1 + b2X2 + b3X3 + b12X1X2 + b13X1X3 + b23X2X3 \tag{3}$$

$$Y = b1X1 + b2X2 + b3X3 + b12X1X2 + b13X1X3 + b23X2X3 + b123X1X2X3 \tag{4}$$

where $b1$, $b2$, $b3$, $b12$, $b13$, $b23$ and $b123$ are the coefficients of equations that are determined according to the model. $X1$, $X2$, $X3$, $X1X2$, $X1X3$, $X2X3$ and $X1X2X3$ are the component proportions in the mixture and Y is the dependent variable (responses). However, for all the responses the quadratic model was chosen to be fit for tensile strength, flexural strength, impact energy, water absorption, thermal expansion and the loss of density. Hence, the quadratic models for three-component mixture design take the final form as given below [39]:

$$\eta = \beta 1x1 + \beta 2x2 + \beta 3x3 + \beta 12x1x2 + \beta 13x1x3 + \beta 23x2x3 \tag{5}$$

where,

η = Response (mechanical, environmental or thermal)

$\beta i's$ = Coefficients of equation, $Xi's$ = Component proportion in mixture.

3.3. Regression Coefficient and Quadratic Model for All Mechanical, Environmental and Thermal Responses

Statistical summaries of model responses and regression coefficients are determined using Minitab software version 19.1.1. The regression equations corresponding to the best fit models were selected for the estimation of the corresponding mechanical property. Regression coefficients with the probabilities are shown in Tables 5–7 for mechanical, environmental and thermal responses respectively. Moreover, quadratic equations were determined from their Regression coefficients as shown below.

$$\eta = 17.35Xp + 19.21Xo - 367.4Xc + 8.53XpXo + 470.4XoXc \tag{6}$$

$$\eta = 33.75Xp + 36.81Xo - 1140Xc + 1854XpXc + 1919XoXc \tag{7}$$

$$\eta = 6.532Xp + 3.905Xo - 59.8Xc - 53.4XpXc - 50.2XoXc \tag{8}$$

$$\eta = 3.35Xp + 7.41Xo + 54.3Xc - 3.53XpXo - 49.21XpXc \tag{9}$$

$$\eta = 1.933Xp + 11.916Xo + 18.8Xc - 8.89XpXo \tag{10}$$

$$\eta = -12.48Xp - 12.08Xo + 970Xc - 1123XpXc - 1096XoXc \tag{11}$$

The terms having not significant probability were eliminated from the final equations. Regression Equations (6)–(8) are for tensile strength, flexural strength, and impact strength respectively. Equations (9)–(11) are used for the estimation of water absorption, loss of density due to weathering, and coefficient of thermal expansion respectively.

Table 5. Analysis of Variance for Tensile strength (MPa), Flexural strength (MPa), and Impact Strength (KJ/m^2).

Source	DF	Seq SS	Adj SS	Adj MS	F-Value	*p*-Value
Tensile strength Regression	5	130.971	130.971	26.194	16.48	0.002
Linear	2	8.081	101.697	50.848	32.00	0.002
Quadratic	3	122.889	122.889	40.963	25.78	0.001
*Xp*Xo*	1	6.537	6.913	6.913	4.35	0.082
*Xp*Xc*	1	26.506	105.793	105.793	66.57	0.000
*Xo*Xc*	1	89.846	89.846	89.846	56.54	0.000
Residual Error	6	9.535	9.535	1.589		
Total	11	140.506				
Flexural strength Regression	5	2099.31	2099.3	419.86	5.31	0.033
Linear	2	282.76	1490.1	745.05	9.43	0.024
Quadratic	3	1816.55	1816.5	605.52	7.66	0.018
*Xp*Xo*	1	312.47	322.4	322.42	4.08	0.090
*Xp*Xc*	1	9.61	1395.8	1395.85	17.66	0.006
*Xo*Xc*	1	1494.47	1494.5	1494.47	18.91	0.005
Residual Error	6	474.13	474.1	79.02		
Total	11	2573.44				
Impact strength Regression	5	28.1916	28.1916	5.63833	55.56	0.000
Linear	2	26.8809	6.2415	3.12073	30.75	0.081
Quadratic	3	1.3107	1.3107	0.43692	4.31	0.061
*Xp*Xo*	1	0.0949	0.0902	0.09024	0.89	0.382
*Xp*Xc*	1	0.1915	1.1597	1.15968	11.43	0.015
*Xo*Xc*	1	1.0243	1.0243	1.02428	10.09	0.019
Residual Error	6	0.6089	0.6089	0.10148		
Total	11	28.8005				

Table 6. Analysis of Variance (component proportions) for water absorption (%) and decrease in density (%).

	Source	DF	Seq SS	Adj SS	Adj MS	F-Value	*p*-Value
Water absorption (%)	**Regression**	5	58.1772	58.1772	11.6354	62.11	0.000
	Linear	2	55.2076	11.7230	5.8615	31.29	0.081
	Quadratic	3	2.9696	2.9696	0.9899	5.28	0.040
	*Xp*Xo*	1	1.2039	1.2183	1.2183	6.50	0.043
	*Xp*Xc*	1	1.1209	0.9798	0.9798	5.23	0.062
	*Xo*Xc*	1	0.6448	0.6448	0.6448	3.44	0.113
	Residual Error	6	1.1241	1.1241	0.1873		
	Total	11	59.3013				
Loss in density (%)	**Regression**	5	222.356	222.356	44.4711	110.49	0.000
	Linear	2	208.552	61.435	30.7177	76.32	0.014
	Quadratic	3	13.803	13.803	4.6012	11.43	0.007
	*Xp*Xo*	1	7.493	7.497	7.4972	18.63	0.005
	*Xp*Xc*	1	6.295	0.126	0.1259	0.31	0.596
	*Xo*Xc*	1	0.016	0.016	0.0157	0.04	0.850
	Residual Error	6	2.415	2.415	0.4025		
	Total	11	224.770				

Table 7. Analysis of variance for the coefficient of thermal expansion.

	Source	DF	Seq SS	Adj SS	Adj MS	F-Value	*p*-Value
Thermal expansion	**Regression**	5	2199.86	2199.9	439.97	5.57	0.030
	Linear	2	1573.42	645.9	322.96	4.09	0.176
	Quadratic	3	626.43	626.4	208.81	2.64	0.144
	*Xp*Xo*	1	113.86	117.4	117.39	1.49	0.269
	*Xp*Xc*	1	24.87	512.1	512.10	6.48	0.044
	*Xo*Xc*	1	487.71	487.7	487.71	6.17	0.048
	Residual Error	6	474.23	474.2	79.04		
	Total	11	2674.08				

3.4. Analysis of Variance for Mechanical Responses

ANOVA Table 5 shows the analysis of all mechanical responses.

For the tensile strength (MPa) response, the quadratic model is well fitted with a probability $p > 99\%$. Two-component interactions, e.g., $Xp*Xo$, $Xp*Xc$ and $Xo*Xc$ have probability of $p = 91\%$, $p > 99\%$ and $p > 99\%$ respectively. The two-component interactions of $Xp*Xc$ and $Xo*Xc$ significantly affect the tensile strength of hybrid composites with a probability of $p > 99\%$. Still, $Xp*Xo$ component interaction has an insignificant effect with the probability of $p = 91\%$ ($p < 95\%$). The quadratic model also fits the flexural strength (MPa) with a $p > 98\%$ as shown in two-component proportion interaction. Both $Xp*Xc$ and $Xo*Xc$ affect the response significantly with $p > 99\%$ respectively. Therefore, both have the highest degree of influence on flexural strength (MPa). However, $Xp*Xo$ component interaction has not been significantly influential due to lower probability ($p = 91\%$). Analysis of variance shows that the quadratic model is also well-fitted for the impact strength of the hybrid composites because it indicates a significant probability of $p = 94\%$. Two-component proportion interaction of $Xp*Xo$ has an insignificant effect on impact response with the probability of $p = 62\%$. However, other two-component interactions $Xp*Xc$ and $Xo*Xc$ have a strong effect on impact response with the probability of $p > 98\%$ respectively. Therefore, the quadratic model is considered a good fit for the impact strength of hybrid composites.

3.4.1. Normal Probability Plots of Residuals for Mechanical Responses

Normal probability plots of the residuals determine whether the collected data fit the selected models by the plot point distribution and by comparing the distribution of different samples. If the plotted point distribution is close to the fitted line slope, then the model distribution fits the collected data. For all mechanical properties the residual normal probability plots as shown in Figure 4.

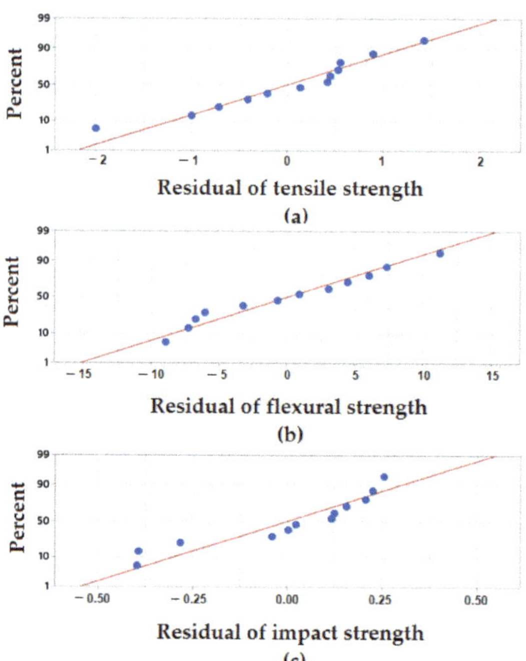

Figure 4. Normal probability plots of residuals (**a**) tensile strength, (**b**) flexural strength and (**c**) impact strength.

Figure 4a,b show the scatter plots for tensile strength and flexural strength respectively. The plots show that the majority of the points are distributed close to the slope (line) and only a few points are distributed far from the line. In the case of flexural strength, the normal probability plot indicates that approximately all the points are very close to the distribution slope as compared to the plot of tensile strength. Figure 4c shows the normal probability plot of impact strength. It can be noted that the points are somewhat clustered with a relatively lower probability of distribution on the slope line. In general, it can be concluded that all the mechanical responses are quite close to the slope of normal distribution.

3.4.2. Response Trace Plots for Mechanical Properties

Responses trace plots are also known as component effect plots. These plots help to analyze the effect of each proportion of individual components in a mixture on a selected response. The effect of varying component proportions along the imaginary line connecting the reference mixture on the vertex is shown by response trace curves. For the mechanical properties, response trace plots are shown in Figure 5a–c.

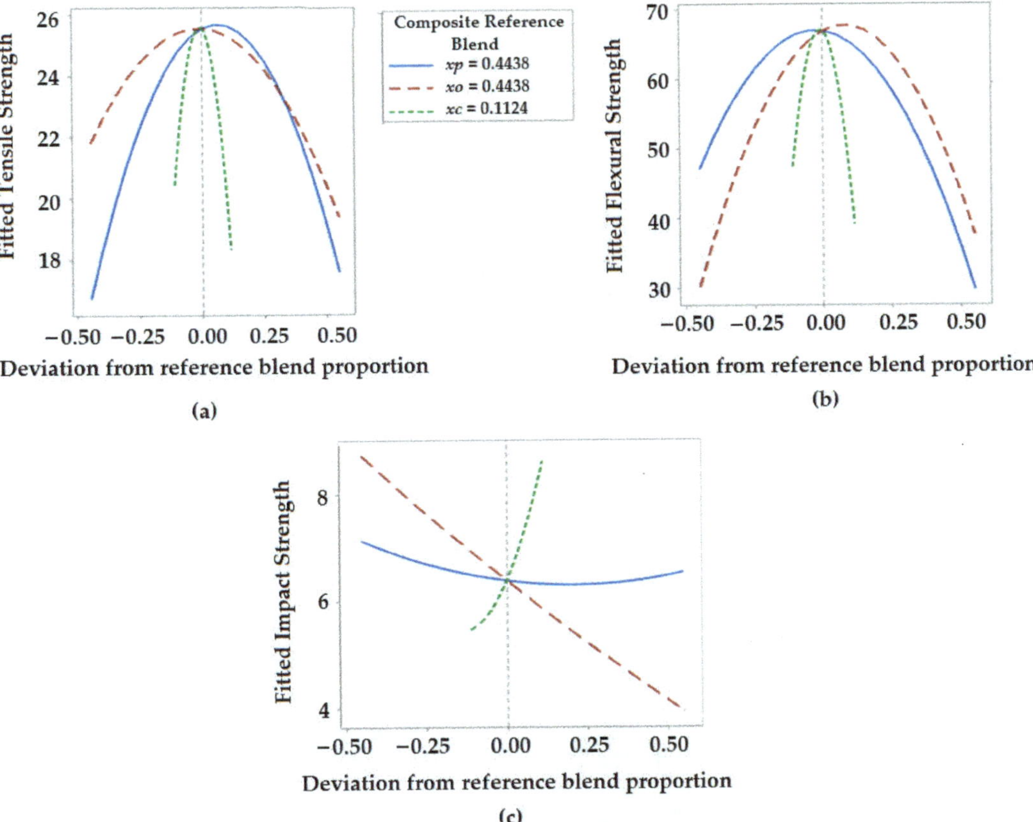

Figure 5. Response trace plots for (**a**) Tensile strength (MPa), (**b**) Flexural strength (MPa), (**c**) Impact strength (KJ/m^2).

As can be seen in Figure 5a, there is an increase in tensile strength to a magnitude of about 27 MPa with an increasing proportion of polyester waste fibers (Xp). The proportion of olive root fibers (Xo) is maintained at an optimum level before the tensile strength starts to decrease. Such findings are in accordance with the mechanical properties of polyester and olive root fibers. The polyester fibers are stronger than the olive root fibers and thus dominate their influence on the tensile strength of the composites. The addition of excess fibers beyond a certain limit leads to improper impregnation and a decrease in strength. It is also visible that the proportion of coir pith micro-particles (Xc) enhances the tensile strength to an optimum extent and further addition of coir pith decreases the tensile strength. micro-particles provide a very high interfacial area for bonding and thus enhance mechanical properties in a composite system. The behavior is reflecting the findings in literature based on Halpin–Tsai equations [32,33].

Figure 5b shows that flexural strength approaches a maximum magnitude of 78 MPa with an increasing proportion of olive root fibers and a decreasing proportion of polyester fiber in the mixture. The flexural strength is dominated by the olive root fibers as they are coarser and show higher stiffness as compared to polyester waste fibers. Moreover, the flexural strength increases to an optimum level with an increase in the proportion of micro-fillers, and subsequently shows a decreasing trend.

The response trace plot in Figure 5c indicates that the impact strength of hybrid composites linearly decreases with an increased proportion of olive root fibers. Polyester fibers are stronger and have a higher shock-absorbing capacity. Moreover, the coir pith micro-particles have a similar effect on the impact properties of hybrid composites. The micro-fillers improve overall bonding with the matrix and thus enhance the bulk mechanical properties, e.g., absorption of impact stresses. The component ratio is observed to have a synergistic effect on the mechanical properties investigated in this research.

3.4.3. Contour Plots for Mechanical Responses

A contour plot is a two-dimensional plot that indicates the effect of varying combinations of mixture components on the magnitude of response. The contour plot shows differently colored regions corresponding to the magnitude of responses. The higher magnitude of response is shown by the highlighted darker-colored regions compared to the lighter-colored regions. The contour plot can calculate the proportions of the components in the mixture and indicate the magnitude of different responses for corresponding color regions. The contour plots for mechanical properties are shown in Figure 6.

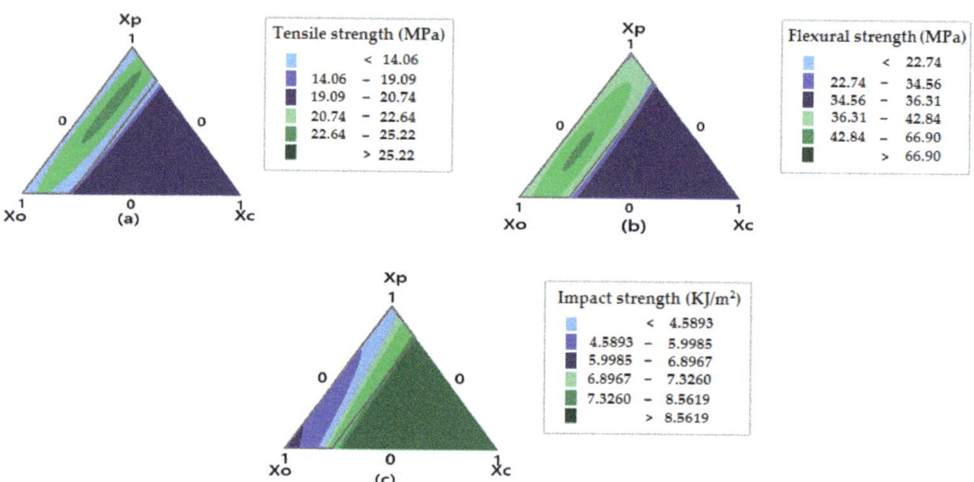

Figure 6. Contour plots for (**a**) Tensile strength, (**b**) Flexural strength, (**c**) Impact strength.

The contour plot shown in Figure 6a indicates the highest value for tensile strength (higher than 27.02 MPa) in the darker green region which is at the edge of Xp (polyester waste fiber) and Xo (olive root fibers). Moving along the contour plot, the highest value for the tensile strength response was 27.02 MPa, as shown in Table 4 for sample C6 (Xp = 58.5, Xo = 28.81, Xc = 12.67). The minimum value for the tensile strength was obtained for the sample C12 (Xp = 0.00, Xo = 77.72, Xc = 22.48) as seen in Table 4. The contour plot of Figure 6b shows the maximum flexural strength which is indicated by the darker green region just like the tensile response plot on the edges of Xo (olive root fiber) and Xp (polyester waste fiber) with optimum content of Xc (coir pith micro-particles). Figure 6c indicates a bigger region of darker green shade which implies that the impact strength is substantially affected by a combination of polyester and olive root fiber proportions. Significant improvement of impact strength can be achieved by increasing the proportion of polyester fibers and coir pith micro-particle content. The results are in agreement with the experimental findings. The trends are similar to theoretical calculation of mechanical properties as reported by other researchers based on Halpin–Tsai models [32,33].

As reported in the literature, remarkable enhancement in mechanical properties was observed in hybrid composites from date palm leaf incorporated recycled poly (ethylene terephthalate) developed by injection molding. Impact strength increased with higher loading of fiber. In addition, tensile strength and flexural strength were enhanced by adding a higher proportion of fiber when compared to the neat matrix. Recycle PET combined with date palm leaf could be a good alternative to obtain eco-friendly products [40]. Hybrid composites fabricated by using wood flour, coir pith powder and corn cob with different ratios in PVC, significantly improved tensile strength, flexural strength and energy absorption properties. The incorporation of coir pith and corn cob reduces voids and cavities in the hybrid composite resulting in the improvement of mechanical properties [41].

3.5. Analysis of Variance for Weathering Related Properties

ANOVA Table 6 shows the analysis of variance for water absorption response and loss of density after weathering treatment.

It reveals that the quadratic model is a suitable model for the water absorption behavior of hybrid composites since it predicts the response with a probability value of p = 96%. Two-component interactions, e.g., Xp^*Xo, Xp^*Xc and Xo^*Xc have probability of p = 96%, p = 94% and p = 89% respectively. Both Xp^*Xo, Xp^*Xc affect the water absorption properties of composites in a significant manner. The interaction, Xo^*Xc component is insignificant due to its poor probability of p = 89%. This means that the water absorption is critically dependent on the proportion of polyester fibers and olive root fibers. Since the proportion of coir pith micro-particles is relatively smaller, it has an insignificant effect on the overall water absorption capacity.

Variance analysis (Table 6), for the percentage decrease in density, indicates that the quadratic model is a significant predictive model with a probability value of p > 99%. The two-component interactions of Xp^*Xo, Xp^*Xc and Xo^*Xc have probability value of p > 99%, p = 40% and p = 15% respectively. These observations indicate that the loss of density is mainly governed by the proportions of polyester and olive root fibers in the hybrid composite. The coir pith micro-particles have an insignificant influence on the loss of density due to their lower proportion.

3.5.1. Normal Probability Plots of Residuals for Environmental/Weathering Related Properties

The normal probability plots of residuals for environmental properties are shown in Figure 7. The normal probability plot in Figure 7a shows that the points are very close to the slope line. This validates the normal distribution of water absorption behavior observed in hybrid composites. Additionally, from Figure 7b, it can be observed that all the fitted points are closer to the distribution slope in case of loss in the density of the composite samples after weathering treatment. This also confirms a normal distribution of the data obtained through the weathering experiment.

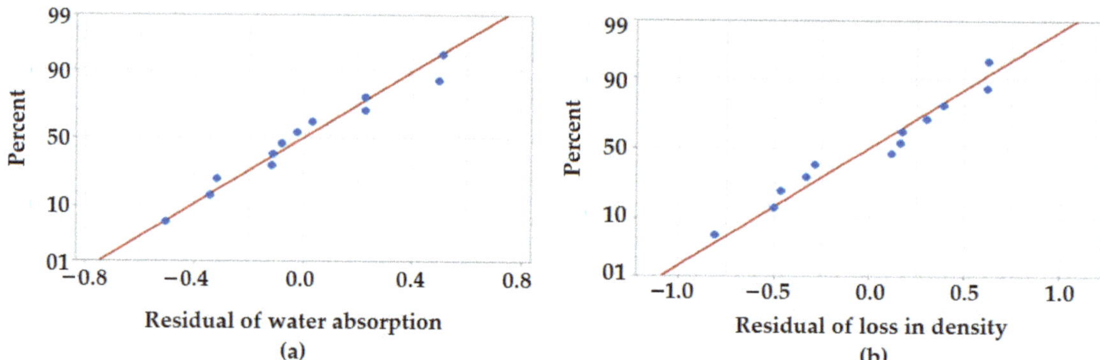

Figure 7. Normal probability plots of residuals for (**a**) water absorption (%), (**b**) density loss percentage.

3.5.2. Response Trace Plot for Environmental Properties

The response trace plot for the water absorption properties of hybrid composites is shown in Figure 8a. It indicates that with the increase in the proportion of polyester fiber, water absorption decreases, and it increases with an increase in the proportion of both olive root fiber and coir pith micro-particles. Moreover, from Figure 8b, it can be concluded that there is a linear decrease in density after exposure to the outdoor environment based on both the proportion of olive root fibers and coir pith micro-particles. However, the proportion of polyester waste fiber shows a negligible effect on the loss of density of hybrid composites when exposing them to the outdoor environment. Generally, the density of each composite sample decreases but with a different percentage after exposure. As the incorporation of polyester waste fiber decreases and the proportion of olive root fiber and coir pith micro-particles increases, there is a substantial decrement of density.

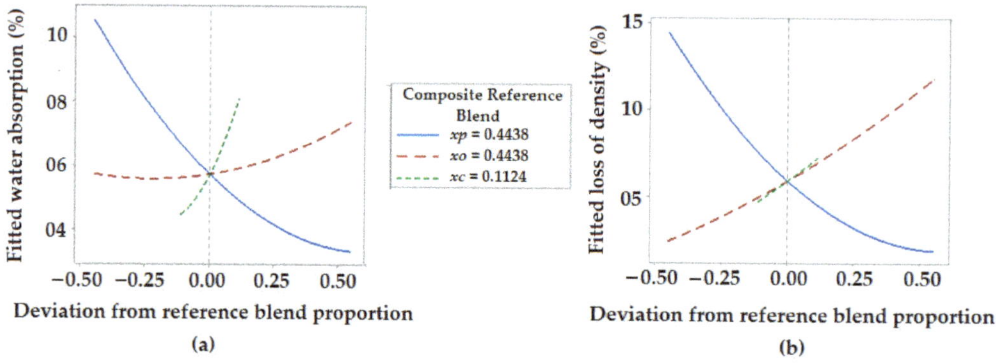

Figure 8. Response trace plot for (**a**) water absorption (%), (**b**) percentage decrease in density.

3.5.3. Contour Plots for Environmental/Weathering Related Properties

Figure 9a shows a contour plot for the water absorption properties of hybrid composites. Moving along the contour plot, it can be observed that the minimum value is indicated in the green region dominated by Xp (polyester waste fiber). When moving from this region towards the Xo (olive root proportion) and Xc (coir pith filler) region, the water absorption increases which is indicated by darker red color. The lowest value as given in Table 4 is about 3.24 (%) for the sample C1 of (Xp = 100, Xo = 0.00, Xc = 0.00). Water absorption behavior increases with the increasing proportion of olive root fiber and coir pith micro-particle filler in the hybrid composite. In practical terms, these natural origin fibers and micro-particles are much more hygroscopic as compared to polyester fibers. The maximum water absorption of hybrid composite was about 10.92% for the sample C12 ($Xp = 0.00$, $Xo = 77.52$, $Xc = 22.48$).

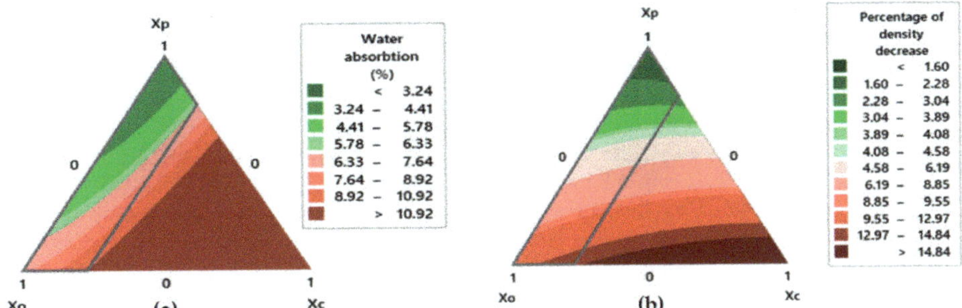

Figure 9. Contour plot for (**a**) water absorption (%), (**b**) percentage loss of density.

The impact of weathering treatment on the percentage loss of density can be noted in Figure 9b. The maximum decrease in the density can be seen in the darker red region and the edge of Xo and Xc contour plot. The indication of lower density loss was given in the darker green region at the edge of Xp. Maximum value for density loss was about 14.84% for the sample C12 ($Xp = 0.00$, $Xo = 77.52$, $Xc = 22.48$) and the minimum value for the density loss was about 1.60% for sample C1 ($Xp = 100$, $Xo = 0.00$, $Xc = 0.00$). The polyester fiber is least degradable which leads to minimum density loss. Biological origin materials, e.g., olive root fibers or coir pith micro-particles are susceptible to degradation and loss of density. Due to weathering, some of the fibers and micro-particles are degraded and this leads to loss of weight. Further, the volume is found to increase causing a loss of density.

Figures 10 and 11 show the variations of water absorption and change in density before and after weathering for all the hybrid composite samples as compared to fiber glass sunshade panels.

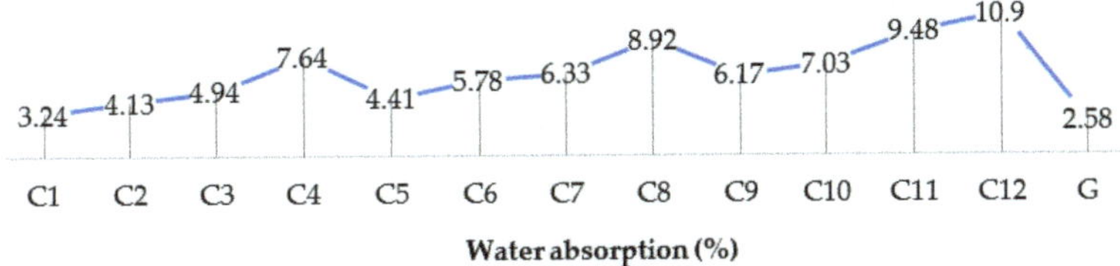

Figure 10. Water absorption (%) of hybrid composites.

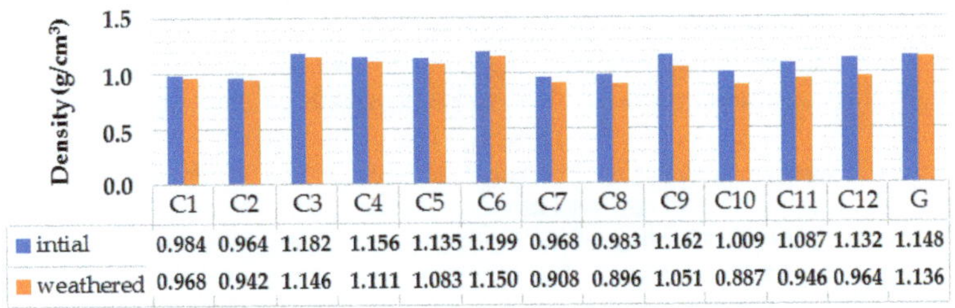

	C1	C2	C3	C4	C5	C6	C7	C8	C9	C10	C11	C12	G
intial	0.984	0.964	1.182	1.156	1.135	1.199	0.968	0.983	1.162	1.009	1.087	1.132	1.148
weathered	0.968	0.942	1.146	1.111	1.083	1.150	0.908	0.896	1.051	0.887	0.946	0.964	1.136

Figure 11. Densities of hybrid composite before and after weathering.

3.5.4. Surface Damage after Weathering Treatment

After weathering treatment, the changes on the surface of each hybrid composite were visually examined. It was observed that the hybrid composite samples having olive root fibers and coir pith micro-particles as major constituents resulted in more surface degradation as compared to composite samples containing polyester waste fibers as a major constituent. Moreover, coir pith micro-particles and olive root fibers protruded on the surface of the composite samples. This is due to the degradation of olive root fibers and coir pith during weathering treatment. Moreover, the observed higher loss in the density of such hybrid composites based on the major proportion of olive root fibers and coir pith micro-particles can be due to the effect of moisture and UV radiation on cellulose, lignin and hemicellulose in these biological origin materials. The hybrid composites rich in cellulosic components degraded due to exposure to a combination of xenon arc radiation and water spray. However, the resistance to environmental degradation of such hybrid composites is significantly improved by increasing the proportion of PET fibers in the composite [23].

3.6. Analysis of Variance for Thermal Properties

ANOVA of the thermal expansion properties in hybrid composites is given in Table 7.

The results indicate that the quadratic model is a suitable model to fit the thermal expansion behavior of hybrid composites. It predicts the thermal expansion response with a probability value of $p = 86\%$. Two-component interaction $Xp*Xo$ has a probability of $p = 74\%$ and thus this interaction has a relatively insignificant effect on the thermal response. While two-component interactions of $Xp*Xc$ and $Xo*Xc$ show a significant influence on thermal expansion properties of hybrid composite with a probability value of $p = 96\%$ and $p = 94\%$ respectively. Therefore, the quadratic model is a relatively good choice to fit the coefficient of thermal expansion as compared to the linear model. The thermal expansion in a composite system is governed by the Halpin–Tsai models and the hybrid composites in the present investigation follow a similar trend [32,33]. The matrix constitutes the majority component in the composites and thus the thermal properties are dominated by the matrix thermal properties. The addition of micro-particles restricts the thermal deformations in the hybrid composites. Since polyester fiber is thermoplastic in nature, a higher proportion always leads to higher deformation in response to thermal conditions. On the other hand, the presence of a higher proportion of cellulosic fiber like olive root fibers restricts thermal deformation.

3.6.1. Normal Probability Plot of Residuals for Thermal Properties

A residual normal probability plot for the coefficient of thermal expansion (CTE), as seen in Figure 12, indicates that five data points out of twelve data points fall very close to the fitted slope line. The remaining points are distributed in groups very close to each other and also close to the probability line. Residual normal probability plot thus validates the distribution of coefficient of a thermal expansion near to normal distribution.

3.6.2. Response Trace Plot for Thermal Properties

As shown in Figure 13, the response trace plot for the thermal expansion of hybrid composites concludes that the two-component interaction of polyester waste fiber with olive root fiber $(Xp*Xo)$ affects a linear decrement in the thermal expansion behavior of samples to an optimum level. After that the hybrid composites start to expand. Moreover, the addition of coir pith micro-particles enables a linear decrement of thermal expansion coefficient in composites. Both the reinforcing fibers as well as the micro-particle filler provide restrictions to the expansion of the composite by offering resistance at the interfacial region with the matrix. Such behavior is well established by theoretical models using the rule of the mixture as well as modified Halpin–Tsai equations [32,33].

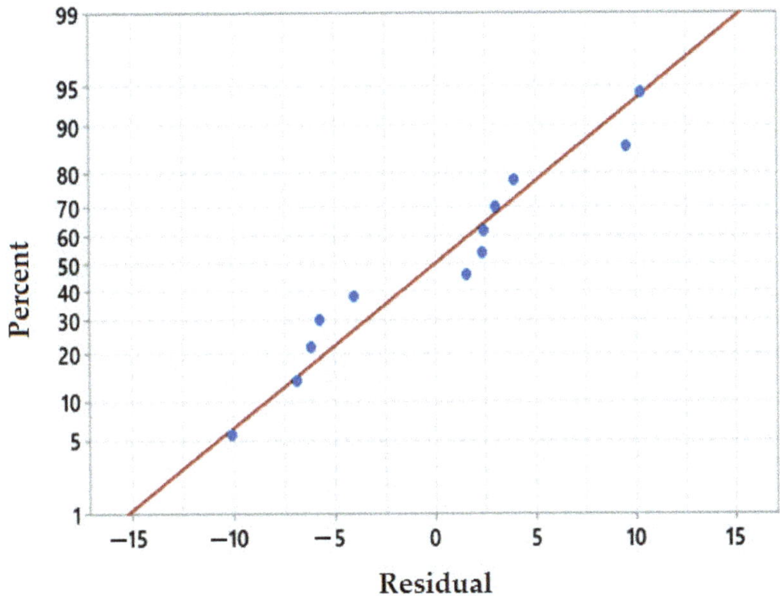

Figure 12. Residual normal probability plot of CTE.

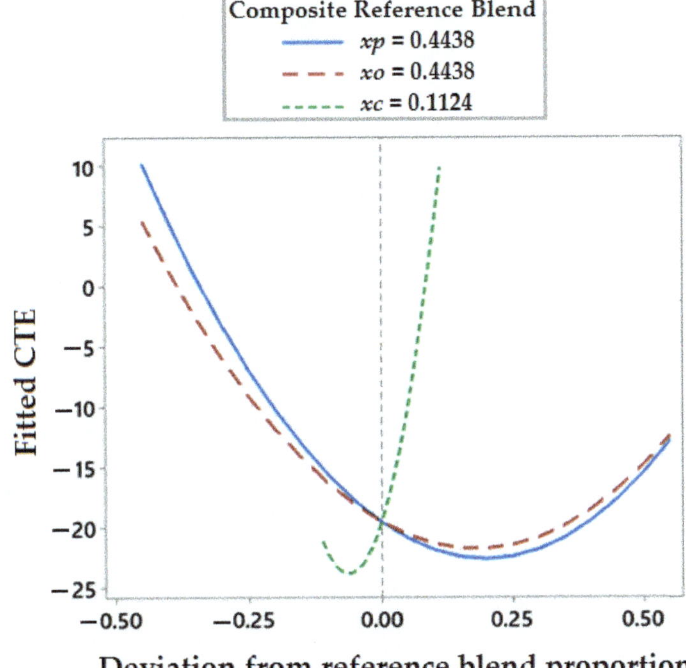

Figure 13. Response trace plot of CTE.

3.6.3. Contour Plot for Thermal Response

For the thermal expansion behavior, the contour plot is shown in Figure 14.

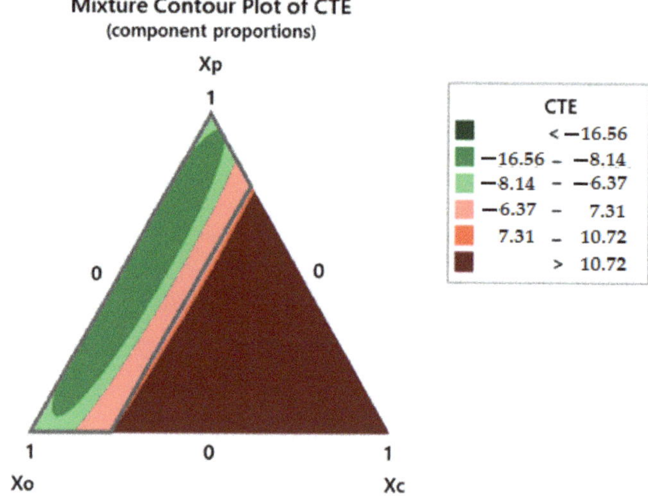

Figure 14. Coefficient of thermal expansion for hybrid composites.

A higher proportion of Xp (polyester waste fiber) offers a negative coefficient of thermal expansion. This tendency of negative thermal expansion was observed for the samples C1 to C8 as shown in Tables 3 and 4. The minimum negative thermal expansion value was observed for sample C8. These trends indicate that reinforcing fibers actually cause a thermal shrinkage. The shrinkage is reduced as the polyester fiber proportion is reduced. As polyester is a thermoplastic fiber material, it might deform during the increase in temperature and cause this shrinkage. On the other hand, olive root fibers are plant-origin cellulosic materials. They do not tend to deform as the temperature rises. The hybrid composites containing polyester waste fibers, exhibit expansion in linear trend from 25–75 °C. As the temperature further increases, the rate of contraction in the composite increases constantly up to 150 °C and the coefficient of thermal expansion turns negative. However, the hybrid composites containing a major proportion of olive root fibers and coir pith micro-particles in the mixture expand linearly up to 150 °C and show relatively lower coefficients of thermal expansion. These observations indicate that biological origin (cellulosic) reinforcements and fillers provide thermal stability and shape retention to a hybrid composite as opposed to thermoplastic fibers like polyester. Guangfa et.al also observed that polyester fiber starts to contract and the coefficient of thermal expansion turns into a negative value as the temperature increased above 220 °C [42].

3.7. Thermogravimetric Analysis of Hybrid Composites and Composites

TGA (Thermo-Gravimetric Analysis) results of the hybrid composites describe the thermal stability at elevated temperatures. It can be observed that thermal stability is significantly affected by a mixture of polyester waste fiber, olive root fiber and coir pith micro-particles. The TGA curves shown in Figure 15 explain the weight loss at the onset of thermal degradation.

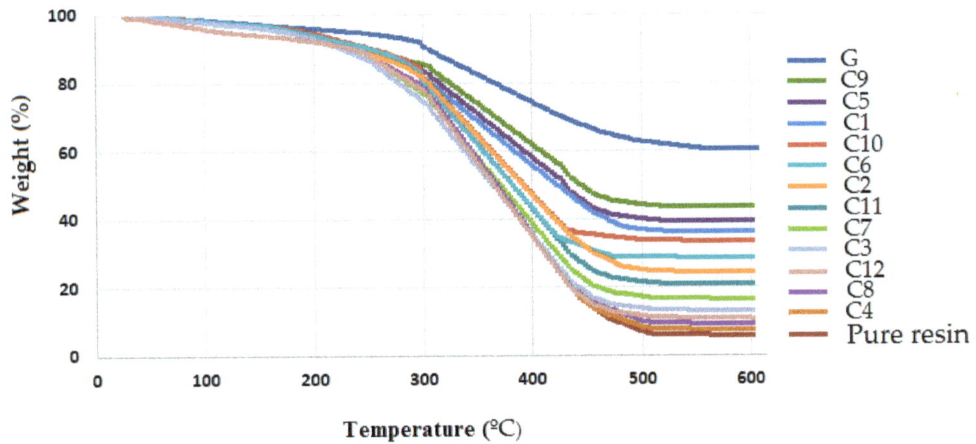

Figure 15. TGA–curves of hybrid composites.

The TGA results of composite samples as shown in Figure 15 indicate weight loss of 0.3% to 6% of the original weight between 100 °C and 200 °C respectively due to elimination of moisture in the matrix. At about 450 °C, the weight loss was between 15% to 57% due to the degradation and volatilization of some components in the matrix. Subsequently, there is no further loss of weight up to 600 °C. This trend is visible for all the hybrid and non-hybrid composites as well as composites in the present study.

It can be stated that the thermal stability of hybrid composites is significantly influenced by the matrix and different reinforcement materials in the component mixture. The thermal decomposition takes place in a three-step process for each composite sample as can be seen in the TGA curves. The decomposition temperature for the olive root fiber-based composite samples was lower than the pure polyester fiber-based samples. This may be due to higher moisture content and volatile substances in natural origin cellulosic fibers as compared to polyester fibers. In the first stage, a small mass loss was observed for all samples between 210 °C to 310 °C due to the degradation of organic elements in the composites. For the second stage of decomposition, in the temperature range from 310 °C to 370 °C, the hemicellulose component in the olive root fiber may be responsible. The third peak at about 420 °C was mainly due to the degradation of the unsaturated polyester matrix.

Table 8 shows the maximum temperature of decomposition, mass loss (%) and char (%) at 600 °C for all the samples.

Table 8. TGA results for hybrid composites.

Samples	T MAX (°C)	Mass Loss (%)	Char (%) at 600 °C
C1	454	62	38
C2	438	75	25
C3	433	82	18
C4	432	90	10
C5	456	58	42
C6	440	69	31
C7	436	80	20
C8	433	86	14
C9	457	55	45
C10	448	65	35
C11	438	78	22
C12	436	84	16
G	467	38	62

The hybrid composites show changes in the process of decomposition due to varying component interactions. The composites with a higher proportion of polyester waste fibers show better thermal stability up to a higher peak temperature of degradation. Hybridization with cellulosic fibers reduces the peak decomposition temperature. On the other hand, the addition of scale fillers in the matrix improves thermal stability. This is mainly because the micro-particles derived from cellulosic fibers, e.g., coir are free from volatile components. These micro-particles are highly crystalline and are free from moisture. The incorporation of scale fillers enables dissipation of heat in the composite system and the degradation is prohibited up to a higher temperature level. When the proportion of micro-fillers was increased from 5% to 10%, a further increase in peak temperature was observed.

The mass loss was observed to be reduced by the hybridization of polyester and olive root fibers. Mass loss in polyester-rich composites was lower as compared to olive root fiber-rich samples. The combination of PET and cellulosic materials in a hybrid composite works in a synergistic manner and they compensate and complement the weaknesses of each other. The use of micro-particles from cellulose origin materials proves to be beneficial in order to improve thermal stability and reduce overall mass loss at elevated temperatures.

The mass loss (%) of the hybrid composites, as well as composites, was higher as compared to commercial fiber glass sunshade panels though the glass panel exhibits a slightly higher peak temperature of degradation.

3.8. Optimization for all Responses

The optimal mechanical properties (tensile strength, flexural strength and impact strength), environmental/weathering properties (water absorption and density loss) and the thermal properties (coefficient of thermal expansion) of the hybrid composites were estimated using an optimization approach for each dependent response. The mechanical responses were maximized, environmental properties and thermal expansion properties were minimized. By using this approach, the best combination was found to be $Xp = 70.83\%$, $Xo = 15.16\%$ and $Xc = 14.01\%$.

4. Conclusions

In this study, an attempt was made to fabricate hybrid composite materials using polyester waste fibers, olive root fibers and coir pith particles as reinforcement in unsaturated polyester resin. Micro-particles obtained from coir pith by ball milling process were used as filler in the polyester resin. The mechanical, thermal, and weathering-related properties were determined for all fabricated hybrid and non-hybrid composite samples. These hybrid composites were then compared with commercial glass fiber-based composite panels used as a sunshade in interior applications.

Mixture design analysis was used to optimize the component proportions in the hybrid composite. For all mechanical responses (tensile strength, flexural strength, and impact strength) environmental properties (water absorption and density loss after weathering) and for the coefficient of thermal expansion, statistical models were developed which show good validity with experimental results. The best formulation was determined at 70.83 wt%, 15.16 wt% and 14.01 wt% of polyester waste fibers, olive root fibers and coir pith fillers, respectively.

Among all the hybrid composites, sample C6 showed the maximum tensile strength and maximum flexural strength. In comparison with commercial glass fiber-based composite, sample C6 reached 30.42% of its tensile strength and almost half of its flexural strength. Maximum tensile strength, flexural strength and impact strength in the developed hybrid composite can be achieved by optimum mixture ratio of polyester waste fiber, olive root fiber and coir pith micro-particles.

The maximum impact strength of the developed hybrid composites was obtained in samples C9 and C10. However, by comparison with commercially available glass fiber composite, C9 composite showed almost 15% of its impact strength. The impact strength increased as the mixture proportion of polyester waste fiber and coir pith micro-particles was increased.

With respect to the environmental/weathering-related properties, C1, C5, and C9 showed the minimum water absorption% as well as density loss which is almost nearer to the commercially available glass fiber composite. Comparing the water absorbency of commercially available glass fiber composite and the developed composite sample (C1), it is found that there is a minute difference in their water absorption capacity. The commercially available glass fiber panel loses 1.2% of its density after weathering while the developed composite C1 lost 1.6% of its density. From the results, it was found that the developed hybrid composite containing a major proportion of polyester waste fiber along with other cellulosic components can be preferred for outdoor applications.

At 150 °C, sample C12 showed the minimum value of thermal expansion. The hybrid composite sample C9 containing a major proportion of polyester waste fiber and coir pith micro-particles in the mixture results in a minimum weight loss (%) and a maximum char (%) at 600 °C as per the TGA results. It resulted in 45% residual char as against the commercial glass fiber composite giving about 62% char. Based on mechanical, thermal and weathering properties, the hybrid composites from polyester waste fiber, olive root fiber and coir pith micro-particles can be categorized separately for outdoor (sunshade) and indoor (interior) applications. These can be used in construction and structural applications at the same time addressing issues of waste management and eco-friendliness.

Author Contributions: M.R.T., H.J., R.M., U.H., M.T. and M.M. experiment design; M.R.T., H.J., R.M., U.H. and M.T. testing of mechanical properties and data analysis; M.R.T., H.J., R.M., U.H., M.T. and M.M. wrote and edited the paper; H.J., R.M. and M.M. language correction; R.M. communication with the editors; H.J., R.M., M.T. and M.M. administration and funding; All authors have read and agreed to the published version of the manuscript.

Funding: The work was supported by the internal grant agency of the Faculty of Engineering, Czech University of Life Sciences Prague (No. 2021:31140/1312/3108).

Institutional Review Board Statement: Not applicable.

Informed Consent Statement: Not applicable.

Data Availability Statement: Not applicable.

Conflicts of Interest: The authors declare no conflict of interest.

References

1. Belgacem, M.N.; Gandini, A. The surface modification of cellulose fibres for use as reinforcing elements in composite materials. *Compos. Interfaces* **2005**, *12*, 41–75. [CrossRef]
2. Masuelli, M. *Fiber Reinforced Polymers: The Technology Applied for Concrete Repair*; Intech Open Limited: London, UK, 2013.
3. Valadez-Gonzalez, A.J.; Olayo, C.; Herrera-Franco, P. Effect of fiber surface treatment on the fiber–matrix bond strength of natural fiber reinforced composites. *Compos. Part B Eng.* **1999**, *30*, 309–320. [CrossRef]
4. Gibson, R.F. *Principle of Composite Material Mechanics*, 3rd ed.; CRC Press: Boca Raton, FL, USA, 2011.
5. Joshi, S.V.; Drzal, L.; Mohanty, A.; Arora, S. Are natural fiber composites environmentally superior to glass fiber reinforced composites? *Compos. Part A Appl. Sci.* **2004**, *35*, 371–376. [CrossRef]
6. Mohanty, A.K.; Misra, M.; Drzal, L.T. *Natural Fibers, Biopolymers, and Biocomposites*; CRC Press Taylor and Francis: London, UK, 2005.
7. Faruk, O.; Bledzki, A.K.; Fink, H.P.; Sain, M. Progress report on natural fiber reinforced composites. *Macromol. Mater. Eng.* **2014**, *299*, 9–26. [CrossRef]
8. Saheb, D.N.; Jog, J.P. Natural fiber polymer composites: A review. *Adv. Polym. Tech. J. Polym. Proc. Inst.* **1999**, *18*, 351–363. [CrossRef]
9. Mohanty, A.; Misra, M.; Drzal, L.T. Surface modifications of natural fibers and performance of the resulting biocomposites: An overview. *Compos. Interfaces* **2001**, *8*, 313–343. [CrossRef]
10. Mishra, R.; Gupta, N.; Pachauri, R.; Behera, B.K. Modelling and simulation of earthquake resistant 3D woven textile structural concrete composites. *Compos. Part B Eng.* **2015**, *81*, 91–97. [CrossRef]
11. Dashtizadeh, Z.; Abdan, K.; Jawaid, M.; Khan, M.; Behmanesh, A.M.; Dashtizadeh, M.; Cardona, F.; Ishak, M. Mechanical and thermal properties of natural fibre based hybrid composites: A review. *Pertanika J. Sci. Technol.* **2017**, *25*, 1103–1122.
12. Gururaja, M.; Rao, A.H. A review on recent applications and future prospectus of hybrid composites. *Int. J. Soft Comput. Eng.* **2012**, *1*, 352–355.

13. Franciszczak, P.; Piesowicz, E.; Kalniņš, K. Manufacturing and properties of r-PETG/PET fibre composite–Novel approach for recycling of PETG plastic scrap into engineering compound for injection moulding. *Compos. Part B Eng.* **2018**, *154*, 430–438. [CrossRef]
14. Wu, C.; Hsieh, W.; Su, S.; Cheng, K.; Lee, K.; Lai, C. Study of braiding commingled self-reinforced PET composites. *Int. J. Mod. Phy. B* **2018**, *32*, 1840086. [CrossRef]
15. Ahmad, M.; Majid, M.; Ridzuan, M.; Firdaus, A.; Amin, N. Tensile properties of interwoven hemp/PET (Polyethylene Terephthalate) epoxy hybrid composites. *J. Phy. Conf. Series.* **2017**, *908*, 012011. [CrossRef]
16. Dan-mallam, Y.; Abdullah, M.Z.; Yusoff, P. Mechanical properties of recycled kenaf/polyethylene terephthalate (PET) fiber reinforced polyoxymethylene (POM) hybrid composite. *J. Appl. Polym. Sci.* **2014**, *131*. [CrossRef]
17. Wu, C.; Lin, P.; Murakami, R. Long-term creep behavior of self-reinforced PET composites. *Exp. Polym. Lett.* **2017**, *11*. [CrossRef]
18. Narendar, R.; Dasan, K.P. Development of coir pith based hybrid composite panels with enhanced water resistant behavior. *Environ. Prog. Sust. Energy* **2015**, *34*, 1481–1487. [CrossRef]
19. Essabir, H.; Bensalah, M.; Rodrigue, D.; Bouhfid, R.; Qaiss, A. Structural, mechanical and thermal properties of bio-based hybrid composites from waste coir residues: Fibers and shell micro-particles. *Mech. Mater.* **2016**, *93*, 134–144. [CrossRef]
20. Islam, M.T.; Das, S.C.; Saha, J.; Paul, D.; Islam, M.T.; Rahman, M.; Khan, M.A. Effect of coconut shell powder as filler on the mechanical properties of coir-polyester composites. *Chem. Mate.r Eng.* **2017**, *5*, 75–82. [CrossRef]
21. Narendar, R.; Dasan, K.P. Effect of chemical treatment on the mechanical and water absorption properties of coir pith/nylon/epoxy sandwich composites. *Int. J. Polym. Anal. Char.* **2013**, *18*, 369–376. [CrossRef]
22. Jamshaid, H.; Mishra, R.; Pechociakova, M.; Noman, M.T. Mechanical, thermal and interfacial properties of green composites from basalt and hybrid woven fabrics. *Fiber Polym.* **2016**, *17*, 1675–1686. [CrossRef]
23. Prithivirajan, R.; Jayabal, S.; Sundaram, S.K.; Kumar, A.P. Hybrid bio composites from agricultural residues: Mechanical and thickness swelling behavior. *Int. J. Chem. Tech. Res.* **2016**, *9*, 609–615.
24. Mishra, R.; Huang, J.; Kale, B.; Zhu, G.; Wang, Y. The production, characterization and applications of nanoparticles in the textile industry. *Textile Prog.* **2014**, *46*, 133–226. [CrossRef]
25. Dehas, W.; Guessoum, M.; Douibi, A.; Jofre-Reche, J.A.; Martin-Martinez, J.M. Thermal, mechanical, and viscoelastic properties of recycled poly (ethylene terephthalate) fiber-reinforced unsaturated polyester composites. *Polym. Compos.* **2018**, *39*, 1682–1693. [CrossRef]
26. Abdullah, M.Z.; Dan-mallam, Y.; Yusoff, P. Effect of environmental degradation on mechanical properties of kenaf/polyethylene terephthalate fiber reinforced polyoxymethylene hybrid composite. *Adv. Mater. Sci. Eng.* **2013**, *2013*, 671481. [CrossRef]
27. Dan-Mallam, Y.; Abdullah, M.Z.; Yosuff, P. Impact strength, microstructure, and water absorption properties of kenaf/polyethylene terephthalate (PET) fiber-reinforced polyoxymethylene (POM) hybrid composites. *J. Mater. Res.* **2013**, *28*, 2142–2146. [CrossRef]
28. Amir, M.; Irmawaty, R.; Hustim, M.; Rahim, I. Tensile strength of glass fiber-reinforced waste PET and Kenauf hybrid composites. In Proceedings of the IOP Conference Series Earth and Environmental Science, Bali, Indonesia, 29–30 August 2020; Volume 419, p. 012061. [CrossRef]
29. Ayrilmis, N.; Buyuksari, U. Utilization of olive mill sludge in manufacture of lignocellulosic/polypropylene composite. *J. Mater. Sci.* **2010**, *45*, 1336–1342. [CrossRef]
30. Raghavarao, D.; Wiley, J.B.; Chitturi, P. *Choice-Based Conjoint Analysis: Models and Designs*; CRC Press: Boca Raton, FL, USA, 2010.
31. Zhang, C.; Wong, W.K. Optimal designs for mixture models with amount constraints. *Stat. Prob. Lett.* **2013**, *83*, 196–202. [CrossRef]
32. Tucker, C.L.; Liang, E. Stiffness predictions for unidirectional short-fiber composites: Review and evaluation. *Compos. Sci. Technol.* **1999**, *59*, 655–671. [CrossRef]
33. Halpin Affdl, J.C.; Kardos, J.L. The Halpin-Tsai equations: A review. *Polym. Eng. Sci.* **1976**, *16*, 344–352. [CrossRef]
34. Makarian, K.; Santhanam, S.; Wing, Z.N. Coefficient of thermal expansion of particulate composites with ceramic inclusions. *Ceram. Int.* **2016**, *42*, 17659–17665. [CrossRef]
35. Karch, C. Micromechanical analysis of thermal expansion coefficients. *Model. Numer. Simul. Mater. Sci.* **2014**, *4*, 104–119. [CrossRef]
36. Safi, M.; Hassanzadeh-Aghdam, M.K.; Mahmoodi, M.J. Effects of nano-sized ceramic particles on the coefficients of thermal expansion of short SiC fiber-aluminum hybrid composites. *J. Alloys Compd.* **2019**, *803*, 554–564. [CrossRef]
37. Cornell, J.A. *Experiments with Mixtures: Designs, Models, and the Analysis of Mixture Data*; John Wiley & Sons: Hoboken, NJ, USA, 2011.
38. Oehlert, G.W. *A First Course in Design and Analysis of Experiments*; University of Minnesota: Minneapolis, MN, USA, 2010.
39. Rostamiyan, Y.; Mashhadzadeh, A.H.; Khani, A.S. Optimization of mechanical properties of epoxy-based hybrid composite: Effect of using silica and high-impact polystyrene by mixture design approach. *Mater. Des.* **2014**, *56*, 1068–1077. [CrossRef]
40. Dehghani, A.; Ardekani, S.M.; Al-Maadeed, M.A.; Hassan, A.; Wahit, M.U. Mechanical and thermal properties of date palm leaf fiber reinforced recycled poly (ethylene terephthalate) composites. *Mater. Des.* **2013**, *52*, 841–848. [CrossRef]
41. Nagaraj, C.; Mishra, D.; Reddy, J.D. Estimation of tensile properties of fabricated multi layered natural jute fiber reinforced E-glass composite material. *Mater. Today Proc.* **2020**, *27*, 1443–1448. [CrossRef]
42. Guangfa, G.; Shujie, Y.; Ruiyuan, H.; Yongchi, L. Experimental investigation on thermal physical properties of an advanced polyester material. *Phy. Proc.* **2012**, *25*, 333–338. [CrossRef]

MDPI

St. Alban-Anlage 66

4052 Basel

Switzerland

Tel. +41 61 683 77 34

Fax +41 61 302 89 18

www.mdpi.com

Polymers Editorial Office

E-mail: polymers@mdpi.com

www.mdpi.com/journal/polymers